Computational Modelling in Industry 4.0

Irfan Ali · Prasenjit Chatterjee · Ali Akbar Shaikh ·
Neha Gupta · Ali AlArjani
Editors

Computational Modelling in Industry 4.0

A Sustainable Resource Management Perspective

Springer

Editors
Irfan Ali
Department of Statistics and Operations Research
Aligarh Muslim University
Aligarh, India

Ali Akbar Shaikh
Department of Mathematics
University of Burdwan
West Bengal, India

Ali AlArjani
Industrial Engineering Department
Prince Sattam Bin Abdulaziz University
Alkharj, Saudi Arabia

Prasenjit Chatterjee
Department of Mechanical Engineering
MCKV Institute of Engineering
West Bengal, India

Neha Gupta
Amity School of Business
Amity University
Noida, India

ISBN 978-981-16-7722-9 ISBN 978-981-16-7723-6 (eBook)
https://doi.org/10.1007/978-981-16-7723-6

This Springer imprint is published by the registered company Springer Nature Singapore Pte Ltd.
The registered company address is: 152 Beach Road, #21-01/04 Gateway East, Singapore 189721, Singapore

Dr. Irfan Ali would like to dedicate this book to his loving parents, beloved wife Huma Chauhan and son Rayyan Ali. A special dedication owes to his respected teachers Profs. S. U. Khan and Abdul Bari.

Dr. Prasenjit Chatterjee would like to dedicate this book to his grandparents, father Late Dipak Kumar Chatterjee, his mother Mrs. Kalyani Chatterjee, beloved wife Amrita and his little angel Aheli.

Dr. Ali Akbar Shaikh would like to dedicate this book to my father Azahar Ali Shaikh, mother Sahanara Shaikh Bibi, wife Dilnawas Khatun and son Ayaan Ali Shaikh. A special dedication owes to my respected teachers and supervisors (Profs. Manjusri Basu, Asoke Kumar Bhunia and Leopoldo Eduardo Cardenas-Barron).

Dr. Neha Gupta would like to dedicate this book to her beloved family members for their constant support and to her organization Amity University Uttar Pradesh, Noida for the inspiration and motivation.

Dr. Ali AlArjani would like to dedicate the book to his great big family, his Father Mr. Saeed AlArjani, his mother Mrs. Nawyer AlArjani, and his awesome wife, brother and sisters.

Preface

With the evolution of industry 4.0 and from the social, economic, and environmental point of view sustainable management of natural resources, energy, transportation, supply chain, entrepreneurship, agricultural, etc., are required today for the betterment of the future. A combination of business and sustainability is sustainable management. After the evolution of the fourth industrial revolution, several growing organizations from a broad range of industries (e.g., retail, food and beverage, agriculture, tech, insurance, automotive, military, healthcare, utility, and more) are pursuing sustainable management business goals. With the growing businesses concern for the environmental issues such as global warming, government policies, scarcity of resources, etc. getting attention and therefore, the need for sustainability in every business is required. This book presents the recent developments, emerging issues & challenges; opportunities occurred in sustainable manufacturing after the accession of industry 4.0 under one umbrella.

This book presents total 19 chapters which are based on different problems, practices, challenges, opportunities, etc. in sustainable management with the decision-making techniques and are also showcasing the application of modelling in sustainable management and industry 4.0. Chapter 1 discusses the emergence and applications of Industry 4.0 systematically and concludes that robotics would take over manufacturing activities more intelligently, leading to industry 5.0 in the future. Chapter 2 examines opportunities and the challenges facing the circular economy implementation and suggests waste management, waste prevention, and resources efficiency to foster a circular transformation of resources. Chapter 3 identifies the challenges in the virtual workspace by the inception of industry 4.0. Chapter 4 developed the cost-based optimization model of solar power generation whereas Chap. 5 developed the mathematical model of green solid logistics problem and studied the impact of carbon tax policy. Chapter 6 focussed on sustainable connectivity management via e-SIM multi licence-based billing. Chapter 7 presents the sustainable production inventory model. A systematic review on cleaner technology for sustainable industrial practices has been employed in Chap. 8. Chapter 9 applies big data in smart sustainable energy systems & E-mobility.

Chapter 10 presents the pricing and inventory policies for inventory models with lot-size dependent discount for deteriorating items. A fuzzy inventory model has been developed for damageable items in Chap. 11. Chapter 12 shows the impact of blockchain on the retail supply chain. Using hybridized WQPSO in uncertain environment optimization of system reliability has been presented in Chap. 13. Chapter 14 shows the application of mediative fuzzy pythagorean algorithm to multi-criteria decision-making in medical diagnostic. Application of interval type-2 fuzzy numbers in nonlinear programming has been shown in Chap. 15. Chapter 16 presents applications of Deep Neural Network and Support Vector Regression in State of Energy (SoE) Estimation of Li-ion batteries.Chapter 17 developed a framework of parking system for reducing pollution in the campus. Chapter 18 presents the levelized cost of sustainable electricity production and storage. Principal component analysis and analytical hierarchical process has been applied in surface water quality assessment in Chap. 19.

This book will be helpful for academicians, researchers, industrialists, and business people in gaining the understanding of the emerging issues, practices, challenges in sustainable management with the accession of industry 4.0 and their solutions.

Aligarh, India — Irfan Ali
West Bengal, India — Prasenjit Chatterjee
West Bengal, India — Ali Akbar Shaikh
Noida, India — Neha Gupta
Alkharj, Saudi Arabia — Ali AlArjani

Acknowledgements

The editors wish to express their hearty gratitude to those who provided inputs, support, suggestions, and comments and assisted in editing and proofreading of this book.

This book would not have been possible without the valuable scholarly contributions of authors across the globe; the editors, therefore, acknowledge these efforts.

The editors acknowledge the endless support and motivation from their family members and friends.

The editors are very grateful to members who served in the editorial and review process of the book.

The editors are very much thankful to the entire Springer publishing team for keeping faith in them and showing the right path to accomplish this book.

Last but not the least, the editors take this opportunity to thank all the readers and expect that this book will continue to inspire and guide them for their future endeavours.

Irfan Ali
Prasenjit Chatterjee
Ali Akbar Shaikh
Neha Gupta
Ali AlArjani

Contents

About the Editors

Dr. Irfan Ali received B.Sc., M.Sc., M.Phil., and Ph.D. degrees from Aligarh Muslim University. He is currently working as an Assistant Professor with the Department of Statistics and Operations Research, Aligarh Muslim University. He received the Post Graduate Merit Scholarship Award during M.Sc. (statistics) and the UGC-BSR Scholarship awarded during the Ph.D. (statistics) programs in 2013. His research interests include applied statistics, survey sampling, supply chain networks and management, mathematical programming, and multiobjective optimization. He has supervised M.Sc., M.Phil., and Ph.D. students in operations research. He has completed a research project UGC–Start-Up Grant Project, UGC, New Delhi, India. He has published more than 75 research articles in reputed journals and serves as a Reviewer for several journals. He is currently editing two books to be published by Taylor France and Springer Nature. He is a Lifetime Member of various professional societies: Operational Research Society of India, Indian Society for Probability and Statistics, Indian Mathematical Society, and The Indian Science Congress Association. He delivered invited talks in several universities and institutions. He also serves as an Associate Editor for some journals.

Dr. Prasenjit Chatterjee is currently the Dean (Research and Consultancy) at MCKV Institute of Engineering, West Bengal, India. He has over 100 research papers in various international journals and peer-reviewed conferences including International Journal of Production Research, International Journal of Intelligent Systems, Expert Systems with Applications, Operations Management Research, Applied Soft Computing, Computers and Industrial Engineering, Socio-Economic Planning Sciences, Management Decision, Clean Technologies and Environmental Policy, Journal of Cleaner Production, Journal of Natural Fibers, Benchmarking: an International Journal, OPSEARCH, International Journal of Advanced Manufacturing Technology, Materials and Design, Robotics and Computer Integrated Manufacturing to name a few. He has authored and edited more than 15 books on intelligent decision-making, supply chain management, optimization techniques, risk and sustainability modelling. He has received numerous awards including Best Track Paper Award, Outstanding Reviewer Award, Best Paper Award, Outstanding

Researcher Award and University Gold Medal. Dr. Chatterjee is the Editor-in-Chief of Journal of Decision Analytics and Intelligent Computing. He has also been the Guest Editor of several special issues in different SCIE / Scopus / ESCI (Clarivate Analytics) indexed journals. He is the Lead Series Editor of "Disruptive Technologies and Digital Transformations for Society 5.0", Springer. He is also the Lead Series Editor of "Concise Introductions to AI and Data Science", Scrivener - Wiley; AAP Research Notes on Optimization and Decision Making Theories; Frontiers of Mechanical and Industrial Engineering, Apple Academic Press, co-published with CRC Press, Taylor and Francis Group and "River Publishers Series in Industrial Manufacturing and Systems Engineering". Dr. Chatterjee is one of the developers of two multiple-criteria decision-making methods called Measurement of Alternatives and Ranking according to COmpromise Solution (MARCOS) and Ranking of Alternatives through Functional mapping of criterion sub-intervals into a Single Interval (RAFSI).

Dr. Ali Akbar Shaikh is an Assistant Professor of Mathematics at The University of Burdwan, West Bengal, India. Earlier, he was a Postdoctoral Fellow at the School of Engineering and Sciences of Tecnológico de Monterrey, México. He has obtained the award SNI of level 1 (out of 0–3) presented by the National System of Researchers of México from Government of México in 2017. He obtained his PhD and MPhil in Mathematics from The University of Burdwan, and MSc in Applied Mathematics from University of Kalyani, India. He has published more than 55 research papers in different international journals of repute. His research interests include inventory control, interval optimisation, and particle swarm optimisation.

Dr. Neha Gupta is presently working as an Assistant Professor at Amity School of Business, AUUP. She obtained her M.Sc. (Operations Research), M.Phil. and PhD (Operations Research) from Aligarh Muslim University, Aligarh (INDIA). Her broad research area includes optimization and decision sciences. She is an Editorial Member of International Journal of Mathematics and Systems Science and International Journal of Data Mining, Modelling & Management, Inderscience Publications. Recently She has edited a special issue for Journal of Revenue and Pricing Management, Springer, and authored two books for international publishers like Taylor and Francis. She is a life member of the Operational Research Society of India and has more than 30 publications in journals of national and international repute. She has participated in several conferences of national and international level.

Dr. Ali AlArjani is an Assistant Professor at the Industrial Engineering Department at Prince Sattam bin Abdulaziz University, Alkharj, KSA. He did a Bachelor of Science degree in Mechanical Engineering from King Fahad University for Petroleum and Minerals, and master and PhD in Industrial Engineering from the USA. Dr. AlArjani's research background is in industrial optimization, mathematical modelling, energy, data clustering analysis and machine learning classification and prediction applications.

Chapter 1
A Systematic Review on the Emergence and Applications of Industry 4.0

Umar Muhammad Modibbo, Neha Gupta, Prasenjit Chatterjee, and Irfan Ali

Abstract The transition from traditional manufacturing to intelligent manufacturing is evolving as a result of technological advancement. Industry 4.0 has a broader spectrum of the domain and has brought a radical shift to innovative activities, though it is still at an early stage of maturity. The need to embrace emerging techniques and technology is imperative more than ever. This article presents a review on the emerging industrial fourth revolution called Industry 4.0 (I4.0). The research noticed no consensus among researchers on the number of technologies responsible for the radical industrial transformation. However, it highlights and discusses some unique enabling technologies of Industry 4.0 with applications areas. The study realized that in the future, robotics would take over the activities of manufacturing more intelligently and thus, leading to Industry 5.0. The article contributes to the bank of literature on Industry 4.0

Keywords Industry 4.0 · Smart manufacturing · Cyber-physical systems · Big data analytics · Industrial internet of things · Industrial robotics

U. M. Modibbo (✉)
Department of Statistics and Operations Research, Modibbo Adama University, PMB 2076, Yola, Nigeria
e-mail: umarmodibbo@mautech.edu.ng

N. Gupta
Amity Business School, Amity University Uttar Pradesh, Noida, India

P. Chatterjee
Department of Mechanical Engineering, MCKV Institute of Engineering, Howrah, West Bengal, India

I. Ali
Department of Statistics and Operations Research, Aligarh Muslim University, Aligarh 202002, India

I. Ali et al. (eds.), *Computational Modelling in Industry 4.0*,
https://doi.org/10.1007/978-981-16-7723-6_1

1 Introduction

Industry 4.0 is the radical change in goods and services due to technology referred to as the fourth industrial revolution. It emerges from market expansion, globalization, and competitiveness growth (Piccarozzi et al. 2018). It is, therefore, a technology-driven revolution for smart production, automation in business communications, and manufacturing (Zheng et al. 2021). Information technology (IT) advancement will lead the fourth industrial revolution to integrate manufacturing and service delivery in an automated and digital fashion (Rüßmann et al. 2015). Industry 4.0, in other words, is an intelligent production/manufacturing consisting of the Internet of things (IoT), cloud computing, and cyber-physical systems (CPS). In manufacturing, for instance, the whole system is adaptable via the CPS. In CPS, the communication is between the processing unit and the computer serving as a head connected as a network. Any change in the input parameter will adjust to the output parameter, thereby creating a feedback loop in the decision-making process (Schwab 2016; Kamble et al. 2018). The second part of the industry 14.0 component is cloud computing related to IT service delivery such as storage, databases, and servers located remotely and easily accessible from everywhere within a short possible time. The IoT is entirely human-free operations. It is a system or machine that connects and interacts without third-party intervention (Yu et al. 2017). Each machine has its unique code identifier connecting the other device and controls remotely at a higher speed and efficiency, making the entire operation bright and more intelligent, leading to the fourth industrial revolution (Lee et al. 2017).

Industry 4.0 brings new industrial perspectives on adding value chain products in the manufacturing system via different technologies and using little resources to realize maximum output. The innovations of industry 4.0 mainly affected the growth and employment sector (Piccarozzi et al. 2018). The application of technology in industries will improve and enhance productivity in the manufacturing, investment, revenue generation, and employment sectors (Rüßmann et al. 2015, Kamble et al. 2018). Therefore, industry 4.0 is a revolutionary shift from the traditional manufacturing norms (machine dominant) to the technological-based practice (intelligent manufacturing) (Oztemel and Gursev 2020).

One of the essential parts of any research is the literature review, through which a research gap can be identified and possibly bridge. Research is a continuous process; a thorough review of existing works, new ideas emerge and widen the research cycle's scope. Many authors studied industry 4.0 from different dimensions and perspectives. A corporate social responsibility studied industry 4.0 and the youth generation via a sustainable human resources management framework (Scavarda et al. 2019). This chapter seeks to review the literature extensively due to the growing interest surrounding industry 4.0. the paper will retrospect the kind of industry 4.0 revolutions such as virtual manufacturing, artificial intelligence (AI), cybernetics (man–machine interactions), machine interactions, industry 4.0 technologies, sustainability, blockchain, and Big data analytics.

2 Industry 4.0 Technologies

As mentioned earlier, technology is the back-born of industry 4.0. It is an essential aspect of the fourth industrial revolution. Nowadays, combining digital and manufacturing technologies integrates organizations' system vertically, horizontally, and end to end across the value chain (Klingenberg et al. 2019). The enabling industry 4.0 technologies are enormous, and there is no consensus on the number among researchers as it varies from one domain to another (Fettermann et al. 2018; Riel and Flatscher 2017). The enabling technologies include but are not limited to Cyber-physical systems, IoT, big data analytics, cloud servers, enterprises resource planning software, blockchain technology, visualization technology, artificial intelligence, and modelling and simulation, automation, and industrial robot system. The applications of these technology enablers for industry 4.0 has been studied extensively (Piccarozzi et al. 2018; Zheng et al. 2021; Rüßmann et al, 2015; Schwab 2016; Kamble et al. 2018; Yu et al. 2017; Lee et al. 2017; Oztemel and Gursev 2020; Scavarda et al. 2019; Klingenberg et al. 2019; Fettermann et al. 2018; Riel and Flatscher 2017; Nascimento et al. 2019; Alguliyev et al. 2018; Monostori et al. 2016; Nagy et al. 2018; Wong and Kim 2017; Yang et al. 2017; Li et al. 2017; Rao and Prasad 2018).

2.1 Cyber-Physical Systems

The CPS is the underlying technology representing Industry 4.0. It is a system that integrates the aspect of computation, automation, networking, and processing. It serves as a modular for manufacturing systems and can massively manufacture highly customized products (Nascimento et al. 2019). It encompasses mobile device technologies, robotics, AI, IoT, 3D printing, and cyber-security (Piccarozzi et al. 2018). It enables the fusion between the virtual and physical world via IoT connectivity. It has been applied in new smart product development and sensing (Alguliyev et al. 2018; Monostori et al. 2016). Additionally, it captures and exchanges data in real time. The companies employing CPS tend to satisfy customers and partners more efficiently (Nagy et al. 2018).

2.2 Internet of Things

Internet of Things improves efficiency and productivity through the combination of predictive analytics and intelligent machines. It can disseminate data and information significantly faster and accurately and provide sensing ability (Wong and Kim 2017). Data are collected, analyzed, and managed using IoT. It is also helpful in the manufacturing system as a control process (Yang et al. 2017). It helps synchronize

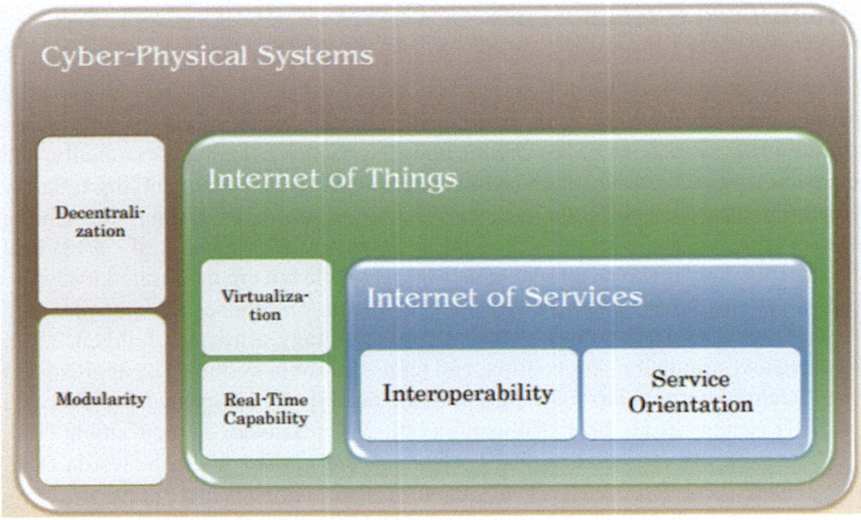

Fig. 1 Industry 4.0 framework (adapted from Oztemel and Gursev 2020)

and coordinate product and information flow effectively. The advantages of IoT in Industry 4.0 is enormous. Using RFID, the IoT provides more flexibility in logistics processes (Fig. 1).

2.3 Big Data and Analytics

Big data and analytics (BDA) is another technology enabler for Industry 4.0. The BDA technology collects and analyzes many available data with several filtering, capturing, and insights reporting techniques. It processes voluminous data with higher velocity and variability (Zheng et al. 2021). For any digital transformation in manufacturing sectors, powerful data analytics is imperative; therefore, expert skills to analyze massively available data are mandatory (Lee et al. 2017). The BDA technologies are real-time data collection tools; they support real-time decision-making, improving product quality, manufacturing flexibility and overall, predictive maintainability of equipment (Li et al. 2017). It has been used widely for finding fault and monitoring processes in manufacturing industries. Manufacturing agility can be derived from the big data intelligence exploitation and the high data quality and analytics expertise required to fulfil the task. However, there is a challenge of data confidentiality and consistency over the complex supply chain in the long run.

2.4 *Artificial Intelligence*

Artificial intelligence is a system that thinks like a human being. It rationally behaves under neural language processing, knowledge-based representation, machine learning, automated reasoning, robotics, and computer vision. Artificial intelligence technologies are helpful in product design, planning, control, scheduling, and quality improvement, among others (Rao and Prasad 2018). It has been used in the manufacturing industry for unmanned training, unmanned vehicles, aeroplanes, developing intelligent factories, and automatic prediction. Nowadays, AI system based is taking part in the enterprise decision-making process as robots, solving healthcare human resource crisis. Typical AI technologies include fuzzy logic, genetic algorithms, neural networks, random forest, expert systems, etc.

2.5 *Cybernetics and Human–Machine Interactions*

Cybernetics or man–machine interactions is an old interdisciplinary science. It studies control and communication concepts in living organisms, machines, and organizations, including self-organization (Novikov 2016). Its inherent feature focuses on how mechanical, biological, or digital systems process information and respond to it or change it based on feedback for better functioning. The feature is typical in industry 4.0 nowadays as modern manufacturing comprises the human component, the CPS, and cyber. IoT integrates all the parts as a connecting element. The human-cyber-physical interactions are regarded as one of the tools for intelligent manufacturing implementation (Qian et al. 2017). The cybernetic interactions with different systems and their originated year depicted in Fig. 2. The significant challenges of deploying IoT-based CPS in an organization is the fear of cyber-attacks and data theft due to malware that could defect the product quality. The cyber-attacks issues need to be addressed to improve the systems' trustworthiness and acceptability and guarantee industrial data safety.

2.6 *Blockchain Technology*

A blockchain is a database with a distributed and tamper-proof digital ledger of transactions based on a chain of linked blocks, including timestamps maintained by participating nodes that enable sharing of information among peers and provide a double-spending problem solution (Zheng et al. 2021; Ghadimi et al. 2019). Blockchain technology is applied in an integrated supply chain effectively with real-time material identification and tracking system. It automatically enables cross-

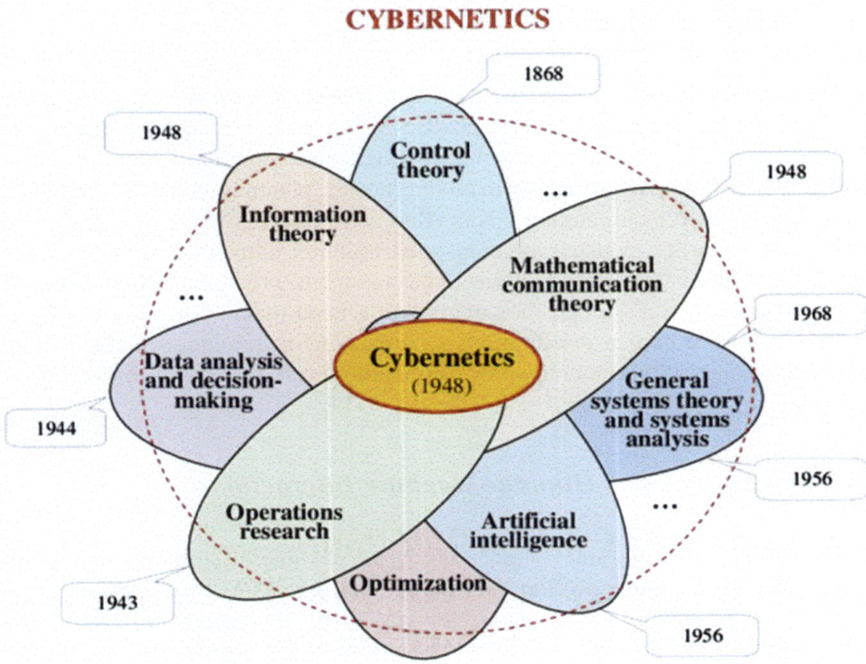

Fig. 2 The structural composition of cybernetics system (adapted from Novikov 2016)

organizational collaboration among stakeholders (Zheng et al. 2021; Ghadimi et al. 2019). Also, it is applicable in smart purchasing and supply chain management, smart contracts in negotiating energy supply agreement among enterprises (Fraga-Lamas and Fernández-Caramés 2019; Mohamed et al. 2019).

2.7 *Visualization Technology*

Visualization Technology (VT) is an indispensable part of automation. Softwares such as computer-aided design (CAD) are used in printing a three-dimensional object in a 3D printer. Visualizing a production process in 3D gives a clear picture or image of the process and makes understanding better. It is known as additive manufacturing, unlike the traditional manufacturing of drilling, cutting, grinding, etc. The VT is automated. The VT can be either augmented reality (AR) or virtual reality (VR). The AR is a set of humanly innovative techniques to interact with a computer capable of embedding virtual objects to interact and co-exist in the natural environment. The AR helps visualize the end product, thereby giving proper assessment and supporting the developmental product process.

On the other hand, VR is an interactive world created by a computer technology application. The user can control the virtual object and the entire virtual scene in real time (Zheng et al. 2021; Cohen et al. 2019). The VT has been applied in intelligent procurement and supply management, shop floor visualization, automated manual tasks guidance for operators, staff training, products alignment, etc.

2.8 Automation and Industrial Robot System

The operational process in manufacturing systems is mainly automated. Humans and machines share similar learning environment. For instance, robotics collaborate and learn from human. It has been said, in the future industrial robotics and robots will take over most if not all of the human activities. Modern robots exhibit some level of flexibility and autonomy, and they can interact with one another and human beings. Industrial robots reduce the cost of operations and are faster and intelligently do factory work. With an increase in programming knowledge, several researchers proposed different robots for industry 4.0 (Wang et al. 2016).

3 Conclusion

This article discusses and investigated the industry 4.0 current state-of-art—the primary industry 4.0 enabling technologies identified and discussed briefly to understand the subject matter. Applications areas of these technologies also elaborated. The study presents the scope of industry 4.0 in recent years. The study discovered no consensus on the different technologies enabling the fourth industrial revolution. However, it found the major players in the era of intelligent manufacturing and the digital economy. It understands that robotics will take over the performance of activities more intelligently shortly. The paper is limited because the number of growing research papers in industry 4.0 has not reported due to its diverse approaches and concept used by several authors in the literature. In future, more robust statistical tools will be employed with graphs and figures in illustrating the trend and the kind of research conducted on Industry 4.0.

References

Alguliyev R, Imamverdiyev Y, Sukhostat L (2018) Cyber-physical systems and their security issues. Comput Ind 100:212–223

Cohen Y, Naseraldin H, Chaudhuri A, Pilati F (2019) Assembly systems in Industry 4.0 era: a road map to understand Assembly 4.0. The Int J Adv Manuf Technol 105(9):4037–4054

Fettermann DC, Cavalcante CGS, Almeida TDD, Tortorella GL (2018) How does Industry 4.0 contribute to operations management? J Indus Prod Eng 35(4):255–268

Fraga-Lamas P, Fernández-Caramés TM (2019) A review on blockchain technologies for an advanced and cyber-resilient automotive industry. IEEE Access 7:17578–17598

Ghadimi P, Wang C, Lim MK, Heavey C (2019) Intelligent sustainable supplier selection using multi-agent technology: theory and application for Industry 4.0 supply chains. Comput Ind Eng 127:588–600

Kamble SS, Gunasekaran A, Gawankar SA (2018) Sustainable Industry 4.0 framework: a systematic literature review identifying the current trends and future perspectives. Process Saf Environ Prot 117:408–425

Klingenberg CO, Borges MAV, Antunes Jr JAV (2019) Industry 4.0 as a data-driven paradigm: a systematic literature review on technologies. J Manuf Technol Manage 32(3):570–592. https://doi.org/10.1108/JMTM-09-2018-0325

Lee MX, Lee YC, Chou CJ (2017) Essential implications of the digital transformation in Industry 4.0. J Sci Indus Res 76(8):465–467

Li D, Tang H, Wang S, Liu C (2017) A big data enabled load-balancing control for smart manufacturing of Industry 4.0. Cluster Comput 20(2):1855–1864

Mohamed N, Al-Jaroodi J, Lazarova-Molnar S (2019) Leveraging the capabilities of Industry 4.0 for improving energy efficiency in smart factories. IEEE Access 7:18008–18020

Monostori L, Kádár B, Bauernhansl T, Kondoh S, Kumara S, Reinhart G, Sauer O, Schuh G, Sihn W, Ueda K (2016) Cyber-physical systems in manufacturing. CIRP Ann 65(2):621–641

Nascimento DLM, Alencastro V, Quelhas OLG, Caiado RGG, Garza-Reyes JA, Rocha-Lona L, Tortorella G (2019) Exploring Industry 4.0 technologies to enable circular economy practices in a manufacturing context. J Manuf Technol Manage 0(3):607–627. https://doi.org/10.1108/JMTM-03-2018-0071

Nagy J, Oláh J, Erdei E, Máté D, Popp J (2018) The role and impact of Industry 4.0 and the Internet of Things on the business strategy of the value chain—the case of Hungary. Sustainability 10(10):3491

Novikov DA (2016) Systems theory and systems analysis. Systems engineering. In: Cybernetics. Springer, Cham, pp 39–44

Oztemel E, Gursev S (2020) Literature review of Industry 4.0 and related technologies. J Intell Manuf 31(1):127–182

Piccarozzi M, Aquilani B, Gatti C (2018) Industry 4.0 in management studies: a systematic literature review. Sustainability 10(10):3821

Qian F, Zhong W, Du W (2017) Fundamental theories and key technologies for smart and optimal manufacturing in the process industry. Engineering 3(2):154–160

Rao SK, Prasad R (2018) Impact of 5G technologies on industry 4.0. Wirel Pers Commun 100(1):145–159

Riel A, Flatscher M (2017) A design process approach to strategic production planning for Industry 4.0. In: European conference on software process improvement. Springer, Cham, pp 323–333

Rüßmann M, Lorenz M, Gerbert P, Waldner M, Justus J, Engel P, Harnisch M (2015) Industry 4.0: the future of productivity and growth in manufacturing industries. Boston Consult Group 9(1):54–89

Schwab K (2016) The fourth industrial revolution; world economic forum: Geneva, Switzerland. ISBN 9781944835002

Scavarda A, Daú G, Scavarda LF, Goyannes Gusmão Caiado R (2019) An analysis of the corporate social responsibility and the Industry 4.0 with focus on the youth generation: a sustainable human resource management framework. Sustainability 11(18):5130

Wang W, Zhu X, Wang L, Qiu Q, Cao Q (2016) Ubiquitous robotic technology for smart manufacturing system. Comput Intell Neurosci

Wong KS, Kim MH (2017) Privacy protection for data-driven smart manufacturing systems. Int J Web Serv Res (IJWSR) 14(3):17–32

Yang C, Lan S, Shen W, Huang GQ, Wang X, Lin T (2017) Towards product customization and personalization in IoT-enabled cloud manufacturing. Cluster Comput 20(2):1717–1730

Yu Z, Ouyang J, Li S, Peng X (2017) Formal modelling and control of cyber-physical manufacturing systems. Adv Mech Eng 9(10):1687814017725472
Zheng T, Ardolino M, Bacchetti A, Perona M (2021) The applications of Industry 4.0 technologies in manufacturing context: a systematic literature review. Int J Prod Res 59(6):1922–1954

Chapter 2
The Implementation Challenges to Circular Economy Via-Sectoral Exploration

Umar Muhammad Modibbo, Idiano D'Adamo, Piergiuseppe Morone, and Irfan Ali

Abstract A circular economy (CE) is one of the economic systems that uses a systemic approach to eliminate waste and enhance resources utilization through redefining growth for sustainable development of businesses, society, and the environment. It closes the gap between production and the lifecycle of the ecosystems. This study examines and reviews the challenges facing the CE implementation and its associated opportunities through sectoral exploration. The chapter highlights the progress and challenges of implementing CE strategies in the sectors related to food, chemicals, metals and minerals, electronics, and building and infrastructure. The study generates and analyses an association map that indicates four significant clusters in the CE literature over the years. It has established further that much attention had not given to cross-case comparison in the literature of implementing CE models. Also, the future reliable and robust CE strategy is shifting to ICT and digital economy of scope rather than scale. This research is implicative in terms of fundamental research and investment. Industrial actors can achieve these through collaboration and efficient communication with stakeholders and access to data availability.

U. M. Modibbo
Department of Statistics and Operations Research, Modibbo Adama University of Technology, Yola PMB 2076, Yola, Nigeria
e-mail: umarmodibbo@mautech.edu.ng

I. D'Adamo (✉)
Department of Computer, Control and Management Engineering, Sapienza University of Rome, Via Ariosto 25, 00185 Roma, Italy
e-mail: Idiano.dadamo@uniroma1.it

P. Morone
Department of Law and Economics, UnitelmaSapienza—University of Rome, Viale Regina Elena 295, 00161 Roma, Italy
e-mail: piergiuseppe.morone@unitelmasapienza.it

I. Ali
Department of Statistics and Operations Research, Aligarh Muslim University, Aligarh 202002, India
e-mail: irfii.st@amu.ac.in

I. Ali et al. (eds.), *Computational Modelling in Industry 4.0*,
https://doi.org/10.1007/978-981-16-7723-6_2

Keywords Circular economy · Economic models · Implementation challenges · Innovation · Opportunities · Technology

1 Introduction

Circular economy (CE) is gaining widespread momentum in sustainable economic growth achievement from the extraction and generation of waste and resource Kalmykova et al. (2018). Several actors engage in exploiting the unique opportunities created by the CE paradigm. They include businesses, policymakers, and the academic community. Recently, the coordinators of CE launched special issues termed 'Advances in the CE' They sought to study the progress through business models, policies, and industrial innovations brought by the CE concepts (Ghisellini et al. 2016; Singh and Ordoñez 2016; Singh et al. 2019). A study recently conducted to highlight the importance of CE applications and establish a possible future direction of research in this domain (Babbitt et al. 2018).

This article aims to contribute within the broader CE context focussing mainly on specific sector-wise challenges and cross-cutting themes. Recently, it has been established that the manufacturing sector, business, construction, and waste electric and electronic equipment have faced challenges (Acerbi and Taisch 2020; Pizzi et al. 2021; Osobajo et al. 2020; Bressanelli et al. 2020). Therefore, a unified CE research is proposed based on these challenges (Borrello et al. 2020; Principato et al. 2019). However, the existing literature neither thoroughly integrates the challenges learned from those sectors nor explores gaps in CE analysis methodology critically. This study aims to bridge these gaps by critically evaluating challenges, assessing existing intersections among them and prioritizing future advancement in research and technology. This work's essential contribution is creating and bringing to the fore the barriers and opportunities for CE implementation across sectors via critically reviewing existing knowledge.

2 Mapping Approach and Themes Identification

This section explores the critical scoring analysis, the opportunities and challenges facing CE themes, and its current trend. The review is focussing on the implementation of CE and its applications in modern society. The terms 'application', 'implementation', 'case study', 'sector', 'operation' are searched in association with 'circular economy' using AND Boolean operator in the Web of Science. The search returned over 300 results. The software for visualizing networks is called VOSviewer version 1.6.16 was used to analyze the downloaded reference terms via the association of keywords. For consistency, similar terms synchronized—the critical challenges and opportunities identified and integrated for evaluating the application and implementation of CE outcomes. The generated association map indicates four significant clusters in the CE literature:

- The industrial symbiosis, ecology, and evolution cluster represented by a yellow colour,
- CE business models, sustainability, supply chain, logistics, barriers, etc., by red colour,
- Bioenergy, bio-economy, renewable energy, biomass, and carbon footprint, etc. denoted by blue colour,
- Metals, minerals, construction, resource recovery, recycling, and other infrastructural systems are shown in green colour (see Fig. 1).

The studies focussed mostly on reuse and recycling, material flow analysis, and cradle-to-cradle perspectives. The blue cluster studies are closely linked with technology to recover the bio-based energy system (anaerobic digestion) and environmental assessment. The fragment in the sectoral studies informed that much attention had not given to cross-case comparison in the literature of implementing CE visa-vee the challenges and opportunities identified.

If critically observing the yellow clusters, more needs to be done in industrial ecologies to link CE research with the existing business models. Industrial

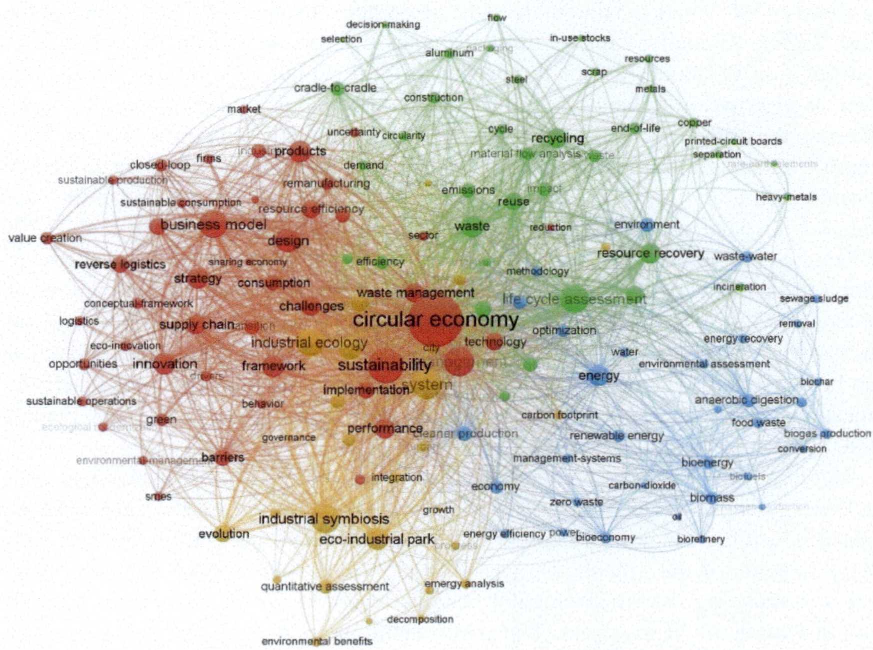

Fig. 1 Keyword association map for circular economy implementation studies

ecology can be seen closely linked to the CE, which is logical considering their standard conceptual bases and the overlapping assessment methodology in that domain. Several industrial ecologies focused on CE studies are revolving around 'Eco-Industrial Parks' and symbiosis from the business perception. Based on the close connectivity of industrial ecology and the assessment methodology such as life cycle assessment, it motivates to analyze key specific sectors like food and its waste, buildings and infrastructure, electronics and associated e-waste, chemicals, metals, and minerals. On the other hand, four primary themes, such as data, models, stakeholder engagement, and business and innovation, will be analyzed regarding the cross-cutting themes.

3 Food Sector and Its Associated Waste

The system of food is central to the studies of CE for two significant reasons. Firstly, the growing global population cannot do without food security. Hence it is critical to living things and economic vitality (Singh et al. 2021). Secondly, because of resources scarcity, they face formidable challenges. Its supply chains consume energy significantly (Pimentel et al. 2008; Del Borghi et al. 2020). The excess nutrients released by freshwater resources contribute to anthropogenic greenhouse gas releases by closely 15%, which is vulnerable to the ecosystems (Singh et al. 2021; Del Borghi et al. 2020). Annually, over 1.3 billion tonnes of food wasted due to the lack of consumption of about 40% of the produced food (Singh et al. 2021). Most food waste is disposed of in landfills, creating environmental degradation and ultimately affecting the climate (Lundie and Peters 2005). CE strategies can transform the food supply chain via the transformation of waste with technological advancement. A considerable literature focuses on a closed-loop supply chain to minimize the loss and maximize potential energy and nutrients (Borrello et al. 2017; Barros et al. 2020; D'Adamo et al. 2020, 2021). Some challenges of the food sector are the inadequate adoption of recovery technologies, stakeholders' disconnection, and streaming of materials. Multi-value streams can be generated by integrating biorefineries as an alternative means; this is an opportunity to explore more in the food sector. Food waste can be converted to biodiesels and biomethane through fermentation and anaerobic digestions (Singh et al. 2021, D'Adamo et al. 2021, Mantalovas et al. 2019, Ferella et al. 2019; Farooq et al. 2021).

The key benefit of environmentally transformed food waste is displacing fossil fuel energy carriers, electricity generation, and synthetic fertilizers. The displacement of fossil fuel, avoiding releases of attendant methane, and landfilling will result in the benefits of the greenhouse gas lifecycle. Stakeholders must show commitments to realizing the environmental benefits of food systems circulation through adopting modern CE strategies. The government, on its part, should provide viable policies to enable stakeholders to overcome the barriers of CE. Certain competing and conflicting objectives have to optimize holistically to pave ways for eco-friendly and economically viable technology-based CE research.

4 Chemicals Sector

The two essential domain of CE implementation in the chemicals sector is the pre and pos consumers (Singh et al. 2021). In the former, feedstock materials are recovered and reused. At the same time, in the latter, the recycling loop is limited to specific non-hazardous waste such as plastics and textile due to challenges facing CE practices (Farooq et al. 2021; Garcia and Robertson 2017). In Europe alone in 2016, the production of plastic materials amounted to about 60 million tonnes annually, and there is a short and midterm increase expectation (Somoza-Tornos et al. 2020). Therefore, on a massive-bases, plastics represent the primary product in the chemical industry. Plastics are versatile, and the most widely demanded type is polyethylene representing 30% of the total production (Somoza-Tornos et al. 2020). The packaging is in the form of bags, bottles, and films. The high percentage of unrecovered plastic packaging renders severe environmental problems. With modern machines and factories, CE practices will enjoy continuity. Practically, the chemical process of conversion gives birth to by-products, side reactions, etc., as wastes during the pre-consumer face. These wastes can be recycled and reused via separation processes such as distillation in petro and biochemical plants (Ioannidou et al. 2020; Kayode and Hart 2019).

In a substantial chemical installation, the linked value chains enable chemical companies to run synthesis processes. One process can be used directly in another. What BASF- a giant chemical company in Germany, calls the '*Verbund* concept'. For instance, in the recycling process of some plastics, it melts and degrades, which can only be reused for lower-valued applications such as building materials, carpets, and clothing. It is challenging to use them in producing new packaging items. Alternatively, incineration can be used as energy valorization even though CO_2 is lost in the volume (Lewtas 2007).

One of the exciting chemical development processes and critical research areas on pre-consumer practices of CE is the valorization of by-products outside its industry. It can be done by industrial symbiosis or eco-industrial parks for environmental and mutual benefits (Chen et al. 2020; Guo et al. 2016). On the other hand, post-consumer most important CE practices barriers are the chemical contamination and its associated risk of life safety (Singh et al. 2021). Contaminants are added by design to enhance product performance but, they inhibit the recycling downstream (Clark et al. 2016). Chemical engineering innovations can be used in chemical industries to promote CE practices, such as valuable metals recovery from e-waste and 'metallurgical waste' and the nutrients recovery from wastewater treatment (Singh et al. 2021; Kalmykova et al. 2018; Zimmerman et al. 2020).

There are four primary challenges to the implementation of CE strategies in the chemical sector. They include chemical mixtures, missing performance and composition data, chemical durability, which is not a desired trait, and hazardous chemical additives that require linear disposal. On the other hand, the opportunities are building on existing matrices and MFG practices for CE, the critical enabler for reusing material in other sectors and green chemistry (Singh et al. 2021).

5 Metals and Minerals Sector

Over the last 50 years, there is a resource-intense decline in the potential for the materials, minerals, and metals industries to move towards a CE—a fewer input with massive production output (Cucciniello and Cespi 2018; Worrell et al. 1997; Cleveland and Ruth 1998). However, the main driver in this sector is economics; as pointed in the literature, dissociating economic growth from resource extraction is fundamental to the foundation of the CE (Behrens et al. 2007; Krausmann et al. 2018; Dorninger et al. 2021; Haberl et al. 2020; Schandl et al. 2020). However, there is increased pressure in this sector due to decreased ore grades and increased total consumption. The focus in the literature for the opportunities in this sector include remanufacturing, recycling, reuse, product redesign, additive manufacturing, and material selection, among others (Singh and Ordoñez 2016; Colorado et al. 2020; Shanmugam et al. 2020; Mesa et al. 2020; Martínez Leal et al. 2020; Dey et al. 2020; Ferrari et al. 2021; D'Adamo and Rosa 2019). However, one of the critical challenges is translating these practices from theoretical to actual production and manufacturing applications (Babbitt et al. 2018; Singh et al. 2021). Recovery rates remain low for most materials; however, recycling remains one of the prime potential areas. Thus, the materials sector can serve as a sink for end-of-life resources. There are three fundamental challenges to the implementation of CE strategies in the metals and minerals sector. They are compositional variability in the stream of recovered resources, inadequate recycling technology, and fluctuation in secondary markets for materials recovered. In comparison, the opportunities can be seen as creating novel by-product streams (Sun et al. 2017; Neves et al. 2020; Henriques et al. 2021).

6 Electronics Sector and Its E-Waste

The electronics sector is an emerging CE case studies related to non-traditional items such as mobile, IoT, clothing, jewellery, toys, household gadgets and appliances, and, of course, healthcare system monitoring tools (Ryen et al. 2015; Hischier et al. 2020; Kasulaitis et al. 2020). Collectively, innovative design and technology strategies influence the strategies of CE in the electronics sector. For example, toxic or emerging containments can be easily eliminated via strategy enhancement. New biodegradable and non-toxic materials such as carbon and pyrene can be integrated to serve as the primary opportunities towards pushing the industry to realize zero-waste sustainable management (D'Adamo et al. 2020; Awasthi et al. 2019). However, the success of these CE strategies in the electronics sector depends on the behaviour and decisions of end-users heavily as a critical stakeholder group. Most attention is given to product lifespan extension and the reuse option in the literature (Coughlan et al. 2018; Wever 2012). Other studies emphasized the link between existing e-waste management and CE strategies in the electronics sector (Gollakota et al. 2020; Kumar et al. 2017). This sector's significant challenges include the evolution of various products and materials, the materials uncertainty and products short lifespan, limited recovery of materials, discourage reusing materials, the lag between waste management infrastructure and existing policies, and the existence of rebound effects.

On the other hand, opportunities in this sector include the availability of data and technology that enhances easy access to information sharing and communication. The uncertainty from the material stream eases with the availability of decision tools. There is a possibility of models sharing in the sector (Singh et al. 2021).

7 Buildings and Infrastructure Sector

Construction sectors consume mega resources, and demolition of the building creates waste in the environment. Heavy-duty construction companies extract minerals exceeding 10 billion metric tonnes annually. It is the only sector with the fastest growth rate during the past century (Cucurachi et al. 2018). Numerous frameworks for CE strategies have been proposed in the built environment. The proposals are with the purview of the Ellen MacArthur Foundation that includes six ways to apply circularity: regenerate, share, optimize, loop, virtualize, and exchange (Foster 2020). Built environment-specific strategies include: reducing construction and design waste and maximizing its value, adaptability design for long life, and component reuse designing. Digital technology can enable construction and design practices in CE.

Research on CE strategies for the built environment is typically focussed on a single strategy, such as using recycled content in new materials, reuse of components, or modularization. New business models will also be required to implement CE strategies in the marketplace. Using the ReSOLVE framework for buildings and infrastructure in new and effective ways will be challenging. However, quantitative assessments of CE strategies' life cycle environmental impacts will be a crucial component of their implementation. The challenges include long life, size and complexity, and a trade-off between CE strategies and cost. At the same time, the opportunities are innovative building materials, new designs and business solutions, the evaluation of life cycle impacts, and analyzing the system implications.

8 Conclusion

There is an increasingly growing interest in CE studies. The literature on this subject matter is extremely significant in terms of works published. For this reason, the proposed critical analysis of CE's impact in several analysis fields is timely and much needed. This study tries to bring some implementation challenges central to the CE topic and opportunities in some sectors implementing CE. It has been established that most at times, the knowledge learnt from one sector do not tend to be useful in another. The literature is relatively fragmented due to a lack of validation in real-life cases studies. In the existing literature of the definition of the CE, few single sectors enjoy the opportunities and CE implementation strategies. Clean technology is expected to facilitate the shift towards the transition to a CE. There is an inadequate methodology for the transition to the CE. This research is implicative in terms of fundamental research and investment. The digital economy can support the development of the CE. It should try to achieve economic scope rather than scale. CE challenges can be addressed through enhanced collaboration with stakeholders and access to data availability.

The results highlight some aspects:

- the point of connection between social change and technological development;
- the need for studies that demonstrate the actual circularity and not the favouring of circular rebound phenomena;
- analysis of consumers to test their willingness to pay for these sustainable products;
- changes in industrial processes not only in developed countries but also in developing countries;
- the definition of indicators that can support decision-makers.

Waste management, waste prevention, and resources efficiency is the direction to implement to foster a circular transformation of resources where political support is required to mitigate the higher costs and encourage implementing these practices to make them competitive in the future.

References

Acerbi F, Taisch M (2020) A literature review on circular economy adoption in the manufacturing sector. J Clean Prod 123086

Awasthi AK, Li J, Koh L, Ogunseitan OA (2019) Circular economy and electronic waste. Nat Electron 2(3):86–89

Babbitt CW, Gaustad G, Fisher A, Chen WQ, Liu G (2018) Closing the loop on circular economy research: from theory to practice and back again. Resour Conserv Recycl 135:1–2

Barros MV, Salvador R, De Francisco AC, Piekarski CM (2020) Mapping of research lines on circular economy practices in agriculture: from waste to energy. Renew Sustain Energy Rev 131:109958

Behrens A, Giljum S, Kovanda J, Niza S (2007) The material basis of the global economy: worldwide patterns of natural resource extraction and their implications for sustainable resource use policies. Ecol Econ 64(2):444–453

Borrello M, Caracciolo F, Lombardi A, Pascucci S, Cembalo L (2017) Consumers' perspective on circular economy strategy for reducing food waste. Sustainability 9(1):141

Borrello M, Pascucci S, Cembalo L (2020) Three propositions to unify circular economy research: a review. Sustainability 12(10):4069

Bressanelli G, Saccani N, Pigosso DC, Perona M (2020) Circular economy in the WEEE industry: a systematic literature review and a research agenda. Sustain Prod Consumption 23:174–188. https://doi.org/10.1016/j.spc.2020.05.007

Chen TL, Kim H, Pan SY, Tseng PC, Lin YP, Chiang PC (2020) Implementation of green chemistry principles in circular economy system towards sustainable development goals: challenges and perspectives. Sci Total Environ 716:136998

Clark JH, Farmer TJ, Herrero-Davila L, Sherwood J (2016) Circular economy design considerations for research and process development in the chemical sciences. Green Chem 18(14):3914–3934

Cleveland CJ, Ruth M (1998) Indicators of dematerialization and the materials intensity of use. J Ind Ecol 2(3):15–50

Colorado HA, Velásquez EIG, Monteiro SN (2020) Sustainability of additive manufacturing: the circular economy of materials and environmental perspectives. J Market Res 9(4):8221–8234

Coughlan D, Fitzpatrick C, McMahon M (2018) Repurposing end of life notebook computers from consumer WEEE as thin client computers—a hybrid end of life strategy for the circular economy in electronics. J Clean Prod 192:809–820

Cucciniello R, Cespi D (2018) Recycling within the chemical industry: the circular economy era. Recycling 3(2):22

Cucurachi S, van der Giesen C, Guinée J (2018) Ex-ante LCA of emerging technologies. Procedia CIRP 69:463–468

D'Adamo I, Rosa P (2019) A structured literature review on obsolete electric vehicles management practices. Sustainability 11(23):6876

D'Adamo I, Falcone PM, Huisingh D, Morone P (2021) A circular economy model based on biomethane: what are the opportunities for the municipality of Rome and beyond? Renew Energy 163:1660–1672

D'Adamo I, Falcone PM, Imbert E, Morone P (2020) A socio-economic Indicator for EoL strategies for bio-based products. Ecol Econ 178:106794

Del Borghi A, Moreschi L, Gallo M (2020) Circular economy approach to reduce water–energy–food nexus. Curr Opin Environ Sci Health 13:23–28

Dey PK, Malesios C, De D, Budhwar P, Chowdhury S, Cheffi W (2020) Circular economy to enhance sustainability of small and medium-sized enterprises. Bus Strateg Environ 29(6):2145–2169

Dorninger C, Hornborg A, Abson DJ, Von Wehrden H, Schaffartzik A, Giljum S, Wieland H (2021) Global patterns of ecologically unequal exchange: implications for sustainability in the 21st century. Ecol Econ 179:106824

Farooq A, Haputta P, Silalertruksa T, Gheewala SH (2021) A framework for the selection of suitable waste to energy technologies for a sustainable municipal solid waste management system. Front Sustain 2:27

Ferella F, Cucchiella F, D'Adamo I, Gallucci K (2019) A techno-economic assessment of biogas upgrading in a developed market. J Clean Prod 210:945–957

Ferrari AM, Volpi L, Settembre-Blundo D, García-Muiña FE (2021) Dynamic life cycle assessment (LCA) integrating life cycle inventory (LCI) and Enterprise resource planning (ERP) in an industry 4.0 environment. J Clean Prod 286:125314

Foster G (2020) Circular economy strategies for adaptive reuse of cultural heritage buildings to reduce environmental impacts. Resour Conserv Recycl 152:104507

Garcia JM, Robertson ML (2017) The future of plastics recycling. Science 358(6365):870–872

Ghisellini P, Cialani C, Ulgiati S (2016) A review on circular economy: the expected transition to a balanced interplay of environmental and economic systems. J Clean Prod 114:11–32

Gollakota AR, Gautam S, Shu CM (2020) Inconsistencies of E-waste management in developing nations—facts and plausible solutions. J Environ Manag 261:110234

Guo B, Geng Y, Sterr T, Dong L, Liu Y (2016) Evaluation of promoting industrial symbiosis in a chemical industrial park: a case of Midong. J Clean Prod 135:995–1008

Haberl H, Wiedenhofer D, Virág D, Kalt G, Plank B, Brockway P, Creutzig F (2020) A systematic review of the evidence on decoupling of GDP, resource use and GHG emissions, part II: synthesizing the insights. Environ Res Lett 15(6):065003

Henriques J, Ferrão P, Castro R, Azevedo J (2021) Industrial symbiosis: a sectoral analysis on enablers and barriers. Sustainability 13(4):1723

Hischier R, Reale F, Castellani V, Sala S (2020) Environmental impacts of household appliances in Europe and scenarios for their impact reduction. J Clean Prod 267:121952
Ioannidou SM, Pateraki C, Ladakis D, Papapostolou H, Tsakona M, Vlysidis A, Koutinas A (2020) Sustainable production of bio-based chemicals and polymers via integrated biomass refining and bioprocessing in a circular bioeconomy context. Bioresour Technol 307:123093
Kalmykova Y, Sadagopan M, Rosado L (2018) Circular economy—from review of theories and practices to development of implementation tools. Resour Conserv Recycl 135:190–201
Kasulaitis B, Babbitt CW, Tyler AC (2020) The role of consumer preferences in reducing material intensity of electronic products. J Ind Ecol
Kayode B, Hart A (2019) An overview of transesterification methods for producing biodiesel from waste vegetable oils. Biofuels 10(3):419–437
Krausmann F, Lauk C, Haas W, Wiedenhofer D (2018) From resource extraction to outflows of wastes and emissions: the socio-economic metabolism of the global economy, 1900–2015. Glob Environ Chang 52:131–140
Kumar A, Holuszko M, Espinosa DCR (2017) E-waste: an overview on generation, collection, legislation and recycling practices. Resour Conserv Recycl 122:32–42
Lewtas J (2007) Air pollution combustion emissions: characterization of causative agents and mechanisms associated with cancer, reproductive, and cardiovascular effects. Mutat Res/Rev Mutat Res 636(1–3):95–133
Lundie S, Peters GM (2005) Life cycle assessment of food waste management options. J Clean Prod 13(3):275–286
Mantalovas K, Di Mino G (2019) The sustainability of reclaimed asphalt as a resource for road pavement management through a circular economic model. Sustainability 11(8):2234
Martínez Leal J, Pompidou S, Charbuillet C, Perry N (2020) Design for and from recycling: a circular ecodesign approach to improve the circular economy. Sustainability 12(23):9861
Mesa J, González-Quiroga A, Maury H (2020) Developing an indicator for material selection based on durability and environmental footprint: a circular economy perspective. Resour, Conserv Recycl 160:104887
Neves A, Godina R, Azevedo SG, Matias JC (2020) Carbon dioxide recovery through industrial and urban symbiosis. In: Proceedings of the 2020 9th international conference on industrial technology and management (ICITM). IEEE, pp 171–175
Osobajo OA, Oke A, Omotayo T, Obi LI (2020) A systematic review of circular economy research in the construction industry. Smart Sustain Built Environ. https://doi.org/10.1108/SASBE-04-2020-0034
Pimentel D, Williamson S, Alexander CE, Gonzalez-Pagan O, Kontak C, Mulkey SE (2008) Reducing energy inputs in the US food system. Hum Ecol 36(4):459–471
Pizzi S, Corbo L, Caputo A (2021) Fintech and SMEs sustainable business models: reflections and considerations for a circular economy. J Clean Prod 281:125217
Principato L, Ruini L, Guidi M, Secondi L (2019) Adopting the circular economy approach on food loss and waste: the case of Italian pasta production. Resour Conserv Recycl 144:82–89
Ryen EG, Babbitt CW, Williams E (2015) Consumption-weighted life cycle assessment of a consumer electronic product community. Environ Sci Technol 49(4):2549–2559
Schandl H, Lu Y, Che N, Newth D, West J, Frank S, Hatfield-Dodds S (2020) Shared socio-economic pathways and their implications for global materials use. Resour Conserv Recycl 160:104866
Shanmugam V, Das O, Neisiany RE, Babu K, Singh S, Hedenqvist MS, Ramakrishna S (2020) Polymer recycling in additive manufacturing: an opportunity for the circular economy. Mater Circ Econ 2(1):1–11
Singh J, Sung K, Cooper T, West K, Mont O (2019) Challenges and opportunities for scaling up upcycling businesses—the case of textile and wood upcycling businesses in the UK. Resour Conservation Recycl 150:104439
Singh S, Babbitt C, Gaustad G, Eckelman MJ, Gregory J, Ryen E, Seager T (2021) Thematic exploration of sectoral and cross-cutting challenges to circular economy implementation. Clean Technol Environ Policy 1–22

Singh J, Ordoñez I (2016) Resource recovery from post-consumer waste: important lessons for the upcoming circular economy. J Clean Prod 134:342–353
Somoza-Tornos A, Gonzalez-Garay A, Pozo C, Graells M, Espuña A, Guillén-Gosálbez G (2020) Realizing the potential high benefits of circular economy in the chemical industry: ethylene monomer recovery via polyethylene pyrolysis. ACS Sustain Chem Eng 8(9):3561–3572
Sun L, Li H, Dong L, Fang K, Ren J, Geng Y, Liu Z (2017) Eco-benefits assessment on urban industrial symbiosis based on material flows analysis and emergy evaluation approach: a case of Liuzhou city, China. Resour Conserv Recycl 119:78–88
Wever R (2012) Design research for sustainable behaviour. J Des Res 10(1–2):2012. Post-print version
Worrell E, Price L, Martin N, Farla J, Schaeffer R (1997) Energy intensity in the iron and steel industry: a comparison of physical and economic indicators. Energy Policy 25(7–9):727–744
Zimmerman JB, Anastas PT, Erythropel HC, Leitner W (2020) Designing for a green chemistry future. Science 367(6476):397–400

Chapter 3
Challenges in the Virtual Workspace by the Inception of Industry 4.0: A Conceptual Frame Work

Aakash Khindri and Samridhi Tanwar

Abstract With the advent of the fourth industrial revolution, Industry 4.0, there has been an increasing interest in the study of the virtual workspace. The concepts of virtual workplace and teams are quite revolutionary, offering various benefits over traditional way of work. However, it is not free from challenges and problems, many of them have risen to the surface in the recent pandemic situation. The purpose of this paper is to develop a conceptual framework highlighting the challenges of the same. In doing so, we qualitatively review the empirical work related to the virtual work setting and associated issues. Finally, important areas for future research in the field of the virtual workplace are addressed as well.

Keywords Virtual teams · Virtual leadership · Challenges · Teamwork · Performance management

1 Introduction

A new team phenomenon known as virtual teams emerged during the 1990s (Furst et al. 2004). But this novice concept has evolved at a slow pace till the mid-90s. Today, due to the unprecedented technological developments in telecommunication technology and the advent of the internet, virtual teams are in vogue. Mainly, virtual teams consist of a group of people working together that are distributed across national boundaries and connected through advanced information technology such as e-mail, instant messaging, and videoconferencing (Scott and Wildman 2015). The current COVID-19 pandemic situation made it necessary to form and operationalize the virtual work environment to survive. The organizations that have adapted rapidly to this requirement have secured a competitive edge over others (Bergiel et al. 2008). An example of this can be observed in the experience of considerable growth of

A. Khindri
Department of Management Studies, Indian Institute of Technology, Roorkee, Uttarakhand, India

S. Tanwar (✉)
The Technological Institute of Textile and Sciences, Bhiwani, Haryana, India

I. Ali et al. (eds.), *Computational Modelling in Industry 4.0*,
https://doi.org/10.1007/978-981-16-7723-6_3

productivity in major companies like Hewlett-Packard (HP) and General Electric due to an early adoption of the concept of virtual teams (Zuofa and Ochieng 2017).

Virtual team members participate in a variety of collaborative activities, both formal and informal using technology and videoconferencing platforms such as Zoom, Skype, Cisco WebEx, Google Meet, and Microsoft Teams (Smith and Ruiz 2020). All these platforms are continuously updating their software packages to provide a real-time, face-to-face meeting experience while maintaining the user-friendliness of the interface. In China, almost 90% of the workforce relies on WeChat as an important tool for workplace communication (Chadee et al. 2021). Virtual teams allow people to come together from different corners of the globe, IT provides the platform, but ultimately it is the people that make the projects work (Vakola and Wilson 2004). The concept of working alone at home is quite fascinating to some individuals. They like the flexibility of working in comfortable clothes, surrounded by their family members and pets. Along with the benefits of knowledge management and flexibility the idea of telework also eliminates the requirement of a cubicle, office, or parking space (Ferreira et al. 2021). It can also contribute to lesser air pollution and traffic congestion, but some people might feel lonely and isolated in such arrangements. The lack of interaction with colleagues and reduced privacy bothers them (Hunsaker and Hunsaker 2008).

Since the inception of the concept, numerous studies have explored and highlighted the advantages and disadvantages of virtual teams over co-located teams. Although there are certain potential pitfalls for virtual teams, the research data shows that the benefits of the virtual team process eventually outweigh the pitfalls (Robertson 2006). But during the current pandemic situation the idea of 'work from home' and virtual teams has been bolstered (Chadee et al. 2021), and various difficulties in the virtual work environment have also surfaced since then. Despite a few earlier studies examining the issues and challenges faced by virtual teams, there has been limited work in exploring the difficulties of virtual teams due to the scant adoption of this mode of work (Smith and Ruiz 2020). A recent study by Ferreira et al. (2021) highlighted the fact that most of the research associated with telework tends to be reporting findings on a single vector of analysis. In the present times, it is of utmost importance to explore these challenges in-depth so that the effectiveness of virtual teams can be improved further by finding the right solutions.

This paper has two goals: (1) To examine the challenges faced by all virtual teams and (2) Segregate these challenges into a few broad categories based on their relatedness as demonstrated in earlier studies so that different types of challenges can be identified clearly. A qualitative literature review was performed to achieve our goals, including 40 relevant studies from the last two decades. The decision regarding the inclusion of studies in review was made through consensus among the authors. The number of studies included in the review was restricted to 40 because a theoretical saturation of broad categories of challenges has been reached, and no new category was observed to emerge beyond that point. To ensure a comprehensive review, studies based on both quantitative and qualitative approaches were included, along with ten review-based studies. In addition to this, the results from this literature review also help to identify the gaps and opportunities for future research.

2 Challenges

Virtual teams are affected by a wide range of different factors. Based on the studies included in this literature review, some factors are clubbed together on the basis of their interrelatedness as established by studies and various categories emerged such as geographic factors, temporal factors, cultural factors, communication factors, technological factors, leadership factors, and those factors which are not adequately related to any of the above-mentioned categories are put under other team-related factors. The subsequent sections of the paper discuss these factors in detail, along with the relevant studies to that factor.

2.1 Geographic Factors

The basic definition of virtual teams highlights the geographically dispersed nature of the team members. It can be commonly stated that virtual teams have discontinuities of several kinds and spatial discontinuity is one of them (Zuofa and Ochieng 2017). The geographic distance among teammates and collaborators can pose several challenges for virtual teams. A study argues that the lack of face-to-face interaction is one of the three main challenges faced by virtual teams (Olariu and Aldea 2014). It has been observed that geographic factors not only act as a challenge in itself but also leads to several other challenges. The presence of other colleagues has a sense of motivation, which is established by 'social facilitation' effects. Also, if some members of the team are working virtually, it not only impacts those employees but also has a bearing over employees working in a physical setup (Lippe and Lippényi 2020). However, these effects are difficult to observe when individuals are working alone, which poses an additional challenge to collaboration (Smith and Ruiz 2020). Moreover, the verbal and nonverbal cues and the synergies that usually build with face-to-face interaction are also missing in virtual work environments (Zuofa and Ochieng 2017).

Lillian (2013) argues that the lack of overlapping working hours can produce a burden related to coordination on the team members as well as on E-leaders. Also, monitoring and maintaining an awareness of members' work progress at remote locations without the facility to have a casual 'look over their shoulders' is another problem for the leaders (Smith and Ruiz 2020). The leader or the facilitator often feels the compulsion to respond quickly to team requests to maintain the level of perceived social support and social closeness that is prevalent in traditional collocated teams (Lillian 2013).

In the absence of face-to-face interaction, the chances of cropping up of defensive routines, such as a human tendency to be unwilling to admit ignorance or failure, can also increase since the members of the team are selected on the basis of their expertise in a particular area. The lack of personal interaction can create an environment that tempts the members to present their knowledge or opinions as certain and unchallengeable (Oakley 1999). And the inability of team members to actually observe the expertise or efforts of others tends to lead to a higher dependence on perceptions and assumptions that could be either biassed or negative (Smith and Ruiz 2020). Such conditions may breed issues related to trust and hamper the team spirit (Zuofa and Ochieng 2017). The virtual work environment is characterized by increased use and higher dependence on machines and information technology that might induce feelings of isolation and purposelessness in individuals. The entire communication facilitated by phones and the internet may cause dehumanizing and alienating experiences in people (Oakley 1999).

2.2 *Temporal Factors*

The challenges of geographic dispersion are usually accompanied by the problems of difference in time zones. Although the two factors are highly related to each other yet (Smith and Ruiz 2020) argue that temporal distance is distinctly different than geographical distance and should be treated as a separate dimension. Factors related to time can be caused by either time-difference in work patterns or differences in time zones (Sarker and Sahay 2004). These difficulties arise partly from a geographical distance where members are working without overlapping working hours and consequently hamper simultaneous work (Lillian 2013). It can also be argued that time-related factors pose a more significant challenge as compared to geographical difficulties because of their impact on coordination (Smith and Ruiz 2020). While the task deadlines and time schedules are challenges for any team, virtual or collocated, e-leaders have to struggle with severe problems while coordinating tasks within virtual teams (Lillian 2013).

The reduction of overlapping working hours is one of the critical disadvantages caused by temporal factors. Ideally, having team members in dispersed locations and different time zone can be viewed as an advantage since at least one or more members would be working at any point of time, thus providing continual progress 24/7, but it is not practically observed. In fact, such time differences usually lead to incongruent schedules and delays in meeting deadlines (See 2018). Furthermore, the problems associated with synchronization also creep into virtual teams due to temporal differences (Sarker and Sahay 2004).

The situations that allow the possibility of informal communication are also eliminated due to temporal distances and thus leading to problems of trust and isolation. It is observed that in large virtual teams, frequent communication remains limited to members of similar time zones since it is convenient to bypass other members. Such a process results in the formation of fragments in teams, and members may harbour jealousy and toxicity among themselves (Smith and Ruiz 2020).

2.3 Cultural Factors

Cultural differences have been studied in different ways by different authors. One of the most popular studies was conducted by Hofstede (1980). He highlighted that culture is the collective programming of the mind which distinguishes the members of one group from another. Generally, the members of a virtual team come from different nations and a variety of cultural backgrounds. Such a heterogeneous composition of the workgroup can pose complications related to mutual understanding and communication (Johnson et al. 2001). According to field-based research by (Kayworth and Leidner 2000), culture can be learned and, therefore, can be manifested in different ways on the basis of nationality, ethnicity, and even organizational settings. Particularly in virtual teams, how individuals perceive, understand, or act upon any information or towards any person is primarily influenced by their cultural backgrounds. The literature related to the impact of cultural factors over the functioning of virtual teams shows ambivalent results. Cultural diversity, along with geographical differences, can be the primary factors that might cause dysfunction in virtual environments. However, if utilized and appropriately managed, cultural diversity can also be the antecedent to team effectiveness (Scott and Wildman 2015).

The presence of members from different cultural backgrounds and experiences brings a variety of perspectives to solve a problem, but it may also create and intensify some communication problems (Kayworth and Leidner 2000). Moreover, the cultural background also influences the behaviour of individuals and decision-making. Dekker et al. (2008) found that teams with a majority of individualistic members tend to have a bearing over the collectivist minority, while the opposite is not observed. Also, individualistic team members tend to be more critical in their remarks towards their co-workers as compared to collectivist members (Scott and Wildman 2015). According to (Paul et al. 2004), the individualistic-collectivistic dimension of culture can be utilized to access the individual's inclination towards teamwork and cooperation.

The most prominent cultural issue is the language barrier. Due to linguistic differences, a significant portion of the information is either lost or distorted, resulting in inefficiency and loss of synergy. It has also been observed that cultural differences may also stem from factors other than nationality and ethnicity. For example, different organizations may have a different work culture that may profoundly affect the expectations and beliefs of the employees. Therefore, it is possible to have two individuals from the same nation and ethnicity yet having different beliefs due to differences in the culture of the organization they have been working with (Kayworth and Leidner 2000).

Besides all these potential pitfalls that may arise due to cultural differences, there are studies that support the conception that cultural diversity in virtual teams works in favour of the team rather than undermining its efficiency. Qureshi and Zigurs (2001) propound those cultural differences are not of much relevance in virtual teams as the virtual environment is more task and work-oriented. The overall effect of culture over virtual teams could be anything on a positive to negative continuum depending upon the management of cultural diversity. In general, the literature seems to agree with the idea that cross-cultural training is vital to leverage the cultural diversity in a virtual team. (Scott and Wildman 2015).

2.4 Communication Factors

Communication problems are one of the most significant issues that have been reported in virtual teams (Johnson et al. 2001). A majority of studies agree on the idea that communication problems primarily occur due to the inability of members to have rich face-to-face communication and due to temporal and geographic dispersion (Kayworth and Leidner 2000; Piccoli et al. 2004; Rosen et al. 2006). In the absence of face-to-face communication, members rely on the internet and information technology to connect and communicate. Since verbal and visual cues are almost absent in virtual communication, it leads to the loss or distortion of various essential components of effective communication. Emotional elements such as hope, anger, affection, or humour may be lost or misinterpreted, giving way to interpersonal conflicts and misunderstandings (Rosen et al. 2006).

Johnson et al. (2001) classify the communication problems into three broad categories: (1) Lack of project visibility: where members remain unclear about the complete progress of the project and their work is going to fit in the big picture. Issues related to accountability and monitoring may also occur under such conditions. (2) Getting in touch with people: where members would face difficulties in getting in touch with teammates, delayed responses, technical glitches, and temporal differences may all exacerbate this issue. (3) Technological constraints: where members of virtual teams would face difficulty in interpreting the meaning and context of messages, especially the text-based messages or e-mails. In such situations, members tend to rely on the symbols and emoticons to convey the contextual meaning, but these can only express very rudimentary feelings (Rosen et al. 2006). And, based on the review study, Bergiel et al. (2008) argue that even if the members possess excellent language and communication skills, they naturally tend to interpret written and verbal communication through the filter of their own culture. In such circumstances, chances of reinforcement of stereotypes based on personal differences multiply that could nullify the relationship-building efforts (Rosen et al. 2006).

Piccoli et al. (2004) suggest that individuals in virtual communication tend to be less attentive and receptive, and sole reliance on technology for communication reduces team effectiveness, which in turn reduces members' satisfaction level. Such factors, along with the impeded communication, can induce power struggles and reduce the identification with the team, leading to the incidence of conflicts (Scott et al. 2015). The virtual nature of communication makes it difficult to manage disagreements. Minute problems that would have come to the leader's attention through a stroll around the office or a casual talk with a member now come to light only if an active conversation is extended from the leader's end (Bergiel et al. 2008).

Although most of the literature emphasizes the absence of physical or verbal cues in virtual communication, and an auto-ethnography study conducted by Depaoli et al. (2014) suggests that within virtual workspaces, physical cues and gestures are closely looked for and interpreted similarly to face-to-face interaction. They further reported that although virtual space does not allow the transmission of information through sensitive, sensuous, physical, or personal cues, yet physical bodies are not entirely absent in the virtual world. They remain present in leadership, interaction, and communication and play an important role. Due to the facilities provided by videoconferencing software where one can see oneself, aesthetic consciousness of self heightens in virtual communication.

It has also been observed in studies that members often introduce some kind of compensatory behaviour while communicating to reduce misinterpretations. Members of virtual teams would tend to enhance their verbal communication by speaking slower and using simpler words and sentences (Scott et al. 2015). Various other techniques and solutions are also suggested by different studies to improve communication effectiveness. An early establishment of the ground rules or norms regarding the content, frequency, and etiquettes of virtual communication can avoid the pitfalls of virtual communication (Malhotra et al. 2007; Scott et al. 2015). Apart from this, training to enhance communication and participation skills along with initial occasional face-to-face interaction can overcome communication barriers (Johnson et al. 2001; Rosen et al. 2006).

2.5 Technological Factors

Technology and ICT are the most vital elements that facilitate the formation of virtual teams (Bergiel et al. 2008). It is quite usual and imperative for virtual teams to embrace the technologies such as telephone, e-mail, videoconferencing, and group decision-making support systems (Rosen et al. 2006; Zuofa and Ochieng 2017). However, using these web-based collaborative tools does not ensure immunity from technological problems (Kayworth and Leidner 2000) since technology can be best defined as a tool that entirely depends upon human input. Regardless of how advanced and sophisticated technology is, it can falter if proper consideration is not given to the perspective and knowledge of humans associated with it (Bergiel et al. 2008). The usage and application of such technologies are highly dependent on multiple factors such as availability of resources, locations of the teammates, and their respective technical awareness. For example, the members of a virtual team from developed parts of the world like the US or Canada may have ready access to such resources while others from countries like Nigeria or Ghana may not (Zuofa and Ochieng 2017).

Furthermore, the selection of an inappropriate or wrong technology can diminish the mutual understanding among members and may lead to ineffectiveness (Rosen et al. 2006). For instance, (Majchrzak et al. 2004) revealed that even e-mail could be a poor way to collaborate for a team as a whole. The one-to-one conversations through e-mails can cause others to feel side-lined, deteriorating the trust and cohesiveness that might lead to an eventual dysfunction. Including every team member in the loop for all the communication has its own problems. Over time, it is complicated to keep track of relevant and prioritized messages, along with the issues associated with storage and spamming. In their field-based research study in 2000, Kayworth and Leidner observed that participants using a web-based collaborative tool encountered technical glitches that removed the participants from the session for no apparent reason. Even sophisticated tools such as videoconferencing are not free from problems. Time delays and difficulties that participants from different time zones face while attending video meetings after regular business hours are quite real (Majchrzak et al. 2004).

Apart from the technological barriers, the individual level of technical knowledge and skill, along with the willingness of employees to go digital, pose some serious challenges (Kayworth and Leidner 2000; Bergiel et al. 2008). Although the coming workforce is tech-savvy and comfortable with technology, yet there might be a considerable number of valuable employees who are technophobic (Johnson et al. 2001). In such circumstances, the idea of virtual teams may actually face a generation gap where the newcomers are keen on working with technology while their seniors may not even have the simple skill of rapidly pointing and clicking (Bergiel et al. 2008). Therefore, a substantial portion of literature seems to agree upon the importance of training in the selection and use of various technological resources could be extremely beneficial for the overall effectiveness and productivity of virtual teams (Johnson et al. 2001; Kayworth and Leidner 2000; Rosen et al. 2006).

2.6 Leadership Factors

Basically, leadership can be defined as 'the process of influencing others to understand and agree about what needs to be done and how it can be done effectively, and the process of facilitating individual and collective efforts to achieve the shared objective' (Sutanto et al. 2011). Leading a virtual team requires a higher level of acumen and skills as compared to leading a collocated team (Malhotra et al. 2007) since the incidence of various challenges due to geographical, temporal, cultural, communication, and technological factors result in some severe leadership challenges. E-leaders may face some unique challenges related to relationship building, goal setting, performance monitoring, recognizing the members' contribution, and conflict management, which may not be experienced in a traditional face-to-face environment (Saafein and Shaykhian 2014).

The problem of providing the members with the recognition they deserve is well acknowledged by leaders in collocated teams. Such difficulties multiply in a virtual environment (Malhotra et al. 2007). The role of a leader is closely related to things like feedback, rewards, motivation, and encouragement. In a traditional work environment, these functions could be carried out through physical presence, positive comments, and a 'slap on the back', but in virtual work settings, a leader has to consciously engage these leadership aspects (Zigurs 2003). A virtual leader must remember to recognize a member's contribution to the achievement of team goals. Mostly, a public acknowledgement could do wonders to ensure member's performance (Bergiel et al. 2008).

According to Sivunen (2008) the four most essential behaviours that members of any virtual team expect in their leader are motivating members to participate, offering support, providing guidance, and setting clear goals. Those who fail to provide enough detail and role clarity are regarded as poor leaders since these factors may increase communication and coordination problems (Piccoli et al. 2004). Saafein and Shaykhian (2014) have reported a positive relationship between leadership performance and communication effectiveness. The leader can act as the central point for the communication network of the team (Oakley 1999). The importance of being a bridge in virtual communication is highlighted by Sutanto et al. (2011) who reported that members of a virtual team perceived a mediator (someone who bridged the interaction between two or more members) as a leader, more than someone who directed or monitored them.

Other factors that might lead to severe leadership issues are associated with monitoring the members' work and providing regular feedback. Bergiel et al. (2008) suggest that virtual team leaders should spare adequate time from their day to connect and monitor members' work and provide constructive feedback to achieve a virtual walk around the entire team. The quality of giving regular feedback and asking the members for their opinions is highly desirable in a virtual leader (Kayworth and Leidner 2000). Additionally, the leader of a virtual team needs to grapple with the problems of member feelings of isolation, team cohesion, earning members' commitment, and relational development in the absence of physical context and cues (Zigurs 2003; Majchrzak et al. 2004; Malhotra et al. 2007). In his longitudinal educational field study, Mitchell (2012) argues that both proactive and reactive interventions could be useful in dealing with leadership challenges. However, collaborative leadership style, excellent communication skills, and leader's ability to remain flexible are some characteristics that could minimize the leadership pitfalls in virtual work environments (Kayworth and Leidner 2000; Malhotra et al. 2007; Oertig and Buergi 2006).

2.7 Team-Related Factors

In addition to the above-mentioned factors, some of the issues inherently associated with the structure and dynamics of a team intersect with other factors posing difficulties for virtual teams (Smith and Ruiz 2020). Diversity among the members of virtual teams is one such factor. The effect of diversity on team effectiveness can be either positive or negative depending upon its management and relevance of diversity to the task (Staples and Zhao 2006). A poorly managed diverse team may affect members' behaviour and therefore complicate communication and execution of work (Lillian 2013). Members of a virtual team could be different in terms of culture, age, experiences, work ethic, and technical expertise. Depaoli et al. (2014) in their auto-ethnography studies revealed that hidden power structures might exist even in a well-managed collegial team. For example, different levels of technological awareness among team members may induce a sense of shame and produce varying power status (Depaoli et al. 2014). Such adverse effects of diversity may cause communication difficulties, misunderstandings, reduced cohesion, and conflict (Staples and Zhao 2006). Diversity on the basis of the age of the members can also be a potential source of tension. Usually, young employees are more computer facile and open to changes as compared to their older colleagues (Vakola and Wilson 2004). Staples and Zhao (2006) reported that in their experimental research that heterogeneous virtual teams were found to be less satisfied as compared to homogenous teams while there was no significant difference in their performance levels. Literature suggests that challenges associated with diversity can be handled if the members have some 'common ground' such as shared experiences, vocabulary, or mental models. Moreover, training in project management and the use of software packages can also be helpful in diversity management (Smith and Ruiz 2020; Vakola and Wilson 2004).

3 Ideas for Future Research

The present literature review can be utilized in multiple ways to explore the concept of virtual teams and associated challenges. Although each type of challenge affects team performance, further research can be conducted to investigate the relative influence of these factors over effectiveness. Some upcoming ideas, such as the use of team-based offices in organizations that mostly work through virtual teams, can be explored (Depaoli et al. 2015). Moreover, an interrelatedness among these categories of factors can be studied for a better understanding.

4 Conclusion

A virtual workspace is one of the vital elements of Industry 4.0 and therefore commands considerable attention. This literature review presents a systemic study of various challenges encountered by virtual teams and classified them into a few categories. Utilizing forty relevant studies, we have categorized the factors into seven broad categories: Geographic factors, temporal factors, cultural factors, communication factors, leadership factors, technological factors, and team-related factors. Finally, popular strategies suggested by researchers to deal with these challenges are also discussed, along with future opportunities for research.

References

Bergiel BJ, Bergiel EB, Balsmeier PW (2008) Nature of virtual teams: a summary of their advantages and disadvantages. Manag Res News 31(2):99–110

Brake T (2006) Leading global virtual teams. Ind Commer Train 38(3):116–121

Chadee D, Ren S, Tang G (2021) Is digital technology the magic bullet for performing work at home? Lessons learned for post COVID-19 recovery in hospitality management. Int J Hosp Manag 92:102718

Davis DD, Bryant JL (2003) Influence at a distance: leadership in global virtual teams. Adv Glob Leadersh 3(1):303–340

De Guinea AO, Webster J, Staples DS (2012) A meta-analysis of the consequences of virtualness on team functioning. Inf Manag 49(6):301–308

De Jong BA, Dirks KT, Gillespie N (2016) Trust and team performance: a meta-analysis of main effects, moderators, and covariates. J Appl Psychol 101(8):1134–1150

De Paoli D, Ropo A (2015) Open plan offices–the response to leadership challenges of virtual project work? J Corp Real Estate 17(1):63–74

De Paoli D, Ropo A, Sauer E (2014) Disappearing bodies in virtual leadership? In: The physicality of leadership: gesture, entanglement, taboo, possibilities, vol 6. Emerald Group Publishing Limited, pp 59–79

Dekker DM, Rutte CG, Van den Berg PT (2008) Cultural differences in the perception of critical interaction behaviors in global virtual teams. Int J Intercult Relat 32(5):441–452

Duran V, Popescu AD (2014) The challenge of multicultural communication in virtual teams. Procedia Soc Behav Sci 109:365–369

Ferreira R, Pereira R, Bianchi IS, da Silva MM (2021) Decision factors for remote work adoption: advantages, disadvantages, driving forces and challenges. J Open Innov Technol Market Complex 7(1):70–94

Gatlin-Watts R, Carson M, Horton J, Maxwell L, Maltby N (2007) A guide to global virtual teaming. Team Perform Manag 13(1–2):47–52

Hunsaker PL, Hunsaker JS (2008) Virtual teams: a leader's guide. Team Perform Manag Int J 14:86–101

Johnson P, Heimann V, O'Neill K (2001) The "wonderland" of virtual teams. J Work Learn 13(1):24–30

Kayworth T, Leidner D (2000) The global virtual manager: a prescription for success. Eur Manag J 18(2):183–194

Kiel L (2003) Experiences in distributed development: a case study. In: Proceedings of international workshop on global software development at ICSE, Oregon, USA, 2003

Lilian SC (2014) Virtual teams: opportunities and challenges for e-leaders. Procedia Soc Behav Sci 110:1251–1261

Majchrzak A, Malhotra A, Stamps J, Lipnack J (2004) Can absence make a team grow stronger? Harv Bus Rev 82(5):131–137

Malhotra A, Majchrzak A, Rosen B (2007) Leading virtual teams. Acad Manag Perspect 21(1):60–70

Mitchell A (2012) Interventions for effectively leading in a virtual setting. Bus Horiz 55(5):431–439

Morris S (2008) Virtual team working: making it happen. Ind Commer Train 40(3):129–133

Morrison-Smith S, Ruiz J (2020) Challenges and barriers in virtual teams: a literature review. SN Appl Sci 2:1–33

Oakley JG (1999) Leadership processes in virtual teams and organizations. J Leadersh Stud 5(3):3–17

Oertig M, Buergi T (2006) The challenges of managing cross-cultural virtual project teams. Team Perform Manag Int J 12(1/2):23–30

Olariu C, Aldea CC (2014) Managing processes for virtual teams—a BPM approach. Procedia Soc Behav Sci 109:380–384

Paul S, Samarah IM, Seetharaman P, Mykytyn PP Jr (2004) An empirical investigation of collaborative conflict management style in group support system-based global virtual teams. J Manag Inf Syst 21(3):185–222
Pauleen DJ (2003) Leadership in a global virtual team: an action learning approach. Leadersh Org Dev J 24(3):153–162
Piccoli G, Powell A, Ives B (2004) Virtual teams: team control structure, work processes, and team effectiveness. Inf Technol People 17(4):359–379
Qureshi S, Zigurs I (2001) Paradoxes and prerogatives in global virtual collaboration. Commun ACM 44(12):85–88
Rosen B, Furst S, Blackburn R (2006) Training for virtual teams: an investigation of current practices and future needs. Hum Resour Manage 45(2):229–247
Saafein O, Shaykhian GA (2014) Factors affecting virtual team performance in telecommunication support environment. Telemat Inform 31(3):459–462
Sarker S, Sahay S (2004) Implications of space and time for distributed work: an interpretive study of US–Norwegian systems development teams. Eur J Inf Syst 13(1):3–20
Sarker S, Ahuja M, Sarker S, Kirkeby S (2011) The role of communication and trust in global virtual teams: a social network perspective. J Manag Inf Syst 28(1):273–310
Scott C P, Wildman J L (2015) Culture, communication, and conflict: a review of the global virtual team literature. Leading Global Teams, Springer, New York
See M (2018) 18 International collaboration: are the challenges worth the benefits? J Anim Sci 96(suppl 3):2–2
Sivunen A (2008) The communication of leaders in virtual teams: expectations and their realisation in leaders' computer-mediated communication. J eWork 2(1):47–60
Staples DS, Zhao L (2006) The effects of cultural diversity in virtual teams versus face-to-face teams. Group Decis Negot 15(4):389–406
Sutanto J, Tan CH, Battistini B, Phang CW (2011) Emergent leadership in virtual collaboration settings: a social network analysis approach. Long Range Plan 44(5–6):421–439
Vakola M, Wilson IE (2004) The challenge of virtual organisation: critical success factors in dealing with constant change. Team Perform Manag Int J 10(5/6):112–120
Van der Lippe T, Lippényi Z (2020) Co-workers working from home and individual and team performance. N Technol Work Employ 35(1):60–79
Weinel M, Bannert M, Zumbach J, Hoppe HU, Malzahn N (2011) A closer look on social presence as a causing factor in computer-mediated collaboration. Comput Hum Behav 27(1):513–521
Zigurs I (2003) Leadership in virtual teams: Oxymoron or opportunity? Organ Dyn 31(4):339–351
Zuofa T, Ochieng EG (2017) Working separately but together: appraising virtual project team challenges. Team Perform Manag Int J 23(5/6):227–242

Chapter 4
A Cost-Based Optimization Modelling of Solar Power Generation in India for Sustainable Development

Kuljeet Jolly, Umar Muhammad Modibbo, Jahangir Chauhan, and Mohd. Shamim Ansari

Abstract This study examines the socio-economic cost of power generation through solar energy sources. It develops a model to optimize its per unit cost and implied revenue while satisfying India's growing demand for power with sustainability. Conversely, complete replacement of thermal energy seems irrelevant in the present scenario for on-grid supply due to solar radiation variation. Alternatively, off-grid supply comes with high storage cost of facilities establishment. This study also attempts to determine the required replacement rate. The ever-rising energy demand can be satisfied efficiently without depriving our future generations of the country's limited exhaustible resources. The study results have implications for the government, corporate and other stakeholders, who have a massive dependence on power. Moreover, it helps in policy planning to achieve Sustainable Development Goals, 2030, particularly India's energy-related goals.

Keywords Electricity demand · Energy planning · Sustainable development goals · Cost–benefit analysis · GHGs emission · Renewable energy

1 Introduction

Sustainability is the process that aims at achieving harmony between the ecosystem and the growing human civilization. Sustainable development is a system-based approach that seeks to meet the current and future human needs and improve life quality whilst endeavouring to conserve natural capital. Its scope is not limited to environmental protection and considers the need fulfilment of communities, social inclusion, and economic justice.

K. Jolly · J. Chauhan (✉) · Mohd. S. Ansari
Department of Commerce, Aligarh Muslim University, Uttar Pradesh, Aligarh, India
e-mail: jchauhan.cm@amu.ac.in

U. M. Modibbo
Department of Statistics and Operations Research, Modibbo Adama University, Yola, Nigeria
e-mail: umarmodibbo@mautech.edu.ng

I. Ali et al. (eds.), *Computational Modelling in Industry 4.0*,
https://doi.org/10.1007/978-981-16-7723-6_4

Sustainable development is a proactive strategy that aims to mitigate climate change by replacing fossil fuels' consumption (responsible for greenhouse gas emissions) with renewable energy sources. This shift requires the necessary infrastructure and technology to boost employment opportunities, reinforcing the economy's stronger financial footing. Population rising at an alarming rate builds pressure on the natural environment to fulfil bare life essentials; therefore, sustainable development should be the pressing priority. Realizing the exigency of addressing this sustainability issue, the Sustainable Development Goals (SDGs) or Global Goals, signed by 193 Member States of the U.N. General Assembly in 2015, were adopted as a sustainable road map for future people and the planet. This ambition was made tangible under the 'Transforming our world: the 2030 Agenda for Sustainable Development', targeted for the next 15 years.[1] SDGs collectively consist of 17 goals covering 169 targets. Each target is measured by 1 to 3 indicators; 232 approved indicators ascertain the appropriate degree of progress.

The SDGs replaced its precursor Millennium Development Goals (MDGs), adopted in 2000 and set for 2015 (Fehling et al. 2013; Kumar et al. 2016; Assefa et al. 2017). As per UNICEF reports, MDGs indicate positive results in specific parameters like poverty eradication, child mortality, access to education, clean drinking water, stunting, etc. However, MDGs suffered from limited and uneven progress across nations. The lack of justification for the objectives adopted (8 goals measured by 18 targets), flaws in its structure as the goals were not interconnected, absence of accountability, lack of legitimacy, and appropriate indicators for measuring progress.

Hence, SDGs were framed with more insight and have evolved over an extensive consultative process. Under the scrutiny of the U.N. General Assembly Open Working Group, established for this specific purpose, it is to overcome MDGs' gaps and shortcomings.[2]

Energy is the critical input for the growth and prosperity of economies across the globe. SDGs address it in its Goal 7, which states, 'Ensure access to affordable, reliable, sustainable and modern energy for all'.[3] Population growth has a dominant role in expedited energy demands, making it essential to improve current energy availability. Non-renewable energy resources such as fossil fuels affect the environment and health negatively. Moreover, the current consumption pattern will exhaust the overall energy-based carbon budget in 20 years. Transformation towards renewable energy sources by investments in imperative infrastructure and modernizing existing technology would improve energy efficiency. It will enable access to clean energy across all nations, bringing us a step closer to a sustainable future.

This goal is affiliated with Goal 13 of SDGs, 'Take urgent action to combat climate change and its impacts by regulating emissions and promoting developments

---

[1] (2015), Transforming our world: the 2030 Agenda for Sustainable Development. Retrived from https://www.un.org/en/development/desa/population/migration/generalassembly/docs/globalcompact/A_RES_70_1_E.pdf.

[2] *Future We Want - Outcome document.:. Sustainable Development Knowledge Platform.* Retrieved November 25,2020, from https://sustainabledevelopment.un.org/futurewewant.html.

[3] Goal 7: Affordable and clean energy | UNDP. Retrieved from https://www.undp.org/content/undp/en/home/sustainable-development-goals/goal-7-affordable-and-clean-energy.html.

in renewable energy'.[4] The current global energy system must undergo a profound transition towards a decarbonized power system to reduce greenhouse gases (GHGs) emissions. Hence, it limits global warming to 1.5-degree celsius above pre-industrial levels (a special report by the International Panel of Climate Change). Goal 13 of the SDGs aims to muster US$100 billion annually by 2020, as reflected in the commitments made by developed countries to the united nations framework convention on climate change (UNFCCC). It aims to cater to developing countries' needs regarding investment in low-carbon development and mitigate climate change.[5]

India, which has a population of 1.37 billion, is one of the world's largest and fastest-growing economies. The energy demand in India is rising at the rate of 3.6 per cent. The country has set colossal renewable energy (RE) targets (i.e. achieving 175GW of renewable energy capacity by 2022). It solicits a clear strategy roadmap, integrated planning, and a whole-of-system approach. Presently India's energy demand is 180 GW and is expected to reach 690 GW by 2036. Although the country is endowed with ample renewable energy resources, they are grossly under-utilized. Since a significant portion of the current power demand has been met from conventional sources, these reserves of these are continually depleting due to massive exploitation. It has expected to exhaust in a few decades to come.

The power demand is also overgrowing, creating an urgent need to replace conventional energy sources with renewable sources to attain the SDGs related to energy sufficiency. The increase in renewable sources in the current power generation mix will reduce Green House Gases, which have far-ranging environmental and health effects. Renewable energy sources offer more benefits from an ecological and social perspective. However, high installation, low capacity utilization, and increased storage and distribution costs lead to higher per unit economic costs than units generated through conventional energy sources. The price per unit of electricity produced through solar energy is higher than the per unit cost of electricity produced using conventional sources such as thermal and nuclear. However, to achieve supply sustainability for meeting the ever-rising power demands, there is a need to optimize solar power generation's production cost. It is the most important and abundant energy source the country has. This study examines the socio-economic cost of power generation through solar energy sources. It develops a model to optimize its per unit cost and implied revenue while satisfying India's growing demand for power.

Conversely, complete replacement of thermal power seems irrelevant in the present scenario for on-grid supply due to solar radiation variation. On the other hand, off-grid supply comes with high storage cost of facilities establishment. This study attempts to determine the required replacement rate to efficiently satisfy the ever-rising energy demand without depriving our future generations of the country's

---

[4] Goal 13: Climate action | UNDP. Retrieved from https://www.undp.org/content/undp/en/home/sustainable-development-goals/goal-13-climate-action.html.

[5] Summary for Policymakers of IPCC Special Report on Global Warming of 1.5 °C approved by governments—IPCC. Retrieved November 25, 2020 from https://www.ipcc.ch/2018/10/08/summary-for-policymakers-of-ipcc-special-report-on-global-warming-of-1-5c-approved-by-governments/.

limited exhaustible resources. The study results have implications for the government, corporate and other stakeholders, who have a massive dependence on power. Moreover, it helps in policy planning to achieve Sustainable Development Goals, 2030, particularly India's energy-related goals (See Aayog NITI (2015)).

2 Literature Review

Sustainability has become a significant target for the different countries worldwide and is making numerous attempts to achieve this goal. One significant factor that can help achieve sustainable development targets is renewable energy sources (Aftab, 2019). Ram et al. (2018) statistically displayed that all the G20 countries have the opportunity to decrease their energy costs significantly between now and 2030. Sustainable development in any country needs effective and strategic policies that fulfil various and overlapping objectives. The multi-objective optimization model is a well-known multi-criteria decision-making method for practical implementation. Many researchers have used this method to arrive at decisions in various fields over the past years. A fuzzy goal programming (FGP) approach was proposed to analyze India's environmental, energy, and sustainability goals by 2030 concerning India's vital economic sectors (Nomani et al. 2017). The model presented analyzes the prospects for change, the need for efforts, and sustainable development plans. A multi-criteria modelling approach is presented using the linear programming problem (LPP) framework to optimize the primary, secondary and tertiary sectors (Gupta et al. 2018). Besides, an FGP model developed to provide an optimum allocation of resources by achieving potential targets for the gross domestic product (GDP), electricity consumption (EC), and greenhouse gas (GHG) emissions. The study extended to consider condensed major vital sectors of India's economy using the FGP concept (Ali et al. 2021). Modibbo et al. (2021) proposed a multi-objective goal programming model to analyze Nigeria's socio-economic, environmental, and energy sector. AlArjani et al. (2021) proposed a new framework for the SDGs of Saudi Arabia in a similar direction.

3 An Overview on Electricity Energy Demand of India

With its multidimensional use, is an essential part of life now. Our energy needs categorized by two types of sources employed in energy generation: conventional and renewable. In contrast, conventional sources for energy generation are denting our existing reserves. It is adding to environmental degradation and renewable energy. On the other hand, policymakers can replenish sources abundantly available in the natural environment over time. It has the edge of having a diversified portfolio: solar, wind power, hydroelectric energy, biogas, geothermal power, etc., thereby reducing

Table 1 Projected electricity demand (19th EPS)

Year	Electrical energy requirement (BU)	Peak electricity demand (GW)
2021–22	1566	225.751
2026–27	2047	298.774
2029–30	2325	339.973

Source Central Electricity Authority, Ministry of Power, G.O.I., (2019). Draft Report on Optimal Generation Capacity Mix 2029–30

the dependence on limited resources such as coal, lignite and natural gas, etc. Renewable power generation projects entail high costs of installation and storage facilities. However, they do not outweigh the benefits of low operating and maintenance costs once the infrastructure has been laid down; demonstrating a job-creation effect that stimulates economic growth. Renewable energy is eco-friendly with a zero to low-carbon footprint than conventional sources that emit GHGs. They are the primary contributors to global warming; however, one renewable resource is an exception to this, i.e. biomass emits carbon and methane in its combustion process.

Presently, the central supply of our energy in general and electricity comes from conventional sources. It is not feasible both socially and economically and hence, needs to shift to renewable sources. However, an immediate shift is impossible due to various factors, but a gradual shift is the government's priority. The switch to renewable sources is the need of the hour. However, it should be viewed as the importance of uninterrupted supply and ever-rising demand for energy.

Demand estimation is a prerequisite in planning for additions to the existing energy generation capacity to meet the future electricity requirements with efficacy. Making use of the 'bottoms-up' approach to estimate demand requirements, 19th EPS (Electricity Power Supply), which constituted by C.E.A. (Central Electricity Authority), projected the probable figures for electrical energy requirements (in Billion Units) and peak electricity demand (in GigaWatt) on the all-India basis for the years 2021–2022, 2026–27 and 2029–30, as shown in Table 1.[6,7] The year 2030 is significant as SDGs intend to achieve its goals by the end of this year.

The total electricity generation in the country from conventional sources and renewable sources of energy during the year 2009–10 was 805.4 BU, as against the generation of 1376.1 BU during the year 2018–19, which shows a growth rate of 70.86 per cent over the decade. The status of the generation of electricity from renewable sources has boosted up by 2.5 times in the past ten years, thereby inching towards achieving the sustainability target of 175 GW[8] of renewable energy by the end of

[6] Central Electricity Authority, Ministry of Power, GOI, (2018), National Electricity Plan (Volume 1).

[7] Central Electricity Authority, Ministry of Power, GOI, (2019), Draft Report on Optimal Generation Capacity Mix 2029–30.

[8] India plans to produce 175 GW of renewable energy by 2022 - United Nations Partnerships for SDGs platform. Retrieved from https://sustainabledevelopment.un.org/partnership/?p=34566.

Table 2 Electricity generation (Billion units)

Year	Generation from conventional sources		Generation from renewable sources		Total generation
	Generation B.U	Percentage Gen	Generation B.U	Percentage Gen	
2009–10	771.6	95.80	36.9	4.58	805.4
2010–11	811.1	95.38	39.2	4.61	850.4
2011–12	876.9	94.48	51.2	5.52	928.1
2012–13	912.0	94.07	57.4	5.92	969.5
2013–14	967.2	94.80	53.1	5.20	1020.2
2014–15	1048.7	94.44	61.7	5.56	1110.4
2015–16	1107.8	94.39	65.8	5.61	1173.6
2016–17	1160.1	93.43	81.5	6.56	1241.7
2017–18	1206.3	92.22	101.8	7.78	1308.1
2018–19	1249.3	90.79	126.8	9.21	1376.1

Source Ministry of Power, G.O.I., Annual Report 2018–19

Table 3 Category wise electricity generation in India during the year 2019–20

Source	Increase/decrease percentage
Thermal	+2.59
Hydro	−15.62
Nuclear	+22.66
Bhutan Import	+31.94
Renewable	+8.75
Overall Growth	+1.04

Source Ministry of Power, G.O.I., Annual Report 2018–19

2022. Table 2 below enunciates the overall electricity generation in power utilities, including Bhutan imports from 2009–10 to 2018–19. Table 3 provides information as per Ministry for the year 2019–20.[9]

The enactment of the Electricity Act, 2003 delicensed the electricity generation, thus enabling the private sector's unrestricted participation in the power projects. This change resulted in additions to the existing electricity generation capacity (coal-based), thereby strengthening its overall energy frontier. Consequently, the amount of deficiency of power supply in comparison to the electrical energy requirements reduced drastically. Table 4 shows the reducing gap between the electricity demand

[9] Power Sector at a Glance ALL INDIA | Government of India | Ministry of Power. Annual Report 2018–19. Retrieved from https://powermin.nic.in/en/content/power-sector-glance-all-india.

Table 4 CO_2 emission factors

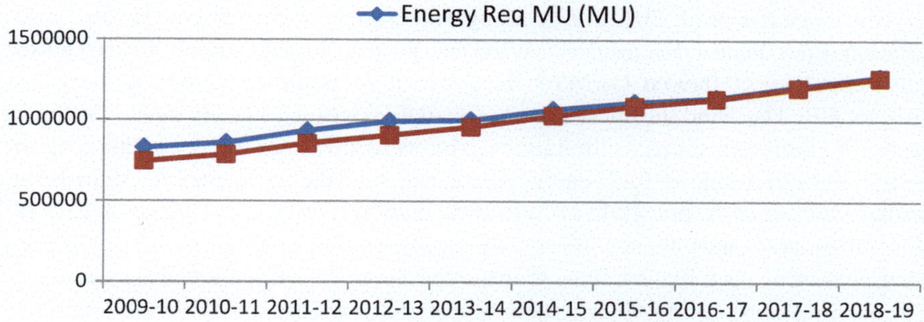

Source Compiled through Ministry of Power, G.O.I., Annual Report 2018–19

Table 5 CO_2 emission factors

Technology	CO_2 Emission Factor (tCO_2/Mwh)
Coal	0.98
Diesel	0.59
Gas	0.45
Lignite	1.38
Hydro	0.011
Nuclear	0.0242
Wind	0.0295
Solar	0.0534

and supply, i.e. 10.1 per cent in the year 2009–2010which lessened to 0.5 per cent in the year 2019–20.[10,11]

The energy sector is responsible for 75% of the global GHGs.[12] With the additions made to the existing electricity generation capacity, there arises a threat of GHGs to the environment as they are significant contributors to climate change. Carbon dioxide, the most prevalent form of GHGs, is taken up for comparison purposes. Although there is no complete wipeout of CO2 from the environment when renewable resources are replaced with non-renewable sources, the former's emissions are minimal compared to later. As shown in Table 5, lignite emits the highest CO_2, i.e. 1.38 tCO_2/Mwh, followed by coal, diesel, and gas. Hydro is responsible for only 0.011 tCO_2/Mwh of carbon emissions, nuclear at 0.0242 tCO_2/Mwh, Wind at 0.0242 tCO_2/Mwh, and Solar at 0.0534 tCO_2/MWh. Hydroelectricity has the least amount of emissions, but it disrupts the aquatic ecosystem. The next best option is

[10] Ministry of Power, GOI, Annual Report 2018–19.

[11] Power Sector at a Glance ALL INDIA | Government of India | Ministry of Power. Retrieved November 25, 2020 from https://powermin.nic.in/en/content/power-sector-glance-all-india.

[12] World Energy Outlook 2015—Executive Summary—English Version. International energy Association 2015. Retrieved November 26, 2020 from www.worldenergyoutlook.org.

nuclear energy which emits negligible carbon emissions but generates high radioactive waste (Harris et al. 2013). Also dangerous for the environment. It can cause radiation if not disposed of properly. Wind energy is a promising alternative to fossil fuels as it releases 0.0295 tCO_2/MWh. However, it has wildlife-related drawbacks as birds get killed by wind turbines. The next available option is chosen due to the three least GHG emission sources' limitations. That is to say, solar energy is taken up to replace non-renewable to fulfil energy needs and save the environment. Apart from fulfilling the social responsibility of lowering carbon footing, solar power is easy to adopt, affordable, and enhances electricity supply resiliency. Small-scale solar plants can use unused space like rooftops to instal the Solar P.V. (Photovoltaic) system.

India aims at replacing 40% of its total installed capacity by 2030 based on non-fossil fuel sources as submitted in Intended Nationally Determined Contributions (INDCs). This transition from non-renewable to renewable energy sources warrants an optimal electricity generation mix that would be cost-effective and environment friendly.

3.1 Electricity Generation Costs and Data

In this section, we will calculate the Levelized costs of coal and solar for producing electricity. Table 6 presents the information regarding the technological parameters of levelized cost and calculates levelized solar and coal costs. Table 7 summarizes the technological parameters of coal and solar in producing electricity.

The other input information summarizes in Table 7, and the information carried out based on the above discussions.

Table 6 Electricity generation costs

Technology →	Coal (X_1)	Solar (X_2)
Plant life (years)	25	25
Discount rate (%)	10	10
Capital cost(Lakh/MW)	760	450
O&M (Lakh/MW)	18	5.62
Fuel price (Rs)	3	0
Specific fuel consumption (Kg/Kwh)	0.627	0
Fuel cost (Lakh/M.W.)	98.86	0
Capacity factor (%)	60	19
Auxiliary consumption (%)	5.25	1
Generation cost (Rs/Kwh)	5.32	3.29

Source Compiled through Ministry of Power, GOI, Annual Report 2018–19

Table 7 Input Parameters Data

Data	Coal (X_1)	Solar (X_2)
Emission (Million tonne)	5154.4	88.94
Current Installed Capacity (MW)	191,092	31,696
Projected Installed Capacity	266,827	300,000

In addition to Table 7, the other information is that the total demand for electrical energy projected for 2029–2030 is 2325000. Furthermore, the estimated peak load is 339973.

4 Methodology

This section presents the techniques for modelling and solving the cost-based optimization problem for solar power generation in the Indian context. Next, the section discusses the levelized cost of the energy formula.

4.1 Levelized Cost of Energy Model

In generating power, the comparison of different technologies is imperative. The total cost of a system (lifetime cost) divided by the total amount of energy production is called **a levelized cost** (**LCOE**). The unit of measurement for the LCOE is either USD/Watt or USD/Kilowatt. The **LCOE formula** helps determine the total cost for generating power and compares different technologies trends (Khan et al. 2021). Based on the available technology and cost parameters, estimations of LCOE can be calculated using the formula as given in Eq. (1)

$$\text{Levelised cost of energy (LCOE)} = \frac{\sum\left[\frac{(I_t+OM_t+F_t)}{(1+r)^t}\right]}{\sum\left[\frac{E_t}{(1+r)^t}\right]} \tag{1}$$

where, I_t is the initial investment (Capital cost), OM_t is the operating and maintenance costs, F_t is the fuel cost, E_t the system energy yield, t is the number of years $\boldsymbol{r}$ is for a discount rate (Lazard, 2016). Next, the section discusses the goal programming approach.

4.2 The Goal Programming Approach

It is easier to solve a single objective programming problem. However, the multi-criteria decision-making problem (MCDM) solution cannot be obtained directly through any algorithm. Therefore some specific algorithms are used to address this issue. The algorithms frequently used for solving the MCDM problems are goal programming (GP). The GP modified the linear programming model, and Charnes et al. (1968) proposed it in the 1960s. It has variants and extensions such as pre-emptive GP, fuzzy GP, interactive fuzzy-GP, intuitionistic FGP, etc. In the literature, these algorithms frequently applied to obtain the compromised solution for the multi-objective decision-making (MODM) problems. In this study, the classical GP approach applied in solving the MODM problem. The mathematical model of the approach is given in Eq. (2).

$$\left.\begin{aligned} & Min\ Z = \sum_{i=1}^{m} \left(d_i^+ + d_i^-\right) \\ & s.t. \\ & \sum_{j=1}^{n} a_{ij}x_j - d_i^+ + d_i^- = b_i,\ i = 1, 2, \ldots, m, \quad (i) \\ & \sum_{j=1}^{n} a_{ij}x_j \begin{bmatrix} \leq \\ = \\ \geq \end{bmatrix} b_i, \quad i = 1, 2, ..., m, \quad (ii) \\ & d_i^+,\ d_i^-,\ x_j \geq 0,\ \ \forall i = 1, 2, ..., m,\ \ \forall j = 1, 2, ..., n.\ (iii), \end{aligned}\right\} \quad (2)$$

Here $d_i^+ + d_i^-$ is the over and under achievements of the goal. Constraint (i) is related to the goal of the decision-maker, constraint (ii) is known as the technological constraints and constraint (iii), the non-negative restrictions on all the variables.

5 Optimization Modelling

This section will model the above-discussed problem of energy production through non-fossil fuel and renewable energy sources. The identified objectives to be optimized include.

(i) Cost and benefits analysis of energy generation sources in the light of SDGs.
(ii) To estimate the optimal electricity generation mix for 2029–3n0, minimizing levelized cost while satisfying different system constraints.
(iii) To find out the optimal electricity generation mix to view the development goals related to reducing greenhouse emissions.

At the same time, the constraints of the model are.

(i) Total projected energy demand in 2029–2030.
(ii) Renewable energy contributes to energy production at least 40% of total energy.
(iii) CO_2 emissions are 35% lesser than that of the 2005 level.
(iv) Coal and solar use capacity in energy at least the current installed capacity.

Based on the above given objective functions and constraints, the mathematical optimization model is as follows:

$$\left.\begin{array}{l} Min \sum_i (LOCE)\,x_i \quad \text{(Levelised Cost)} \\ Min \sum_i e_i x_i \quad \text{(Emissions)} \\ Subject\ to \\ \sum_i p_i\,x_i \geq D \ \text{(Projected Energy Demand in 2029 – 2030)} \\ \text{Renewable energy } \geq 40\% \text{ of total energy} \\ CO_2 \text{ emission } \leq 35\% \text{ lesser than 2005 level} \\ x_i \geq \text{ present instaaled capacity (2019 – 2020)} \\ x_i \leq \text{ Maximum instaaled capacity (2029 – 2030)} \end{array}\right\} \quad (3)$$

where $\boldsymbol{P_i}$ is the electricity produced per unit installed capacity per year and $\boldsymbol{X_i}$ is the decision variable (installed capacity).

Let suppose X_1 is energy production through coal and X_2 is energy production through renewable energy. Putting all input parameter values in Eq. (3) in light of Eq. (1) give the resulting optimization problem of Eq. (4).

$$\left.\begin{array}{l} Min\ Z_1 = \dfrac{2649.8\,x_1 + 540.9\,x_2}{49.8x_1 + 16.5x_2} \quad \text{(levelised cost)} \\ Min\ Z_1 = 5154.4\,x_1 + 88.94\,x_2 \quad \text{(Emissions)} \\ Subject\ to \\ \quad 5259.6\,x_1 + 1665.4\,x_2 \geq 23,25,000 \quad \text{(Energy Demand)} \\ \qquad x_1 + x_2 \geq \ 3,39,973\ (peak\ load) \\ \qquad x_2 \geq 40\% \text{ of total energy} \\ \quad CO_2 \text{ emission } \leq 35\% \text{ lesser than 2005 level} \\ 191092\text{ MW } \leq x_1 \leq 266827 \text{ and } 31696\text{ MW } \leq x_2 \leq \ 300000 \end{array}\right\} \quad (4)$$

6 Results and Discussion

The mathematical model for the cost-based optimization given in Eq. (4) is solved using the Lingo Optimization package in light of the goal programming approach discussed in Sect. 4.2 and the model given in Eq. (2). The levelized cost of energy is optimized to be Rs 4.9/Kwh. At the same time, the carbon dioxide emission level is minimized at 1063.5 Million tons of CO_2. The optimal amount of energy production from coal is 2,03,983.8 MW, and that generated from the solar is 1,35,989.2 MW. It is found from the result that 1.3 per cents will overachieve the goal related to Installed Capacity/PV net electricity generation. Similarly, 8.29 per cent underachievement of the goal related to carbon dioxide emission is evident from the solution. The result is promising; it shows that emission can be reduced by 8.29 million tons from the individual optimum values, whereas the cost will increase by Rs. 0.013/Kwh.

7 Conclusion

This study reviews the power generation of India. The various technologies compared and levelized cost of energy computed. Multi-objective goal programming model formulated for the cost-based power generation problem. The optimal solution results obtained show that the emission can be reduced by 8.29 million tonne from the individual optimum values, whereas the cost will increase Rs. 0.013/Kwh. The optimal solution suggests that, by obtaining the most negligible emissions and low-cost values, the government should generate 60% electricity through thermal-based sources and the remaining from solar. Since thermal sources' current generation capacity is about 65 per cent, there is no need to augment thermal power generation facilities. Thus, the expansion should be made only in solar power generation facilities.

References

Aayog NITI (2015) A report on energy efficiency and energy mix in the indian energy system (2030)–using India energy security scenarios, 2047. Government of India, New Delhi

Aftab I, Kamal M, Irfan A (2020) Mapping Sustainable Development Goal Indicators for Pakistan: Way Forward. JSIS 15(1):10–35

AlArjani A, Modibbo UM, Ali I, Sarkar B (2021) A new framework for the sustainable development goals of Saudi Arabia. J King Saud Univ Sci 33(6):101477. https://doi.org/10.1016/j.jksus.2021.101477

Ali I, Modibbo UM, Chauhan J, Meraj M (2021) An integrated multi-objective optimization modelling for sustainable development goals of India. Environ Dev Sustain 23(3):3811–3831

Assefa Y, Van Damme W, Williams OD, Hill PS (2017) Successes and challenges of the millennium development goals in Ethiopia: lessons for the sustainable development goals. BMJ Global Health 2(2)

Charnes A, Cooper WW, DeVoe JK, Learner DB, Reinecke W (1968) A goal programming model for media planning. Manag Sci 14(8):B-423
Fehling M, Nelson BD, Venkatapuram S (2013) Limitations of the millennium development goals: a literature review. Glob Public Health 8(10):1109–1122
Gupta S, Fügenschuh A, Ali I (2018) A multi-criteria goal programming model to analyze the sustainable goals of India. Sustainability 10(3):778
Harris G, Heptonstall P, Gross R, Handley D (2013) Cost estimates for nuclear power in the UK. Energy Policy 62:431–442
Khan MF, Pervez A, Modibbo UM, Chauhan J, Ali I (2021) Flexible fuzzy goal programming approach in optimal mix of power generation for socio-economic sustainability: A Case Study. Sustainability 13(15):8256
Kumar S, Kumar N, Vivekadhish S (2016) Millennium development goals (MDGS) to sustainable development goals (SDGS): Addressing unfinished agenda and strengthening sustainable development and partnership. Ind J Community Med Offic Publ Ind Assoc Prevent Soc Med 41(1):1
Lazard N (2016) Lazard's Levelized Cost of Energy Analysis–Version 9.0.
Modibbo UM, Ali I, Ahmed A (2021) Multi-objective optimization modelling for analyzing sustainable development goals of Nigeria: Agenda 2030. Environ Dev Sustain 23(6):9529–9563
Nomani MA, Ali I, Fügenschuh A, Ahmed A (2017) A fuzzy goal programming approach to analyze sustainable development goals of India. Appl Econ Lett 24(7):443–447
Ram M, Child M, Aghahosseini A, Bogdanov D, Lohrmann A, Breyer C (2018) A comparative analysis of electricity generation costs from renewable, fossil fuel and nuclear sources in G20 countries for the period 2015-2030. J Clean Prod 199:687– 704.

Chapter 5
The Impact of Carbon Tax Policy in a Multi-Objective Green Solid Logistics Modelling Under Sustainable Development

Soumen Kumar Das, Sankar Kumar Roy, and Gerhard-Wilhelm Weber

Abstract In the last few decades, worldwide atmosphere change has become out to be one of the most significant environmental issues. Carbon emissions and air pollution have motivated a need to design an effective and sustainable logistics network. As a result, environmental policies are included in the transportation system to reconsider and re-structure a greenway distribution network. This chapter addresses a multi-objective optimization problem to design a solid logistics modelling in a green framework. The objectives of the stated problem are as follows: (a) to minimize the total financial costs along with carbon emissions cost, (b) to maximize the customers' satisfaction level simultaneously, and (c) to maximize the sustainable effectiveness conveyances. A multi-objective optimization procedure, namely, global criterion method is introduced to extract a non-dominated solution to the proposed problem. Two numerical examples test the formulated model and solution procedure. A comparative study among the proposed procedure and the other existing relevant procedures is also presented. Concluding remarks are discussed at last.

Keywords Solid logistics modelling · Carbon emissions · Carbon tax policy · Sustainable development · Multi-objective optimization

Abbreviations

The accompanying abbreviations are utilized in this chapter.

DM	Decision-maker,
FP	Fuzzy programming,
GCM	Global criterion method,
SD	Sustainable development,

S. K. Das · S. K. Roy (✉)
Department of Applied Mathematics with Oceanology and Computer Programming, Vidyasagar University, Midnapore, West Bengal 721102, India

G.-W. Weber
Faculty of Engineering Management, Chair of Marketing and Economic Engineering, Poznan University of Technology, Poznan, Poland
e-mail: gerhard.weber@put.poznan.pl

I. Ali et al. (eds.), *Computational Modelling in Industry 4.0*,
https://doi.org/10.1007/978-981-16-7723-6_5

STP	Solid transportation problem,
IFP	Intuitionistic fuzzy programming,
MOSTP	Multi-objective solid transportation problem,
MOGSTP	Multi-objective green solid transportation problem.

1 Introduction and Prior-Related Research

Nowadays, controlling carbon emissions is one of the most attractive and critical issues throughout the world. Figure 1 illustrates the carbon emanations information for various sectors of the world. From Fig. 1, it can be concluded that the carbon emissions have been essentially increased due to logistics network.

Poterba (1991), Chen and Wang (2016) discussed that the increase in carbon outflows is the main reason behind the global warming effect. Thus, governments around the globe have endorsed various policies in which the carbon tax policy (cf. Zhou et al. 2011; Benjaafar et al. 2012; Chen et al. 2013; Konur and Schaefer 2014) is implemented in effective measure to control the global warming effect. As per the carbon charge strategy, the carbon release holders need to pay the carbon charge for every fossil fuel byproduct unit to the policymaker. In the literature, few authors have considered this policy for instance Elhedhli and Merrick (2012), and Turken et al. (2017). This inspired us to formulate green logistics modelling to meet our present needs without polluting the atmosphere. For that reason, the concept of *sustainable development* (SD) (Lélé 1991; Litman and Burwell 2006) has attracted the scientists

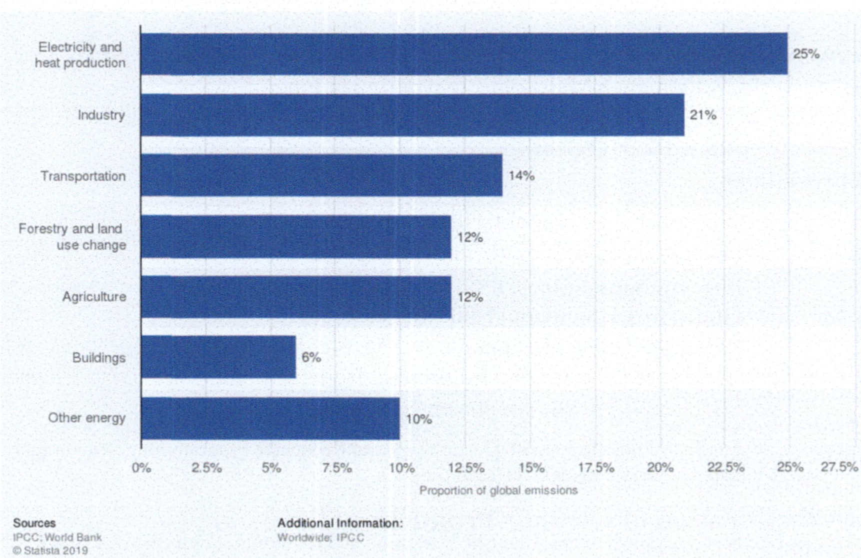

Fig. 1 Distribution of carbon emissions by industry sectors worldwide in 2019

as a field of research. According to the 'World Commission on Environment and Development', SD is defined as 'the concept of meeting the needs of the present without compromising future generations' ability to meet their needs'. In fact, the financial, social, and environmental aspects are considered simultaneously in SD. Nomani et al. (2017b) proposed a multi-objective model to investigate the SD goals of India. Gupta et al. (2018) presented a sustainable logistics model for the mining sector and then they solved it by an integrated multi-objective solution procedure. Maity et al. (2019b) introduced a sustainable environment in a multi-objective logistics network under the time window.

In last few decades, logistics modelling plays a vital role towards globalization. *Solid transportation problem* (STP) is a part of logistics network design. Substantially, STP, an augmentation of the classical transportation model is mainly described by source, demand, and conveyance set of constraints. The main concept of STP is to disseminate homogeneous products from specific sources to a few destinations via different kind of conveyances (e.g., hydrogen vehicle, diesel vehicle, CNG vehicle, bio-diesel vehicle) with the object that the total cost of the conveyance will be diminished. It was first enlightened by Shell (1955), however, the solution procedure was given by Haley (1962). Afterwards, Jiménez and Verdegay (1998) introduced the idea of fuzziness into an STP. They (Jiménez and Verdegay 1999) introduced a parametric approach to solve a fuzzy STP. It is often difficult to handle the real-world problems by a single objective STP. Due to this fact, several authors introduced multi-objective environment on STP, known as *multi-objective solid transportation problem* (MOSTP) from their different points of view. Tao and Xu (2012) presented the concept rough programming into an MOSTP. Liu et al. (2014) studied an STP in which the parameters were taken as type-2 fuzzy variables. Molla-Alizadeh-Zavardehi et al. (2013) described three metaheuristic algorithms for solving an MOSTP. Nomani et al. (2017a) discussed a new solution procedure to solve a multi-objective logistics modelling. Chen et al. (2017) analyzed an uncertain goal programming for solving a bicriteria STP. Mehlawat et al. (2019) incorporated a *data envelopment analysis* (DEA) concept in an STP under sustainable environment. Roy and Midya (2019) studied an MOSTP with product blending under uncertain environment. Biswas et al. (2019) employed a novel NSGA-II algorithm for solving an MOSTP in crisp and interval environments. Roy et al. (2019) discussed the idea of several items into an STP under a hybrid environment.

An exhaustive comparison on different features among the current research and prior associated studies in logistics network is depicted in Table 1. The comparative study delineates the gaps in previous researches and motivation for this investigation.

To resolve all these issues, in this study, we develop a mathematical model, referred as *multi-objective green solid transportation problem* (MOGSTP). To tackle several realistic and practical attributes of the research gaps, three conflicting objectives, transportation modes, vehicles efficiency, carbon emissions reduction policy, and sustainable development are simultaneously considered in the proposed problem. The summary of this chapter is listed as follows:

Table 1 A summary of relevant researches on MOSTP

Literature	TC	TT	CE	CM	VE	Sustainability
Elhedhli and Merrick (2012)	✓		✓			
Paksoy et al. (2012)	✓		✓	✓		
Tao and Xu (2012)	✓	✓		✓		
Chen et al. (2013)	✓		✓			
Molla-Alizadeh-Zavardehi et al. (2013)	✓			✓		
Konur and Schaefer (2014)	✓		✓			
Chen and Wang (2016)	✓		✓	✓		
Nomani et al. (2017a)	✓	✓				
Chen et al. (2017)	✓	✓		✓		
Gupta et al. (2018)	✓		✓	✓	✓	
Roy and Midya (2019)	✓	✓		✓		
Biswas et al. (2019)	✓	✓		✓		
Roy et al. (2019)	✓	✓		✓		
Das et al. (2020a)	✓			✓		
Maity et al. (2019b)	✓	✓				✓
Mehlawat et al. (2019)	✓			✓	✓	✓
Das and Roy (2019)	✓	✓	✓			
Tirkolaee et al. (2020)	✓		✓			✓
Proposed model	✓	✓	✓	✓	✓	✓

TC: Transportation cost; TT: Transportation time; CE: Carbon emissions; CM: Conveyance modes; VE: Vehicles efficiency

- This is the first study to investigate a transportation model to impact the carbon emissions reduction policy under sustainable environment.
- The model gives the decisions regarding the product flow from sources to destinations via sustainable transportation modes.
- The total conveyances cost with maintaining, fixed-charge cost and carbon emissions cost, delivery time, including loading and unloading time, and vehicle efficiency are also studied.
- A solution procedure on multi-objective optimization is described to receive an optimal solution of MOGSTP.
- A comparative study among the proposed procedure and the other existing relevant procedures is investigated to obtain the best Pareto-optimal solution.
- The impact of carbon tax policy in MOGSTP under sustainable environment is also discussed.

This chapter is sorted out as follows: Sect. 2 addresses the problem identification and mathematical model for MOGSTP. The methodology to solve the above model is discussed in Sect. 3. Then, Sect. 4 illustrates the efficiency of the stated problem and procedure with two real-world examples. Section 5 provides the outcomes of the application examples with the impact of carbon tax policy and a comparison with other relevant solution procedures. Finally, concluding remarks are given with future directions of this study in Sect. 6.

2 Mathematical Formulation

Here, initially, we define the proposed problem and its assumptions to formulate the mathematical model. Thereafter, we discuss index sets, decision variables, and parameters. At last, the three objective functions and constraints are described.

2.1 Model Background

The proposed study deals with a multi-objective problem to formulate a sustainable solid logistics network considering carbon emissions. The important objectives are: (O1) to lessen the total shipping cost and time, and to maximize sustainable vehicles efficiency under an emission control policy, and (O2) to seek the optimal amount of transported goods simultaneously. In addition, apart from that, the following factors are described in the model: (I) weights of the vehicles dependent on economical, environmental, and social features; (II) fixed-charge costs such as setup costs underway frameworks, toll charges on an expressway and so on; (III) servicing costs of the conveyances; (IV) loading and unloading time for the transported products, which give more accuracy of the logistic time; (V) follow the carbon tax policy for reducing carbon emissions. This study's main focus is to design a green transportation model by considering outrageous climate occasions for reducing carbon emissions under a sustainable environment. The considered green logistics network as depicted in Fig. 2, is comprised of sources (S_1, S_2, and S_3), destinations (D_1, D_2, D_3 and D_4) and conveyances (T_1, T_2, T_3 and T_4). Moreover, the dotted lines are designated as the product flow from sources to destinations through conveyances under a sustainable environment.

2.2 Nomenclature

The nomenclatures are listed below to construct our model:

Indices

i Index of sources ($i = 1, 2, \ldots, m$);
j Index of destinations ($j = 1, 2, \ldots, n$);
k Index of transportation modes ($k = 1, 2, \ldots, p$).

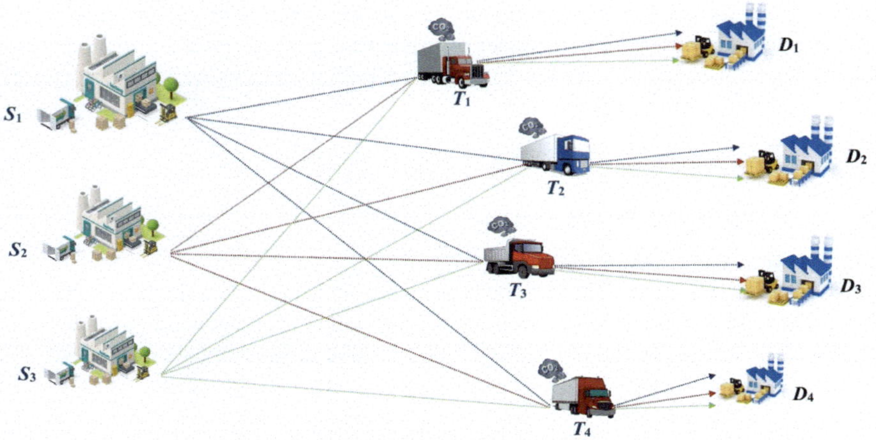

Fig. 2 Pictorial representation of a green solid logistics network

Sets

- X $\{(x_{ijk})$: subject to the constraints $\forall i, j, k\}$: the feasible space;
- F $\{(x^B_{ijk})$: $\forall i, j, k\}$: the optimal feasible set.

Decision variable

- x_{ijk} Quantity to be sent from ith source to jth destination by kth vehicles.

Parameters

- a_i Availability of ith source;
- b_j Demand at jth destination;
- c_k Capacity of kth transportation mode;
- e_{ijk} Unit cost for transporting unit item per unit distance from the ith source to the jth destination by kth vehicle;
- t_{ijk} Transport time for kth vehicle per unit distance from ith source to jth destination;
- l_i Loading time of unit item at ith source;
- l'_j Unloading time of unit quantity at jth destination;
- M_k Servicing cost of kth conveyance;
- f_{ijk} Fixed-charge to distribute products from ith–jth route by kth transportation mode;
- h_k Unit CO_2 mitigation by kth vehicle;
- P_c Carbon tax revenue for each unit of mitigation;
- α_k Effectiveness of kth conveyance in economical aspects;
- β_k Effectiveness of kth conveyance in environmental evaluation;
- γ_k Effectiveness of kth conveyance in social features;
- w_1 Weight for economical aspects;
- w_2 Weight for environmental evaluation;

w_3 Weight for social features.

2.3 Assumption

There are the following assumptions:

- The dispersed product is the homogeneous kind. The idea of transportation modes is heterogeneous. Transportation cost is straightforwardly relative to the unit of delivered products.
- The time is likely to be proportional to the distance that is not dependent on the unit distributed item. CO_2 mitigation is reliant on the distance gone by the movements, fuel consumption, and transported goods.

2.4 Mathematical Model

This section provides a mathematical formulation in the light of solid logistics modelling, carbon tax policy, and sustainable environment. The mathematical model of MOGSTP is addressed as follows:

Model 1

$$\text{minimize } Z_{1(x)} = \sum_{i=1}^{m}\sum_{j=1}^{n}\sum_{k=1}^{p}[e_{ijk}x_{ijk} + M_k d_{ij} y_{ijk} + f_{ijk} y_{ijk}] + P_c \sum_{i=1}^{m}\sum_{j=1}^{n}\sum_{k=1}^{p} h_k d_{ij} x_{ijk} \quad (1)$$

$$\text{minimize } Z_{2(x)} = \sum_{i=1}^{m}\sum_{j=1}^{n}\sum_{k=1}^{p}[t_{ijk} y_{ijk} + (l_i + l'_j) x_{ijk}] \quad (2)$$

$$\text{maximize } Z_{3(x)} = \sum_{i=1}^{m}\sum_{j=1}^{n}\sum_{k=1}^{p}(w_1\alpha_k + w_2\beta_k + w_3\gamma_k) x_{ijk} \quad (3)$$

$$\text{subject to } \sum_{j=1}^{n}\sum_{k=1}^{p} x_{ijk} \le a_i, \quad i = 1, 2, \ldots, m, \quad (4)$$

$$\sum_{i=1}^{m}\sum_{k=1}^{p} x_{ijk} \ge b_j, \quad j = 1, 2, \ldots, n, \quad (5)$$

$$\sum_{i=1}^{m}\sum_{j=1}^{n} x_{ijk} \le c_k, \quad k = 1, 2, \ldots, p, \quad (6)$$

$$y_{ijk} = \begin{cases} 1, \text{ if } x_{ijk} > 0, \\ 0, \text{ otherwise.} \end{cases} \tag{7}$$

$$\sum_{i=1}^{m} a_i \geq \sum_{j=1}^{n} b_j \text{ and } \sum_{k=1}^{p} c_k \geq \sum_{j=1}^{n} b_j, \tag{8}$$

$$x_{ijk} \geq 0, \quad i = 1, 2, \ldots, m; \quad j = 1, 2, \ldots, n; \quad k = 1, 2, \ldots, p. \tag{9}$$

The objective function (1) intends to minimize the total shipping cost along with CO_2 mitigation cost. Terms 1–4 of (1) address the all out transportation cost, servicing cost, fixed-charge cost, and CO_2 mitigation cost separately. The objective function (2) is identified with clients' fulfilment, which expectations to diminish the total shipping time, loading and unloading time from ith source to jth destination via kth conveyance. The objective function (3) is connected with economical, environmental, and social aspects, which indicates choosing conveyances depending on their sustainable effectiveness. Limitation (4) implements that each source's by and the large disseminated amount should be less or equivalent to its ability. Imperative (5) forces that the in generally delivered units of every objective satisfy the interest. Imperative (6) shows that every transportation mode's generally speaking shipped streams can't outperform its capacity. Constraint (7) represents the relationship between continuous and binary variables. Requirements (8) allude to the practical basis of the issue. Eventually, requirement (9) is the non-antagonism conditions.

Definition 1 An ideal solution of the proposed model is the one that optimizes each of the objective independently, i.e., $Z_q(x^*) = \min_{(x)\in F} Z_q(x)$, $q = 1, 2$ and $Z_3(x^*) = \max_{(x)\in F} Z_3(x)$.

Definition 2 A solution $x^A \in F$ of Model 1 is called an anti-ideal if it satisfies the condition $Z_q(x^A) = \max_{(x)\in F} Z_q(x)$, $q = 1, 2$ and $Z_3(x^A) = \min_{(x)\in F} Z_3(x)$.

Definition 3 A solution $x^P \in F$ is a Pareto-optimal solution of Model 1 if there does not exist any other solution $x \in F$ such that $Z_q(x) \leq Z_q(x^P)$ for $1 \leq q \leq 2$ and $Z_3(x) \geq Z_3(x^P)$ with at least one inequality holding as strict inequality.

3 Solution Procedure

In a multi-objective optimization problem, the *decision-maker* (DM) has to minimize/maximize the non-commensurable nature of objective functions all at once. Due to this fact, no single optimum outcome exists that simultaneously optimizes all the objective functions. For that reason, the idea of a Pareto frontis introduced instead of an ideal outcome. Here, we introduce a multi-objective solution procedure, explicitly, a *global criterion method* (GCM) (Shih and Chang 1995) to solve Model 1. An algorithm flowchart is depicted in Fig. 3.

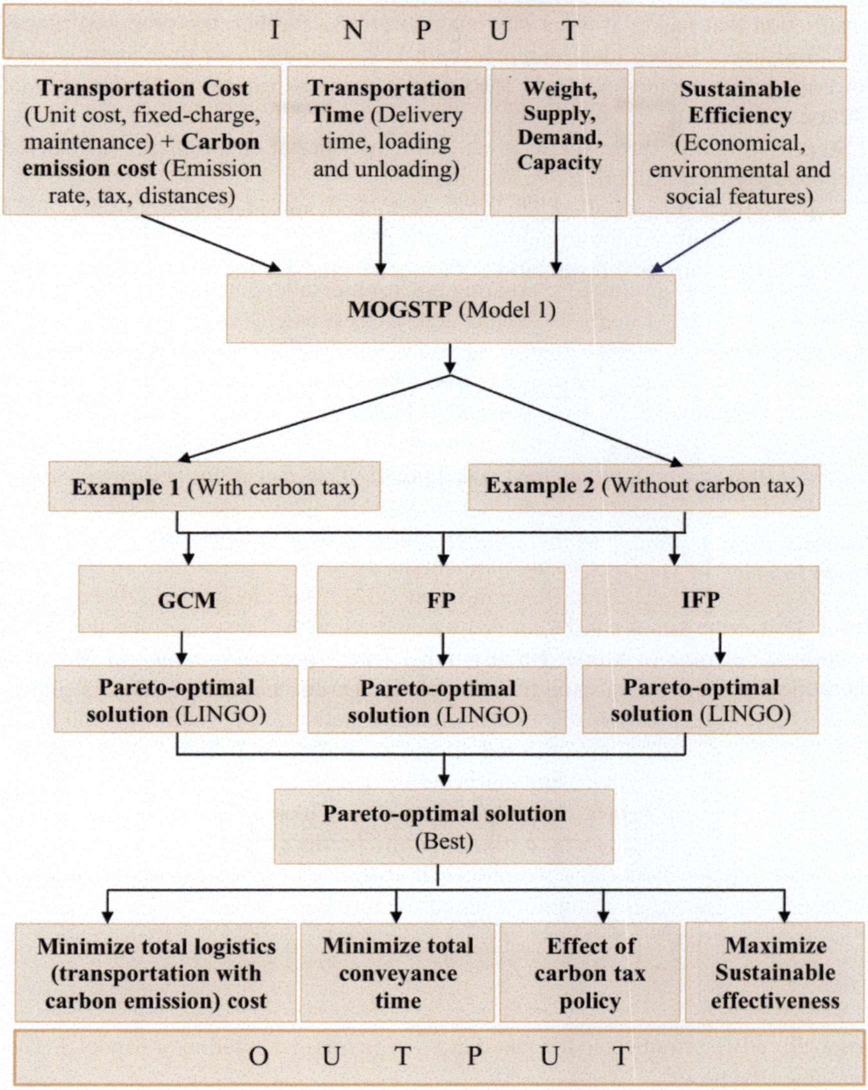

Fig. 3 Algorithm flowchart of the proposed problem

3.1 GCM

GCM uses the possibility of the briefest path from the ideal front to discover a Pareto front; the stated methodology does not need any earlier data (objectives and loads) on target capacities from the DM. Indeed, GCM yields a rudimentary numerical construction that makes it more obvious and utilize. Further, the proposed procedure furnishes a Pareto ideal outcome with less computational time and memory concerning different methodology. The following steps can derive the Pareto-optimal solution of Model 1:

Step 1: Initially, three single green solid transportation problems are solved independently by considering one at a time.

Step 2: Then, the ideal $(Z_1^{\min}, Z_2^{\min}, Z_3^{\max})$ and anti-ideal $(Z_1^{\max}, Z_2^{\max}, Z_3^{\min})$ solutions are calculated employing the results of Step 1.

Step 3: The simplified formulation of Model 1 can be addressed as stated below:

Model 2

$$\text{minimize} \left[\sum_{q=1}^{2} \left(\frac{Z_q(x) - Z_q^{\min}}{Z_q^{\max} - Z_q^{\min}} \right)^2 + \left(\frac{Z_3(x) - Z_3^{\max}}{Z_3^{\max} - Z_3^{\min}} \right)^2 \right]^{\frac{1}{2}}$$

subject to the constraints (4) to (9).

Proposition 1 *Assuming x^P is an optimal outcome of Model 1, then it will be a Pareto front x^P of Model 2.*

Proof This proposition can be proved by contradiction. Let us assume that x^P is an optimal outcome of Model 1 that is not a Pareto-optimal outcome of Model 2. Subsequently, there exists a solution x' such that (x') dominates x^P. This implies:

$$\left[\sum_{q=1}^{2} \left(\frac{Z_q(x') - Z_q^{\min}}{Z_q^{\max} - Z_q^{\min}} \right)^2 + \left(\frac{Z_3(x') - Z_3^{\max}}{Z_3^{\max} - Z_3^{\min}} \right)^2 \right]^{\frac{1}{2}}$$
$$< \left[\sum_{q=1}^{2} \left(\frac{Z_q(x^P) - Z_q^{\min}}{Z_q^{\max} - Z_q^{\min}} \right)^2 + \left(\frac{Z_3(x^P) - Z_3^{\max}}{Z_3^{\max} - Z_3^{\min}} \right)^2 \right]^{\frac{1}{2}},$$

which directly contradicts to the fact that x^P is an optimal solution of Model 2. That completes the proof.

4 Numerical Experiment

In this section, two application examples are illustrated to validate the formulated problem and methodology numerically.

Example 1 A company has four $(m = 4)$ source plants, labelled as S_1, S_2, S_3 and S_4, and three $(n = 3)$ distribution centres like D_1, D_2 and D_3, respectively.

Goods are distributed by two ($p = 2$) transportation modes from source plants to distribution centres. Due to environmental concerns and carbon tax policy, the company chooses two sustainable conveyances: hydrogen fuel-based vehicle (k_1) and bio-diesel based vehicle (k_2). The company's goal is to reduce the total logistics cost and carbon emissions cost, conveyance time, and increase the effectiveness of sustainable vehicles. The distances between sources and destinations are given in Table 2 with their corresponding supplies and demands. The unit logistics cost (INR), delivery time (minutes), and fixed-charge cost (INR) parameters are displayed in Table 3.

The other inputs are as follows:

Capacities of conveyances (KG) $c_1 = 1060$, $c_2 = 1210$; Loading time (minutes) $l_1 = 90$, $l_2 = 120$, $l_3 = 60$, $l_4 = 80$; Carbon emissions tax (INR/KG) $P_c = 0.4$; Unloading time (minutes) $l'_1 = 60$, $l'_2 = 40$, $l'_3 = 30$; Carbon emissions rate (KG/L) $h_1 = 0.3$, $h_2 = 0.5$; Weights of the sustainability aspects $w_1 = 0.3$, $w_2 = 0.5$, $w_3 = 0.2$; Maintenance cost (INR) $M_1 = 0.3$, $M_2 = 0.5$; Effectiveness of conveyances $\alpha_1 = 0.9$, $\alpha_2 = 0.75$, $\beta_1 = 0.85$, $\beta_2 = 0.7$, $\gamma_1 = 0.6$, $\gamma_2 = 0.85$.

Table 2 Distances (d_{ij}) (KM) of sources-destinations with supply and demand (KG)

	D_1	D_2	D_3	a_i
S_1	45	60	80	670
S_2	55	40	100	710
S_3	70	90	50	520
S_4	100	50	60	370
b_j	835	740	695	

Table 3 Transportation cost for unit quantity, conveyance time, and fixed-charge cost ($e_{ijk}, t_{ijk}, f_{ijk}$)

Source–destination	Conveyance ($k = 1$)	Conveyance ($k = 2$)
1−1	(16, 30, 220)	(14, 45, 350)
1−2	(22, 45, 145)	(15, 60, 190)
1−3	(26, 60, 300)	(20, 80, 250)
2−1	(16, 41.5, 280)	(14, 55, 390)
2−2	(16, 30, 245)	(11, 40, 500)
2−3	(12, 75, 360)	(11, 100, 500)
3−1	(15, 52.5, 400)	(10, 70, 435)
3−2	(14, 67.5, 450)	(12, 90, 410)
3−3	(10, 37.5, 425)	(90, 50, 420)
4−1	(18, 75, 400)	(13, 100, 600)
4−2	(17, 37.5, 430)	(15, 50, 495)
4−3	(13, 45, 540)	(12, 60, 470)

Example 2 We consider the fossil fuel byproducts charge $P_c = 0$, and the others continue as before as in Example 1.

5 Experimental Result and Exploration

The optimal solutions of the proposed problem are extracted from the GCM. The GCM is carried out employing LINGO 18.0.56 software on an Intel Core i5 processor with 8 GB RAM under Mac OS environment. The 'most preferred' Pareto fronts of both the applications are listed in Tables 4 and 5. The analytical outcomes reveal that the economic, clients' fulfilment, and sustainable objectives are optimized. When the carbon emissions cost is added along with the total logistics cost, then firms' profit will be reduced. Because of this reality, the organization will consistently be worried about fossil fuel byproducts because of the dispersion of merchandise. The company will always select the sustainable conveyances which emit less CO_2. In this regard, the third objective function plays a vital role in choosing conveyances which increases the sustainable efficiency score.

Along these lines, the expressed detailing can handle financial advancement without inconvenience to ecological and natural assets. Thereafter, the carbon tax policy helps the company choose 'most preferred' choices for enhancing their economic and sustainable development and upholds the policymaker for lessening fossil fuel byproducts. The impact of carbon tax policy is illustrated in Fig. 4.

Table 4 The Pareto ideal outcome of Example 1

Methodology	Pareto-front out come	CPU time (s)	Memory (K)
GCM	$\mathbf{Z_1 = 52257.52, Z_2 = 307230.0, Z_3 = 1765.35}$, $x_{111} = 558.94$, $x_{112} = 67.115$, $x_{122} = 43.942$, $x_{212} = 208.942$, $x_{221} = 501.05$, $x_{332} = 520.0$, $x_{422} = 195.0$, $x_{432} = 175.0$ and remaining all $x_{ijk} = 0$	0.18	72
FP	$\lambda = 0.808$, $Z_1 = 52523.96$, $Z_2 = 307230.0$, $Z_3 = 1765.35$, $x_{111} = 568.30$, $x_{122} = 101.69$, $x_{211} = 48.386$, $x_{212} = 218.30$, $x_{221} = 443.30$, $x_{332} = 520.0$, $x_{422} = 195.0$, $x_{432} = 175.0$ and remaining all $x_{ijk} = 0$	0.27	73
IFP	$\lambda = 0.712$, $\theta = 0.287$, $Z_1 = 52523.96$, $Z_2 = 307230.0$, $Z_3 = 1765.35$, $x_{111} = 568.30$, $x_{122} = 101.69$, $x_{211} = 48.386$, $x_{212} = 218.30$, $x_{221} = 443.30$, $x_{332} = 520.0$, $x_{422} = 195.0$, $x_{432} = 175.0$ and remaining all $x_{ijk} = 0$	0.23	74

Bold-front refers the 'most preferred' Pareto front

Table 5 The Pareto-ideal outcome of Example 2

Methodology	Pareto-front out come	CPU time (s)	Memory (K)
GCM	$\mathbf{Z_1 = 34154.44, Z_2 = 307230.0, Z_3 = 1765.35}$, $x_{111} = 523.94, x_{112} = 137.13, x_{122} = 8.937, x_{212} = 173.94, x_{221} = 536.06, x_{332} = 520.0, x_{422} = 195.0, x_{432} = 175.0$ and remaining all $x_{ijk} = 0$	0.18	72
FP	$\lambda = 0.865, Z_1 = 34156.79, Z_2 = 307230.0, Z_3 = 1765.35, x_{111} = 561.28, x_{112} = 62.428, x_{122} = 46.285, x_{212} = 211.28, x_{221} = 498.71, x_{332} = 520.0, x_{422} = 195.0, x_{432} = 175.0$ and remaining all $x_{ijk} = 0$	0.23	73
IFP	$\lambda = 0.865, \theta = 0.135, Z_1 = 34156.79, Z_2 = 307230.0, Z_3 = 1765.35, x_{111} = 561.28, x_{112} = 62.428, x_{122} = 46.285, x_{212} = 211.28, x_{221} = 498.71, x_{332} = 520.0, x_{422} = 195.0, x_{432} = 175.0$ and remaining all $x_{ijk} = 0$	0.27	74

Bold-front refers the 'most preferred' Pareto front

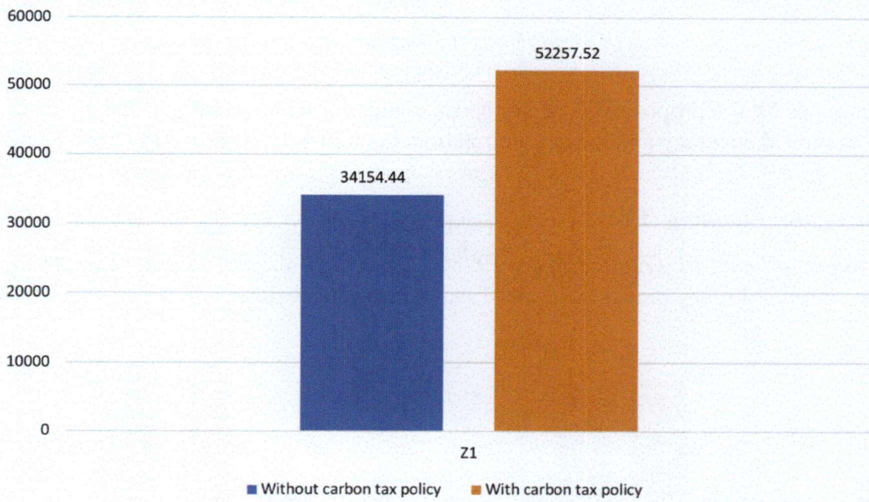

Fig. 4 Effect of carbon tax policy in MOGSTP

5.1 Comparative Study

In this subsection, the aforementioned examples are also resolved with existing relevant two solution procedures, namely: (i) *Fuzzy programming* (FP) and (ii) *Intuitionistic fuzzy programming* (IFP). The obtained outcomes are compared so that the DM has a choice to get the best Pareto-optimal solutions. Table 6 shows that the obtained

Table 6 Outcomes comparison

Example	GCM	FP	IFP
Example 1	$Z_1 =$ 52257.52	$Z_1 =$ 52523.96	$Z_1 =$ 52523.96
	$Z_2 =$ 307230.0	$Z_2 =$ 307230.0	$Z_2 =$ 307230.0
	$Z_3 = 1765.35$	$Z_3 = 1765.35$	$Z_3 = 1765.35$
	CPU time = 0.18 s	CPU time = 0.23 s	CPU time = 0.27 s
	Memory = 72 K	Memory = 73 K	Memory = 74 K
Example 2	$Z_1 =$ 34154.44	$Z_1 =$ 34156.79	$Z_1 =$ 34156.79
	$Z_2 =$ 307230.0	$Z_2 =$ 307230.0	$Z_2 =$ 307230.0
	$Z_3 = 1765.35$	$Z_3 = 1765.35$	$Z_3 = 1765.35$
	CPU time = 0.18 s	CPU time = 0.23 s	CPU time = 0.27 s
	Memory = 72 K	Memory = 73 K	Memory = 74 K

outcomes by the proposed GCM are better compared with the FP and IFP in terms of sustained accuracy and lower computational complexity, displayed in Figs. 5 and 6.

***Fuzzy programming* (FP)**

After employing FP (Zimmermann 1978; Inuiguchi et al. 1990; Li and Lai 2000), the simplified fuzzy model of Model 1 can be stated as below:

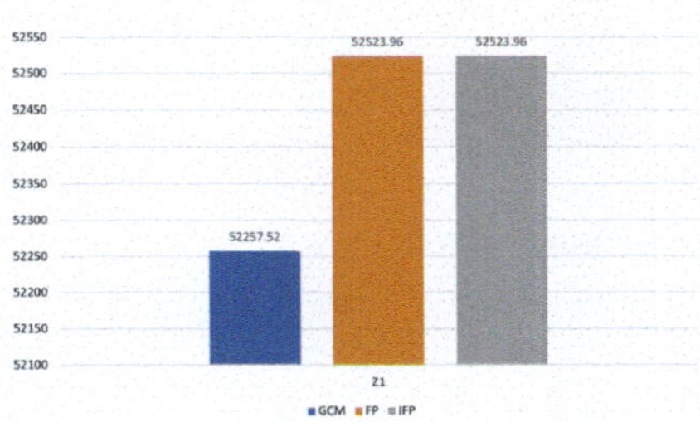

(a) Optimum value of Z_1 by three procedures for Example 1.

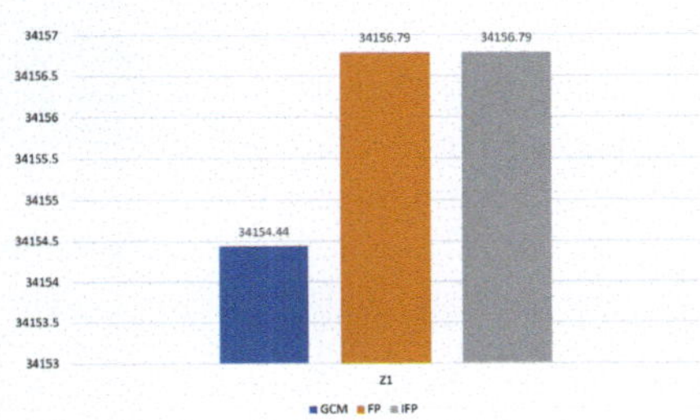

(b) Optimum value of Z_1 by three procedures for Example 2.

Fig. 5 Comparison optimum value of Z_1 by three procedures

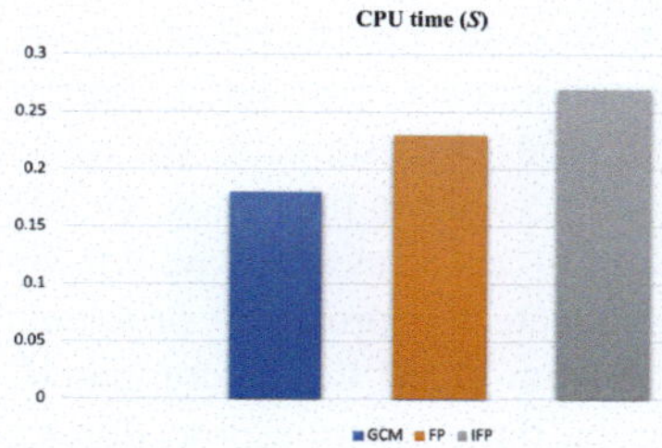

(a) CPU time for the three methods.

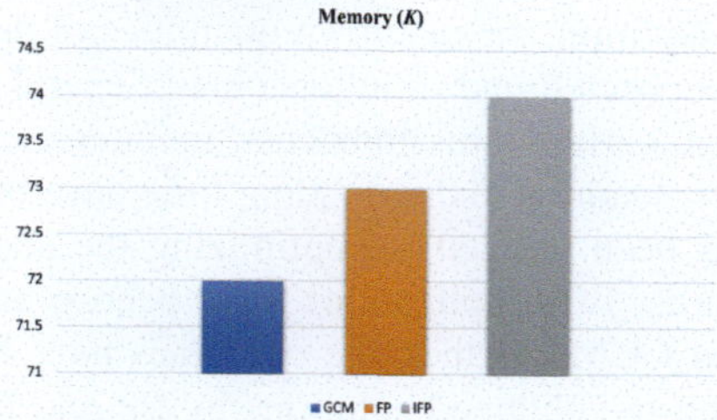

(b) Computational memory by the three methods.

Fig. 6 Comparison lower computational complexity of three approaches

Model 3

$$\begin{aligned}&\text{maximize } \lambda\\&\text{subject to } Z_q(x,) + \lambda(U_q - L_q) \le U_q, q = 1, 2, 3,\\&\text{the constraints (4) to (9)},\\&\lambda \ge 0.\end{aligned}$$

Here, λ is the degree of fulfilment of anoutcome.

***Intuitionistic fuzzy programming* (IFP)**

After employing IFP (Angelov 1997; Roy et al. 2018a; Roy and Midya 2019), the intuitionistic optimization model for Model 1 can be expressed as follows:

Model 4

$$\begin{aligned}&\text{maximize } \theta - \mu\\&\text{subject to } Z_q(x,) + \theta(U_q - L_q) \le U_q, q = 1, 2, 3,\\&Z_q(x,) - \mu(U_q - L_q) \le L_q, q = 1, 2, 3,\\&\text{the constraints (4) to (9)},\\&\theta \ge \mu, \theta + \mu \le 1, \theta, \mu \in [0, 1].\end{aligned}$$

Here, θ and μ indicate the grade of fulfilment and disappointment of anoutcome.

6 Concluding Remark and Outlook

In this chapter, a green solid logistics modelling has been addressed by thinking about the affordable, clients' fulfilment, and ecological goals under sustainable development. To help the choices, the stated model has been detailed with the aforementioned three non-commensurable objectives under a fossil fuel byproducts decrease strategy.

In addition, the unknown quantities of distributed products by various conveyances have been found simultaneously. Several factors like unit logistics cost, fixed-charge cost, carbon emissions cost, maintenance cost, three types of sustainable parameters, loading, and unloading time have been taken into account in this study. GCM has been utilized for optimizing the above objectives in a successful way. Afterwards, two applications have been presented to delineate the proposed problem and GCM. Further, the outcomes have been derived from GCM that is compared with the obtained solutions by FP and IFP. Among the three solution procedures, better Pareto-optimal outcomes have been received from the proposed GCM in terms of sustained accuracy and lower computational complexity. Ultimately, choices in regards to decreasing carbon emanations because of transportation frameworks have been examined, as well. It has been concluded that our formulation can proceed with the economic progress without detriment to ecological and regular assets.

There are a few interesting topics for future research works of this study by considering different kinds of environments such as grey systems (Roy et al. 2017a), robust environment (Goli et al. 2019; Khalilpourazari et al. 2019), stochastic (Mahapatra et al. 2013), neutrosophic set (Das and Roy 2019), dual-hesitant fuzzy (Maity et al. 2019a), interval valued (Roy and Maity 2017), multi-choice type (Roy et al. 2017b), location problem (Das et al. 2020a, b, c, 2021), rough sets (Roy et al. 2018b), type-2 fuzzy set (Roy and Bhaumik 2018; Roy and Maiti 2020), and biofuel supply chain (Paksoy et al. 2013). Further, the several nature inspired optimization algorithms and stochastic methods can be employed to solve the proposed model.

References

Angelov PP (1997) Optimization in an intuitionistic fuzzy environment. Fuzzy Sets Syst 86(3):299–306

Benjaafar S, Li Y, Daskin M (2012) Carbon footprint and the management of supply chains: Insights from simple models. IEEE Trans Autom Sci Eng 10(1):99–116

Biswas A, Shaikh AA, Niaki STA (2019) Multi-objective non-linear fixed charge transportation problem with multiple modes of transportation in crisp and interval environments. Appl Soft Comput 80:628–649

Chen L, Peng J, Zhang B (2017) Uncertain goal programming models for bicriteria solid transportation problem. Appl Soft Comput 51:49–59

Chen X, Benjaafar S, Elomri A (2013) The carbon-constrained EOQ. Oper Res Lett 41(2):172–179

Chen X, Wang X (2016) Effects of carbon emission reduction policies on transportation mode selections with stochastic demand. Transp Res Part E: Logist Transp Rev 90:196–205

Das SK, Roy SK (2019) Effect of variable carbon emission in a multi-objective transportation-p-facility location problem under neutrosophic environment. Comput Ind Eng 132:311–324

Das SK, Roy SK, Weber G-W (2020a) Heuristic approaches for solid transportation-p-facility location problem. CEJOR 28:939–961

Das SK, Roy SK, Weber GW (2020b) An exact and a heuristic approach for the transportation–facility location problem. Comput Manag Sci 17:389–407

Das SK, Roy SK, Weber G-W (2020c) Application of type-2 fuzzy logic to a multi-objective green solid transportation-location problem with dwell time under carbon tax, cap and offset policy: fuzzy vs. non-fuzzy techniques. IEEE Trans Fuzzy Syst 28(11):2711–2725

Das SK, Pervin M, Roy SK, Weber G-W (2021) Multi-objective solid transportation-location problem with variable carbon emission in inventory management: a hybrid approach. Ann Oper Res. https://doi.org/10.1007/s10479-020-03809-z

Elhedhli S, Merrick R (2012) Green supply chain network design to reduce carbon emissions. Transp Res Part D: Transp Environ 17(5):370–379

Goli A, Tirkolaee EB, Malmir B, Bian G-B, Sangaiah AK (2019) A multi-objective invasive weed optimization algorithm for robust aggregate production planning under uncertain seasonal demand. Computing 101(6):499–529

Gupta P, Mehlawat MK, Aggarwal U, Charles V (2018) An integrated ahp-dea multi-objective optimization model for sustainable transportation in mining industry. Resour Policy. https://doi.org/10.1016/j.resourpol.2018.04.007

Haley K (1962) New methods in mathematical programming—the solid transportation problem. Oper Res 10(4):448–463

Inuiguchi M, Ichihashi H, Tanaka H (1990) Fuzzy programming: a survey of recent developments. In: Stochastic versus fuzzy approaches to multiobjective mathematical programming under uncertainty. Springer, pp 45–68

Jiménez F, Verdegay JL (1998) Uncertain solid transportation problems. Fuzzy Sets Syst 100(1–3):45–57

Jiménez F, Verdegay JL (1999) Solving fuzzy solid transportation problems by an evolutionary algorithm based parametric approach. Eur J Oper Res 117(3):485–510

Khalilpourazari S, Mirzazadeh A, Weber G-W, Pasandideh SHR (2019) A robust fuzzy approach for constrained multi-product economic production quantity with imperfect items and rework process. Optimization 69(1):63–90

Konur D, Schaefer B (2014) Integrated inventory control and transportation decisions under carbon emissions regulations: LTL vs. Tl carriers. Transp Res Part E: Logist Transp Rev 68:14–38

Lélé SM (1991) Sustainable development: a critical review. World Dev 19(6):607–621

Li L, Lai KK (2000) A fuzzy approach to the multiobjective transportation problem. Comput Oper Res 27(1):43–57

Litman T, Burwell D (2006) Issues in sustainable transportation. Int J Glob Environ Issues 6(4):331–347

Liu P, Yang L, Wang L, Li S (2014) A solid transportation problem with type-2 fuzzy variables. Appl Soft Comput 24:543–558

Mahapatra DR, Roy SK, Biswal MP (2013) Multi-choice stochastic transportation problem involving extreme value distribution. Appl Math Model 37(4):2230–2240

Maity G, Mardanya D, Roy SK, Weber G-W (2019a) A new approach for solving dual-hesitant fuzzy transportation problem with restrictions. Sādhanā 44(4):75. https://doi.org/10.1007/s12046-018-1045-1

Maity G, Roy SK, Verdegay JL (2019b) Time variant multi-objective interval-valued transportation problem in sustainable development. Sustainability 11(21):6161. https://doi.org/10.3390/su11216161

Mehlawat MK, Kannan D, Gupta P, Aggarwal U (2019) Sustainable transportation planning for a three-stage fixed charge multi-objective transportation problem. Ann Oper Res 1–37. https://doi.org/10.1007/s10479-019-03451-4

Molla-Alizadeh-Zavardehi S, Nezhad SS, Tavakkoli-Moghaddam R, Yazdani M (2013) Solving a fuzzy fixed charge solid transportation problem by metaheuristics. Math Comput Model 57(5–6):1543–1558

Nomani MA, Ali I, Ahmed A (2017a) A new approach for solving multi-objective transportation problems. Int J Manag Sci Eng Manag 12(3):165–173

Nomani MA, Ali I, Fügenschuh A, Ahmed A (2017b) A fuzzy goal programming approach to analyse sustainable development goals of india. Appl Econ Lett 24(7):443–447

Paksoy T, Özceylan E, Weber G-W (2013) Profit oriented supply chain network optimization. CEJOR 21(2):455–478

Paksoy T, Pehlivan NY, Özceylan E (2012) Fuzzy multi-objective optimization of a green supply chain network with risk management that includes environmental hazards. Hum Ecol Risk Assess Int J 18(5):1120–1151

Poterba JM (1991) Tax policy to combat global warming: on designing a carbon tax. Technical report, National Bureau of Economic Research

Roy SK, Bhaumik A (2018) Intelligent water management: a triangular type-2 intuitionistic fuzzy matrix games approach. Water Resour Manag 32(3):949–968

Roy SK, Ebrahimnejad A, Verdegay JL, Das S (2018a) New approach for solving intuitionistic fuzzy multi-objective transportation problem. Sādhanā 43(1):3. https://doi.org/10.1007/s12046-017-0777-7

Roy SK, Midya S, Vincent FY (2018b) Multi-objective fixed-charge transportation problem with random rough variables. Int J Uncertain Fuzz Knowl Based Syst 26(6):971–996

Roy SK, Maiti SK (2020) Reduction methods of type-2 fuzzy variables and their applications to stackelberg game. Appl Intell 50:1398–1415. https://doi.org/10.1007/s10489-019-01578-2

Roy SK, Maity G (2017) Minimizing cost and time through single objective function in multi-choice interval valued transportation problem. J Intell Fuzzy Syst 32(3):1697–1709

Roy SK, Maity G, Weber G-W (2017a) Multi-objective two-stage grey transportation problem using utility function with goals. Cent Eur J Oper Res 25(2):417–439

Roy SK, Maity G, Weber GW, Gök SZA (2017b) Conic scalarization approach to solve multi-choice multi-objective transportation problem with interval goal. Ann Oper Res 253(1):599–620

Roy SK, Midya S (2019) Multi-objective fixed-charge solid transportation problem with product blending under intuitionistic fuzzy environment. Appl Intell 49(10):3524–3538

Roy SK, Midya S, Webe G-W (2019) Multi-objective multi-item fixed-charge solid transportation problem under twofold uncertainty. Neural Comput Appl 31(12):8593–8613

Shell E (1955) Distribution of a product by several properties, directorate of management analysis. In: Proceedings of the second symposium in linear programming, vol 2, pp 615–642

Shih C, Chang C (1995) Pareto optimization of alternative global criterion method for fuzzy structural design. Comput Struct 54(3):455–460

Tao Z, Xu J (2012) A class of rough multiple objective programming and its application to solid transportation problem. Inf Sci 188:215–235

Tirkolaee EB, Mardani A, Dashtian Z, Soltani M, Weber G-W (2020) A novel hybrid method using fuzzy decision making and multi-objective programming for sustainable-reliable supplier selection in two-echelon supply chain design. J Clean Prod 250:119517

Turken N, Carrillo J, Verter V (2017) Facility location and capacity acquisition under carbon tax and emissions limits: to centralize or to decentralize? Int J Prod Econ 187:126–141

Zhou S, Shi M, Li N, Yuan Y (2011) Impacts of carbon tax policy on CO_2 mitigation and economic growth in china. Adv Clim Chang Res 2(3):124–133

Zimmermann H-J (1978) Fuzzy programming and linear programming with several objective functions. Fuzzy Sets Syst 1(1):45–55

Chapter 6
Sustainable Connectivity Management via e-SIM Multi Licence-Based Billing

Atul Narain, Pratibha Garg, and Dilshad Ahmad Ansari

Abstract This paper talks about e-SIM as a software module being used under a licencing model similar to current practices of Software Licencing. The paper talks on the sustainable model of switching to e-SIM and doing away with paper-based physical SIM especially in IoT industry. And the paper talks about how billing would be done in such a scenario of a single profile being used concurrently on multiple devices/subscriptions in different ways. The conventional one-to-one relationship between a device and its associated service profile is broken to allow for plural/concurrent service profiles to be activated and billed for use in connection with a single user profile. A separate billing identification is maintained for each service profile in order to allow for billing of concurrent usage across devices. This would enable concurrent billing across multiple devices having the same IMSI but different IMEI and eUICC ID. The Billing would be based on the combination of several parameters like—number of simultaneous devices that are active, type of device on which the profile is active (based on IMEI) and type of profile being used actively on the device. The selection of a certain profile to be active is made through a combination of IMEI, eUICC ID, and the service profile, which will enable unique billing for concurrent usage of a single user profile on multiple devices while still maintaining the device specific usage statistics.

Keywords eSIM · Billing · eUICC ID · Multi device provisioning

1 Introduction

Today when we buy any software, it comes with licence binding which can be of any of the below types. An enterprise or an individual buying a piece of software would have the flexibility to choose from any of these licencing model.

A. Narain · P. Garg (✉)
Amity School of Business, Amity University, Uttar Pradesh, Noida, India
e-mail: pgarg1@amity.edu

D. A. Ansari
A.I.D.C, Department of Commerce, Associated Lucknow University, lucknow, India

I. Ali et al. (eds.), *Computational Modelling in Industry 4.0*,
https://doi.org/10.1007/978-981-16-7723-6_6

On the contrary when we look at telecom industry today, although the user has the flexibility to buy a bill plan, however, he is restricted with a single physical SIM card to use at a time in a mobile handset. For use of multiple numbers, either a mobile handset has to support multiple SIM cards (which today is limited to maximum 2 in almost all of handsets) or the user has to do the tedious activity of changing the SIM cards every time.

The projections for IoT/M2M suggest 50 Billion connected devices in the network by 2020 and there is a need for the businesses to design and understand the impact that IoT solutions will have on integrating them into the Business Support Systems.

Now considering the evolution and rise in IoT devices, this current mechanism of physical SIM card management is a nightmare. Today mostly physical SIM is used in devices. There is an option to move towards software-based SIM called e-SIM going ahead in devices. Unlike physical SIM, an e-SIM can have multiple profiles saved in it and at any given time one of the desired profiles can be activated. If an e-SIM is used today it can be used with multiple devices with the service profile that is pre-configured in the HLR against the IMSI and MSISDN. The billing system is configured to process and bill according to the active plan associated with the MSISDN (Koshy and Rao 2018).

This paper explores the conceptual model of introducing the current Software licencing models in provisioning of e-SIM and providing the same for enterprises for the management of their IoT devices. Using an e-SIM is a great example of sustainable data consumption. Compared to cell phones SIM cards, e-SIM cards are more durable and can withstand adverse weather conditions. Some of the current Software licencing models are:

i. **Subscription Software Licence**

An individual or an enterprise can subscriber to a service and get a licence to use the service based on a certain periodic frequency e.g., each month or annual regardless of the use. The subscription is billed as per the frequency opted for. Software as a Service which is billed at a particular frequency is a subscription-based licence. It is one of the most popular approaches to software licences today. With the technology advancement, more and more of services are moving towards cloud hosted platform and subscription-based licencing is primarily available for these cloud hosted applications and services. It is easier to manage and record usage of an application or service and bill accordingly over a cloud hosted environment. Subscription-based licencing enables recurring revenue for service and application providers and as such is a preferred mode for them.

In an analogy to Telecom setup, the service providers are the hosted environment which can provide the e-SIM under subscription model and that would in return establish the recurring revenue stream.

ii. **Perpetual Software Licence**

Buy once and use as much and as long as one wish. The perpetual licence enables a user to use the application or the service forever. Traditionally before cloud hosted environments, application and service providers had only perpetual licencing model only where applications were installed on-premise. Today and enterprise, which has a very large base of employee or software users, prefer to use perpetual licence model, so not to incur an operational recurring cost but rather incur a onetime capital expense to the requirement.

iii. **Consumption or Pay-Per-Use Software Licence**

Internet usage charging is a very relevant example for Consumption or Pay-Per-Use Software licence. It is a licencing model where an enterprise or a user is billed based on the amount of the usage or the number or instances of usage of an application or service. Smaller enterprises may not be able to afford very expensive software's or any organization may not have a consistent need for software usage. As such, paying for how much and what they use is a very convenient model for them, rather than investing on buying expensive applications. On the other hand, software providers tend to move away from this model and rely more of subscription-based mode, as the subscription-based model tends to provide a more recurring and stable revenue. However, for users or enterprises to ger a taste of or an experience of the application or service, the enterprises may still provide this as one-time trial option (Leh 2014).

iv. **Node Locked Licence**

Sometimes the application providers link the application or software to the device on which it is being used. As such in Node Locked Licence as the name suggests, the user is bound to use the application or software on a particular device only.

v. **Floating Licence**

Contrary to Node Locked, if there are multiple devices, the floating licence types enable the application or software use on any of the devices as long as the overall limitation of usage is not exceeded. Each licence can be used on any machine in the network, and when one machine stops using it, it can release the licence to be used by any other device. It is usually controlled by how many licences can run concurrently in a network at any given point of time (Gull and Wehrmann 2009).

Smart Cities, IoT, and other aspects have created the need today for enterprises to manage and provision large number of devices in their network via SIM modules for connectivity via Voice, SMS, and Data. Some of the examples of such cases are below:

- Smart City: An enterprise managing multitude of cameras, AC's, access points, etc.in a building via SIM modules
- Enterprise: Managing inventory and logistics of goods movement
- Enterprise providing Services: A typical example is car manufacturers who need connectivity to vehicles to record and manage car performances etc.
- Smart Farming: Multitude of sensors connected with SIM to network providing data related to soil, weather, temperature, humidity, etc. (Stålbrand 2017).

The paper discusses in detail the conceptual architecture of how a service provider can leverage and use the licencing models for extending provisioning and billing services towards enterprises for their multi SIM requirements (Provisioning 2018).

2 Technical Background

Today mostly physical SIM is used in devices. There is an option to move towards software-based SIM called e-SIM going ahead in devices. Unlike physical SIM, an e-SIM can have multiple profiles saved in it and at any given time one of the desired profiles can be activated. If an e-SIM is used today it can be used with multiple devices with the service profile that is pre-configured in the HLR against the IMSI and MSISDN. The billing system is configured to process and bill according to the active plan associated with the MSISDN.

In the current scenario there is limitation of one-to-one relationship between each profile on the e-SIM and the associated bill plan in the billing system against the profile. This implies that all profiles in the e-SIM are mostly provisioned in the HLR and the Billing and network systems as well and only one of them will be active at any given time.

Also todays system do not differentiate between a normal telecom subscriber and a machine. There is a need to provision and bill based on subscriber categorization e.g., IoT, M2M, Normal User, Enterprise, Critical Services, etc., which would become more of reality with the upcoming 5G network.

The conventional one-to-one relationship between a device and its associated service profile is broken to allow for plural/concurrent service profiles to be activated and billed for use in connection with a single user profile. A separate billing identification is maintained for each service profile in order to allow for billing of concurrent usage across devices. The present paper talks about eSIM as a software module being used under a licencing model similar to current practices of Software Licencing. The paper talks about how billing would be done in such a scenario of single profile being used concurrently on multiple devices/subscriptions in different ways. This would enable concurrent billing across multiple devices having the same IMSI but different IMEI and eUICC ID.

The Billing would be based on the combination of several parameters like—number of simultaneous devices that are active, type of device on which the profile is active (based on IMEI) and type of profile being used actively on the device. The selection of a certain profile to be active is made through a combination of IMEI, eUICC ID, and the service profile, which will enable unique billing for concurrent usage of a single user profile on multiple devices while still maintaining the device specific usage statistics.

Advantages of the proposed solution

The advantages of the stated approach span across different segments—customer or End User, Telecom Service Provider & M2M Service Provider. Advantages for each of them are stated below:

Customer: Assuming a customer owns multiple connections (e.g multiple devices) which can an end user, Enterprise, or an IoT scenario.

- The solution approach would enable the end user to have only a single user profile and activate service profiles in the device of his choice at any given point of time.
- Seamless portability of profiles across devices based on need, location, etc.
- Single Bill generation and single profile maintenance.
- Choice of expansion with licencing at any given time.
- No physical inventory of SIM or damage control.
- Ability of play around with multiple profiles from different Telecom Service Provider enabled on same e-SIM.
- Self-Care Management.

Telecom Service Provider: Assuming TSP is providing multiple service profiles against a given user profile for bulk/multi usage across devices (Abdou 2019).

- Single profile maintenance.
- Move away from individual provisioning and activation to service profiles.
- Move away from 'Always ON' in the network to 'On demand'.
- Move away from physical generation of SIM cards as well as Distributor and Retailer management.
- Monetization using multiple licencing and device to device communication.
- Optimization on HSS, CRM, and Billing systems for data records.
- Empowered customer with Self Care.
- Planning for IoT/User using Meta data driven any industry approach.

M2M Service Provider: Assuming M2M service provider is extending a self-care towards enterprise customer for profile management as well as tie up with Telecom Service Provider for dynamic provisioning and configuration across network (HSS) and BSS (Billing and CRM)

- Move away from Distributor and Retailer management.
- Complete user and service profile Life Cycle Management.
- Extended Self-Care Management with full suite of services.

Monetization using multiple licencing and profile management (Sealy 2019).

3 Solution Approach

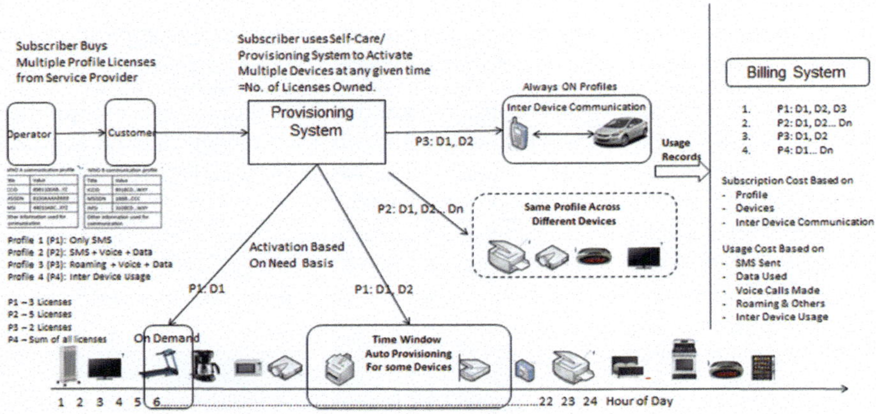

As stated in previous sections, this solution approach makes it possible to have a single bill for multiple devices having multiple profiles owned by a user (Mehmi et al. 2017).

1. A user will have a custom bundle of licence to simultaneously use multiple service profiles across multiple devices.
2. A user procures multiple service profiles associated with a User profile from an operator's /CSP Account salesperson.
3. A service profile can have a bundle of services and each of them can be enabled and activated on eSIM at a time, using self-care based on Customer preference.
4. A cloud-based provisioning system shall be provided as a Mobile (IoS/Android etc.) App or Web Portal, for activation/de-activation of devices and service profiles.
5. The solution enables the user to have multiple choices for associating a service profile with a particular device type.
6. The user can use the same service profile across multiple devices, as well as can use multiple service profiles simultaneously across devices, based on purchased licence bundle.
7. The billing parameters related to usage(SMS, Voice, Data) and licence (Active devices, active profiles) are configured and processed against the unique User profile.
8. The HLR and CRM would be provisioned against the User profile.
9. The purchased licences would be associated with a unique User profile and a single bill will be generated for his complete usage covering different service profiles across all the devices.

10. A single bill generation would happen based on the licence policy defined in the billing system; this will enable the user to pick and choose services across multiple devices concurrently.

The solution focuses on licence-based charging and associating multiple Service Profiles with a single User Profile, enabling the user to activate the same user profile across multiple devices (Vesselkov et al. 2015). Below given are some of the sample use case scenarios for charging as per the solution.

i. **Charging Use Cases:**

Inter Device Usage: This is an example scenario where devices using service profiles under the same user profile communicate to each other using different service type—sms, voice, or data. This is similar to a CUG scenario, and billing configuration can be similar to the way we do it today for CUG subscribers. However, at any given point of time only a specific number of such devices can be made active as per the licence procured by the customer.

Service Profile Licence Usage: This is an example scenario where a customer can configure any number of devices with the same Service profile configuration licences he has bought, however he can only activate a fixed number of devices simultaneously as per the number of licences he has procured against that Service Profile.

Flexible Mode of Billing: The CSP can configure the billing system to enable Billing as per combinations of metering data. E.g. a profile can be activated during night on one device (e.g. TV/AC etc) and during day time on another (e.g. Car etc). The billing in such scenarios can be based on any of the below combination of metering data:

- Kb Data for IP Traffic.
- Voice Minutes.
- SMS Count.
- Active Device ID's.
- Duration of Service Profile Active.
- Service Types Active on a Profile.
- Different Device ID's used against each Service Profile.
- Scheduled Activation of Service Profile on a device.

Site-Based Licencing: The CSP can provide licence based on location or service area defined by BS or BSC/RNC or a building or location. In this scenario the CSP can provide agreed/procured or any number of licences to be configured in the vicinity of the defined location and customized billing (based on location/site) agreements can be made with the customer.

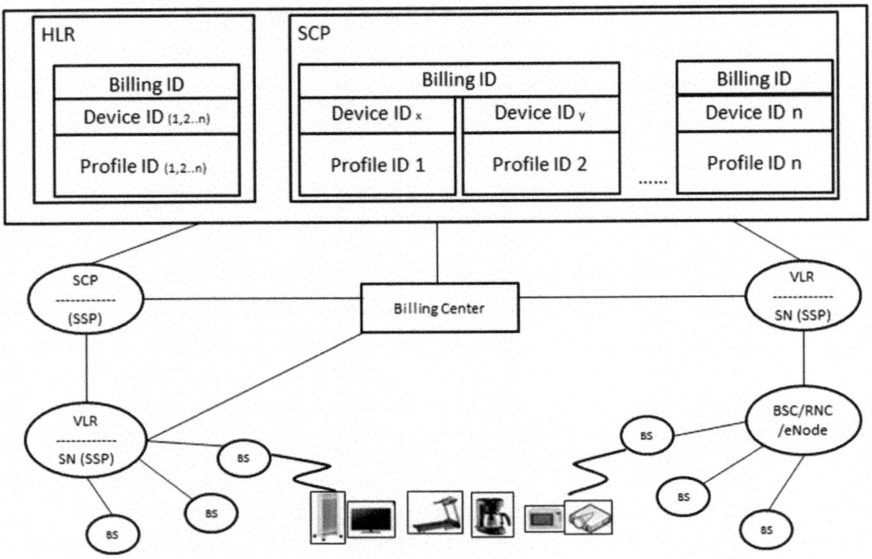

Present solution relates to cellular communication system and in particular to multiple service profile billing for an associated user profile. In a conventional cellular communications system, a one-to-one relationship is typically maintained between the mobile station and the subscription for that mobile station. In this regard the subscription refers to the contract entered by a user with a CSP. The subscription defines not only to whom billing for cellular services should be charged but also set forth the types to cellular calling services and features that are to be provided to the subscriber. Once the subscription to cellular caller services and features has been defined this data is collected in a subscriber service profile that is stored and maintained in a database. The service profile is linked in that database to an identification of the mobile station for the subscriber. Thus the conventional system also maintains a one-to-one relationship between each mobile station and the service profiles that define the cellular services and feature provided to that mobile station.

To address the need of having a single Bill association to multiple devices along with multiple service profiles associated with the user profile and usage of these service profiles concurrently on end device types, a one to many relationship is established between a User Profile ID and its associated Service Profiles and devices.

The Home Location Register (HLR) or HSS of the cellular network stores plural service profiles for such subscripted licences under User Profile ID. Each of the service profile is associated under the same unique User Profile ID to allow singular billing for use of multiple licences across different device for various services.

The user selects one of the service profiles in the Self Care or the Web portal to be activated for a device. At each selection, the service profile is sent to the provisioning system to associate the selected service profile on the e-sim of the selected devices. Once activation of the device is done, the profile would be downloaded from the HRL into the VLR (Park et al. 2013).

Charging on per service profile basis is made for both incoming and outgoing usage (data, sms and voice). In each case metering data is generated which is consumed by the billing system to do the singular charging. These metering records not only identify the service but also the associated devices for which the services are concurrently active. By processing these records, a single bill can be generated against the User Profile ID with bifurcation on device types (Çelik et al. 2018).

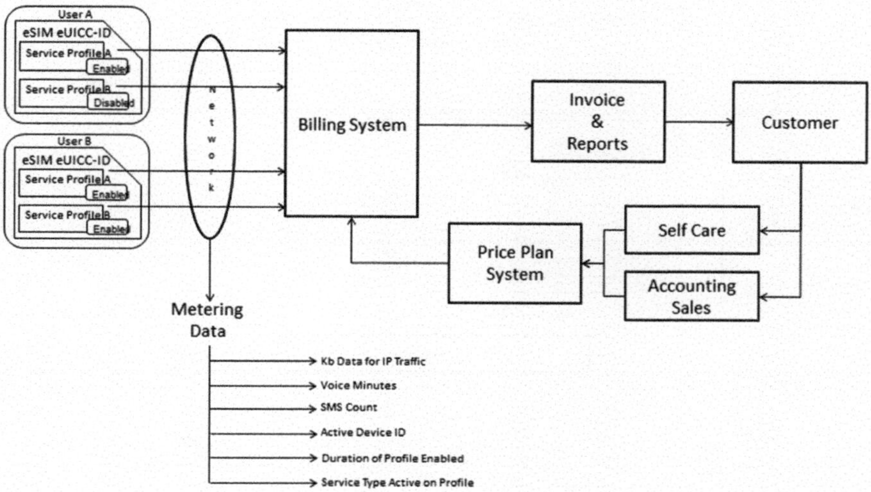

The above figure illustrates the billing scenario for the mentioned solution approach. The components in the architecture are Billing System, Invoice and Reporting, Customer Functions, Accounting and Sales Function, Self-care and a price Plan System.

The Billing system receives the data input from the network, and basis the usage statistics by each device in the network, calculates and generates the bill for the associated device. Post calculation of billing, the information in the process is passed on to the invoicing system, which generates the invoice for the end user.

As part of the customer relationship agreement between the network and customer, the invoicing function generates an invoice for the end user. The invoice contains the information on usage and subscription rental of services. Basis the invoice an end user would make the payment towards the operator.

Every customer is associated with a price plan, which informs him of how he has been charged for usage of various services in the invoice. The customer will also have access of Self-care application (e.g. app running on android or IoS) and also accounting and sales function may help him to update or add or configure service profiles associated with his User profile (Chitroub et al. 2019).

After receiving the information on the active service profile and devices usage the billing system calculates a customer bill based upon the price plan, licences configuration, usage, and time & fixed charges. The calculated bill would then become an input towards the invoicing and reporting function. The customer has visibility on the active service profile, usage, and licences being utilized and can actively modify or configure the service profile using a self-care to make adjustments.

ii. **SIM Provisioning**

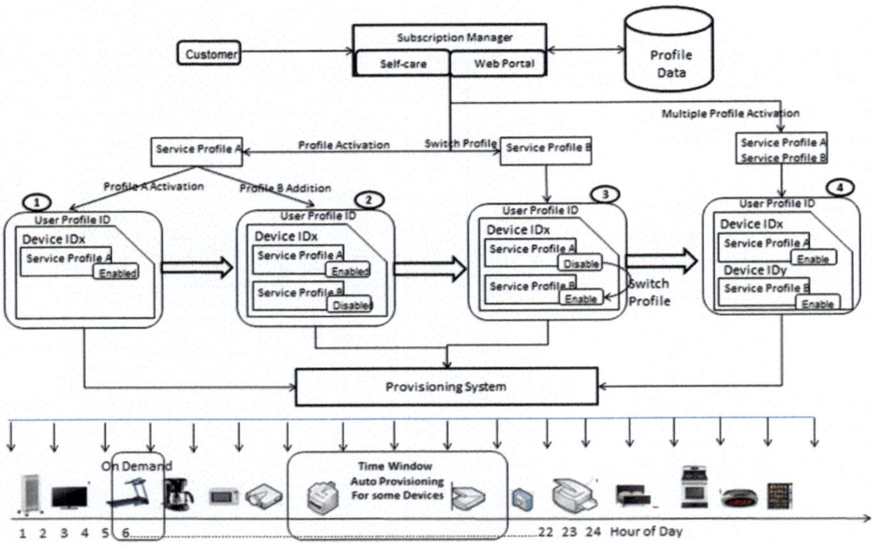

The above figure describes a use case for Service Profile Activation and Provisioning Flow.

a. Customer selects a single Service Profile A associated to his User Profile which the enabled profile for him.
b. Customer acquires a second service profile (B) which is added to his User profile and is in the Disabled state.
c. Customer requests for a profile switch resulting in a Service Profile B to be active and Service Profile A to be disabled.
d. Customer requests for both the profiles (A & B) to be active against his/her User profile.

The customer can select to choose any of the above scenarios of combination of Service Profiles to be activated on the device or multiple devices concurrently using the self-care or any other means of provisioning based on his procured licences (Smith 2014).

4 Conclusion

The concept of Software Licence-based model has multitude of advantages for both Service Provider as well as the Enterprise trying to adopt the same. Some of the advantages are stated below.

a. Move from Silo Individual SIM Card Provisioning to Software Licence based e-SIM Model.
b. One Owner One Bill—There can be a single Bill generated against a single User profile ID which can have multiple Service Profile ID associated with it.
c. The same Service Profile can be used across multiple devices.
d. The solution takes into account e-SIM capability with multi profiling existing in the devices.
e. Scheduled and On Demand plural service profile configuration across devices.
f. Inter Device usage Billing.
g. Metering Data generation for Billing as per Service Profile usage under a single User Profile.
h. Modelling based meta data-driven approach to support any industry.
i. Self-care driven with flexible configuration options to configure need-based service profiles on devices.

We will see disruptive business models which will be enabled by e-SIM adoption and the concept discussed in the paper is one of the means towards that. It's trendy to consume data sustainably. The e-SIM technology is gaining more demand worldwide as more e-SIM compatible devices are.

References

Abdou BA (2019) Commercializing eSIM for Network Operators. In: IEEE 5th World Forum on Internet of Things, WF-IoT 2019—Conference Proceedings, pp 616–621

Çelik A, Yaman H, Turan S, Kara A, Kara F, Zhu B, Dutta D (2018) Millenial politeness in what's up application communication. J Mater Process Technol 1(1):1–8

Chitroub S, Blaid D, Aouadia H, Laouar R (2019) Securing mobile IoT deployment using embedded SIM: concerns and solutions. In: Proceedings *2019* International Conference on Internet of Things, Embedded Systems and Communications, IINTEC 2019 -, pp 75–79

Gull D, Wehrmann A (2009) Optimized software licensing—combining license types in a license portfolio. Proc Bus Inf Syst Eng 1(4):277–288

Koshy DG, Rao SN (2018) Evolution of SIM cards—what's next? Proceedings 2018 International Conference on Advances in Computing, Communications and Informatics, ICACCI 2018, pp 1963–1967

Leh J (2014) 3 Licensing Models of an LMS (Or Any Enterprise Software). https://talentedlearning.com/3-license-models-of-lms/. Accessed 17 April 2021

Mehmi S, Verma HK, Sangal AL (2017) Simulation modeling of cloud computing for smart grid using CloudSim. J Elect Syst Inform Technol 4(1):159–172

Park J, Baek K, Kang C (2013) Secure profile provisioning architecture for embedded UICC. In: Proceedings—2013 International Conference on Availability, Reliability and Security, ARES 2013, pp 297–303

Provisioning RSIM (2018) eSIM Whitepaper The what and how of Remote SIM rovisioning. https://www.gsma.com/esim/wp-content/uploads/2018/12/esim-whitepaper.pdf. Accessed 15 May 2021
Sealy P (2019) The true value proposition. https://go.abiresearch.com/lp-the-true-value-proposition-of-the-esim. Accessed 15 May 2021
Smith I (2014) Embedded SIM remote provisioning architecture. http://www.gsma.com/connectedliving/wp-content/uploads/2014/01/1.-GSMA-Embedded-SIM-Remote-Provisioning-Architecture-Version-1.1.pdf. Accessed 15 May 2021
Stålbrand F (2017) M2M / IoT applications in the agricultural industry. In: Berg Insight, pp 2–5
Vesselkov A, Hämmäinen H, Ikäläinen P (2015) Value networks of embedded SIM-based remote subscription management. In: Proceedings Conference of Telecommunication, Media and Internet Techno-Economics, CTTE 2015

Chapter 7
A Sustainable Production Inventory Model with Variable Demand Dependent on Time, Selling Price, and Electricity Consumption Reduction Level

Amalesh Kumar Manna and Asoke Kumar Bhunia

Abstract Sustainable development in the manufacturing system is the developments that reduce negative environmental impact, save natural resources and energy, and create a safe and economically healthy atmosphere for employees, communities, and consumers in the production of items. On the other hand, customers are always motivated to purchase low electric consumption goods. In this study, a sustainable manufacturing model is developed under the assumption that the manufacturer will use the latest technology to reduce the electric consumption. Here, the electric consumption reduction level and selling price-dependent customers' demand is considered. The manufacturer's social sustainable development cost is also incorporated, which depends on the electric consumption level. This study concludes that the electric consumption reduction level and selling price-dependent customers' demand will help both the manufacturer and customer's environmental and economic benefits.

Keywords Manufacturing · Sustainable development · Promotional demand

1 Introduction

Many manufacturing firms realize that they are responsible for social, economic, and environmental development. This responsibility can make a difference in the sustainable development of human settlements. On the other hand, every developed country aims to reduce the use of natural resources. So, the goal of the sustainable development of the manufacturing system is to maximize the average profit or minimize the average cost. In 2015, the 21st session of the State Party in Paris, France, adopted the UNFC (United Nations Framework Convention) on climate change, which brings the regulations of carbon emission reduction and energy-saving policy by promoting in an alternative way. Sometimes we can say that sustainable development is the transition towards sustainable consumption during production. The

A. K. Manna (✉) · A. K. Bhunia
Department of Mathematics, The University of Burdwan, Burdwan 713104, India

A. K. Bhunia
e-mail: akbhunia@math.buruniv.ac.in

I. Ali et al. (eds.), *Computational Modelling in Industry 4.0*,
https://doi.org/10.1007/978-981-16-7723-6_7

natural resource-saving concept is one kind of sustainable development. Wolf and Chomkhamsri (2015) described the consumption decision affects of the consumer's household resources, space, time etc., in a sustainable production and consumption system. Turkay et al. (2016) studied social and environmental criteria in supply chain management to incorporate attainability. Abreu et al. (2017) studied Lean-Green models for sustainable production. Manna et al. (2019) addressed an imperfect production inventory model with GHG emission control from industrial waste under the fuzzy environment to incorporate sustainable development. Daryanto and Wee (2018) developed a sustainable production model by considering carbon emission from production, holding and waste disposal activities. Shen et al. (2019) addressed a sustainable model on the production process by considering preservation investment for perishable items and carbon tax. Mishra et al. (2020) and Manna et al. (2020a) considered carbon emission cost in their production model to sustain the environment. Lu et al. (2020) also proposed a sustainable development in a production model by reducing carbon emissions. Also, they considered price-dependent demand and solved the problem by the Stackelberg game approach. Tang et al. (2020) investigated sustainable development in a transportation system under a carbon tax policy via a multiplayer dynamic game approach.

Customers' demand plays a vital role in a manufacturing firm to optimize the system's average cost or profit. The management of a manufacturing firm incorporates various types of features for electronic goods such as low radiation level, low electric consumption level, etc., of the produced items to attract the customers. Several researchers reported various types of demand function in their inventory model. You and Hsieh (2007) proposed stock and selling price linked to demand in an EOQ model. Hovelaque and Bironneau (2015) developed a carbon emission dependent demand-based inventory model. Shah (2015) addressed credit-linked and selling price-dependent demand in the formulation of an inventory model. Zerang et al. (2016) suggested marketing effort and selling price based demand in their supply chain model. Bhunia et al. (2017) developed a production model considering variable demand dependent on the item's frequency of advertisement and selling price. Bhunia et al. (2018) assumed advertisement frequency, displayed inventory level and selling price-dependent demand in an inventory model for a perishable item. Kumar and Uthayakumar (2019) proposed price discount and stochastic demand in an inventory model. Manna et al. (2020b) investigated the effects of selling price and warranty policy linked to demand in the production model addressing imperfect items' inspection errors.

This chapter has established a sustainable manufacturing model with reduced electricity consumption of produced electronic goods. It is assumed that the customers will purchase the product verifying the product's electricity consumption level and selling price. Also, the manufacturer's social sustainable development cost has been considered dependent on the electric consumption level. This study's main objective is to determine the optimal selling price and electric consumption reduction level of the manufacturing product, which maximizes the production system's average profit. A numerical experiment has been done by solving a numerical example for testing the validity of the model. Finally, sensitivity analyses have been performed w.r.t. various system parameters.

2 Problem Description

2.1 Notation

$I_m(t)$:	On hand inventory level of the product
P_m:	Manufacturer's production rate
t_m:	Manufacturing time
T_b:	Business period
e_l:	Electricity consumption reduction level of the product
s_r:	Selling price per unit product
$D(e_l, s_r, t)$:	Customers' demand rate
$C(e_l)$:	Unit production cost
h_c:	Holding cost/unit time/unit product
$c_d(e_l)$:	Sustainable development cost/unit product
A_m:	Set-up cost/cycle
$\pi(e_l, s_r, T)$:	Manufacturer's average profit.

2.2 Assumptions

(i) The manufacturing firm produces the non-deteriorating product, and a part of the produced product is defective. The production and defective rates are constants. Also, the demand rate is less than the rate of produced perfect items, i.e. $P > D(e_l, s_r, t)$.

(ii) The customers' demand rate is decreased with the increase of time as well as the selling price of the product. Also, it is increased with the electricity consumption reduction level of the product. Hence the mathematical expression of customers' demand rate is given by

$$\begin{aligned} D(e_l, s_r, t) &= \alpha_0 + \alpha_1 e_l - \alpha_2 s_r - \alpha_3 t \\ &= d(e_l, s_r) - \alpha_3 t, \end{aligned}$$

where $d(e_l, s_r) == \alpha_0 + \alpha_1 e_l - \alpha_2 s_r$ and $\alpha_0, \alpha_1, \alpha_2, \alpha_3$ are positive constants.

(iii) Unit production cost is increased with the increase in electricity consumption reduction level of the product. The mathematical expression of unit production cost is as follows:

$$C(e_l) = c_p e^{\delta e_l}$$

where c_p is the fixed production cost and δ is the control parameter of the electricity consumption reduction level.

(iv) Due to government regulation, the manufacturer is bound to spend an amount of money for society's sustainable development. This amount is proportional to e_l. The sustainable development cost per unit production is given by

$$c_d(e_l) = a - be_l$$

where a is the fixed sustainable development cost and b is sustainable development sensitive cost.

3 Mathematical Representation of the Model

Let us suppose that a manufacturing firm initially starts the production to fulfil the demand of the customers. During the production period $(0,\ t_m)$, the inventory level of perfect item increases at the rate $P_m - D(e_l,\ s_r,\ t)$, whereas it decreases at the rate $D(e_l,\ s_r,\ t)$ and reaches zero at time $t = T_b$. The graphical nature of the production inventory system during $[0,\ T_b]$ is represented in Fig. 1.

The governing differential equations of the inventory level $I_m(t)$ are as follows:

$$\frac{dI_m(t)}{dt} = P_m - d(e_l,\ s_r) + \alpha_3 t, \quad 0 < t \le t_m \tag{1}$$

$$\frac{d\mathrm{I}_m(t)}{dt} = -d(e_l,\ s_r) + \alpha_3 t, \quad t_m < t \le T_b \tag{2}$$

with the conditions $I_m(0) = 0 = I_m(T_b)$. Also $I_m(t)$ is continuous at $t = t_m$.

The solutions of the Eqs. (1) and (2) are as follows:

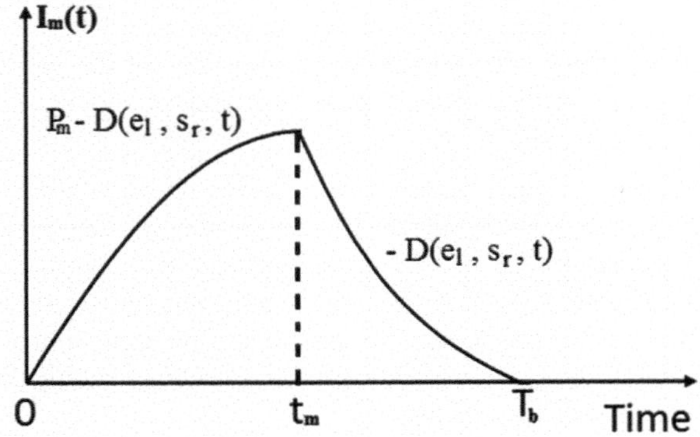

Fig. 1 Schematic diagram of the inventory level of perfect items

$$I_m(t) = \{P_m - d(e_l, s_r)\}t + \frac{\alpha_3}{2}t^2, \quad 0 < t \le t_m \tag{3}$$

$$I_m(t) = (T_b - t)d(e_l, s_r) - \frac{\alpha_3}{2}(T_b^2 - t^2), \quad t_m < t \le T_b \tag{4}$$

where $d(e_l, s_r) == \alpha_0 + \alpha_1 e_l - \alpha_2 s_r$ is independent of t.

From the continuity of $I_m(t)$ at $t = t_m$, we get

$$\begin{aligned}&\{P_m - d(e_l, s_r)\}t_p + \frac{\alpha_3}{2}t_p^2 = (T_b - t_m)d(e_l, s_r) - \frac{\alpha_3}{2}(T_b^2 - t_m^2)\\ &\text{which implies that } t_m = \frac{T_b}{P_m}\left\{d(e_l, s_r) - \frac{\alpha_3}{2}T_b\right\}\end{aligned} \tag{5}$$

The manufacturer's average holding cost (AHC) during $[0, T_b]$ is as follows

$$\begin{aligned}AHC &= \frac{h_c}{T_b}\left\{\int_0^{t_m} I_m(t)dt + \int_{t_m}^{T_b} I_m(t)dt\right\}\\ &= \frac{h_c}{2T_b}\left[P_m t_m^2 + (T_b^2 - 2t_m T_b)d(e_l, s_r) - \frac{\alpha_3}{3}(2T_b^3 - 3t_m T_b^2)\right]\end{aligned}$$

The manufacturer's average production cost (APC) is as follows

$$APC = \frac{1}{T_b}c_p e^{\delta e_l} P_m t_m = c_p\left\{d(e_l, s_r) - \frac{\alpha_3}{2}T_b\right\}e^{\delta e_l}$$

The manufacturer's average sales revenue (ASR) is given by

$$ASR = \frac{s_r}{T_b}\int_0^{T_b}\{d(e_l, s_r) - \alpha_3 t\}dt = s_r\left\{d(e_l, s_r) - \frac{\alpha_3}{2}T_b\right\}$$

The average set-up cost of the manufacturer is $\frac{A_m}{T_b}$.

The average sustainable development cost (ASDC) of the manufacturer is calculated as follows:

$$ASDC = \frac{1}{T_b}(a - be_l)P_m t_m = (a - be_l)\left\{d(e_l, s_r) - \frac{\alpha_3}{2}T_b\right\}$$

Hence the average profit of the manufacturer is calculated as follows

$$\begin{aligned}\pi(e_l, s_r, T_b) &= ASR - APC - ASDC - AHC - \frac{A_m}{T_b}\\ &= s_r\left\{d(e_l, s_r) - \frac{\alpha_3}{2}T_b\right\} - \{(a - be_l) + c_p e^{\delta e_l}\}\left\{d(e_l, s_r) - \frac{\alpha_3}{2}T_b\right\}\end{aligned}$$

$$-\frac{h_c}{2T_b}\Big[P_m t_m^2 + (T_b^2 - 2t_m T_b)d(e_l, s_r) - \frac{\alpha_3}{3}(2T_b^3 - 3t_m T_b^2)\Big] - \frac{A_m}{T_b} \tag{6}$$

Now the problem is to find the optimal values of e_l (electricity consumption reduction level of the product), s_r (selling price/unit product), and T_b (business period), which give the maximum average profit $\pi(e_l, s_r, T_b)$ of the manufacturer.

Hence the corresponding maximization problem is as follows:

$$\begin{aligned}&\textbf{Maximize } \pi(e_l, s_r, T_b)\\ &\textbf{subjectto } e_l > 0,\ \ s_r > 0,\ \ T_b > t_m > 0)\end{aligned} \tag{7}$$

4 Numerical Experiment

To verify the model, the following numerical example is considered.

Example: Let us suppose that a manufacturing firm produces an electrical product (say, AC) at the rate of $P_m = 600$ units per month and fulfils the customers' demand. According to the assumptions, the demand parameters are $a_0 = 380$ units, $a_1 = 25$ units, $a_2 = 2.1$ units, $a_3 = 0.18$ units. Also, the unit production cost parameters are assumed as $c_p = 35$ units and $\delta = 0.34$. The manufacture is bound to spend any amount for sustainable development of the environment and society. It is assumed that the sustainable development cost parameters are $a = 3$ units and $b = 1.2$ units. Again, the set-up and holding costs are considered as $A_m = 300$unit and $h_c = 3$ unit, respectively. The manufacturing authority wishes to find the extreme values of e_l, s_r, and $T_b$by maximizing the manufacturer's average profit.

Solution: Using the mentioned parameters values in the optimization problem (7) and using MATHEMATICA, we get the optimal values of e_l, s_r and T_b and $\pi(e_l, s_r, T_b)$. The optimal solution of the example is shown in Table 1.

The concavities of the average profit w.r.t. e_l, s_r and T_b are shown in Figs. 2, 3, 4, 5, 6 and 7, which are plotted by MATHEMATICA software.

Table 1 Optimal result of the example

Electricity consumption reduction level of the produced item (e_l)	Selling price of the produced item (s_r)	Production period (t_m)	Business period (T_b)	Manufacturer's average profit $\pi(e_l, s_r, T_b)$
0.2836 units	113.23 units	0.3261 units	1.3115 units	10296.50 units

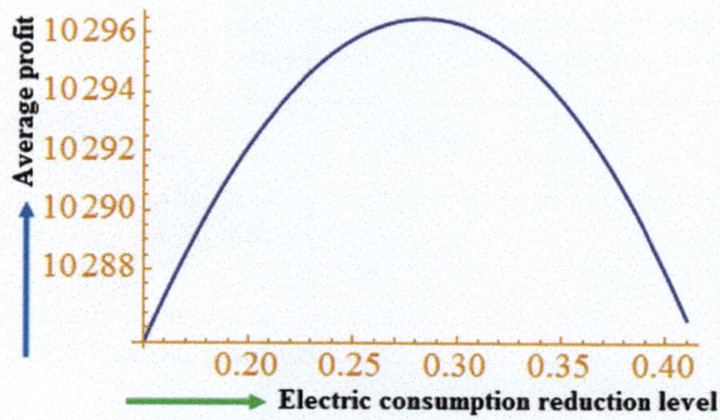

Fig. 2 Concavity of $\pi(e_l, s_r, T_b)$ w.r.t. e_l

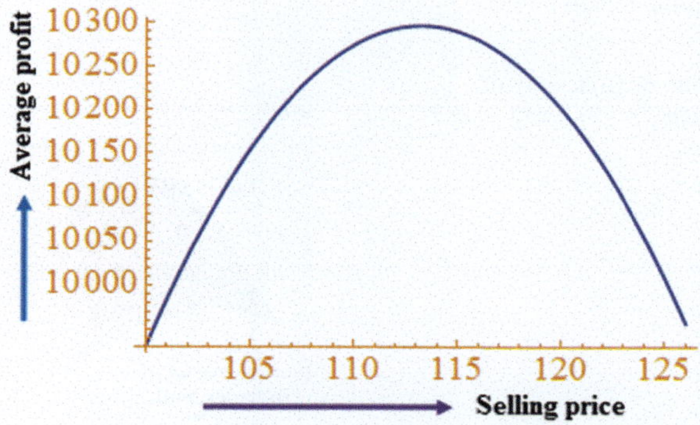

Fig. 3 Concavity of $\pi(e_l, s_r, T_b)$ w.r.t s_r

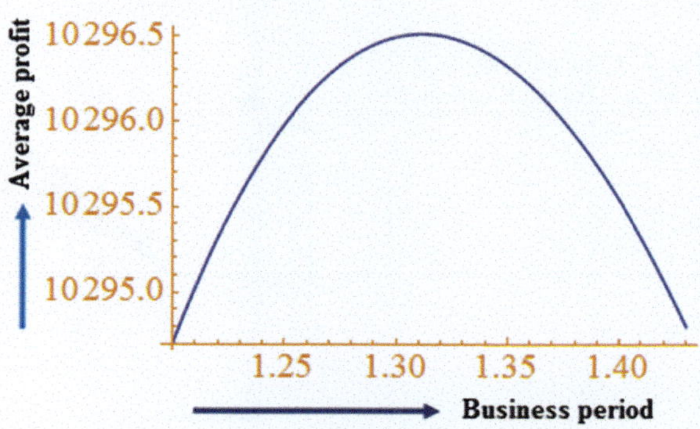

Fig. 4 Concavity of $\pi(e_l, s_r, T_b)$ w.r.t. T_b

5 Sensitivity Analyses and Result Discussion

In this subsection, the post optimality analyses are performed w.r.t. different system parameters like P_m, α_0, α_1, α_2, α_3, c_p, c_h, and A_m by changing the values from -10% to $+10\%$ to investigate the effect of the values' optimum values of the decision variables. The detailed results are shown in Table 2.

Fig. 5 Concavity of $\pi(e_l, s_r, T_b)$ w.r.t. s_r & T_b

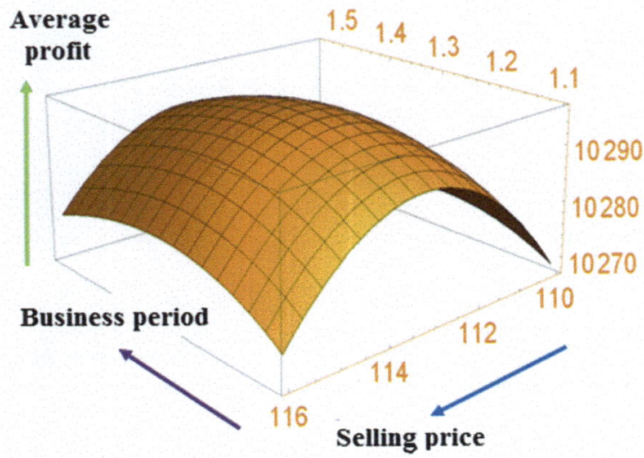

Fig. 6 Concavity of $\pi(e_l, s_r, T_b)$ w.r.t. e_l & T_b

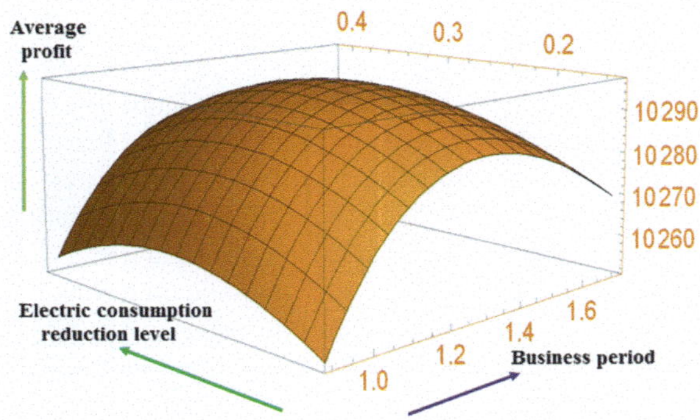

Fig. 7 Concavity of $\pi(e_l, s_r, T_b)$ w.r.t. s_r & e_l

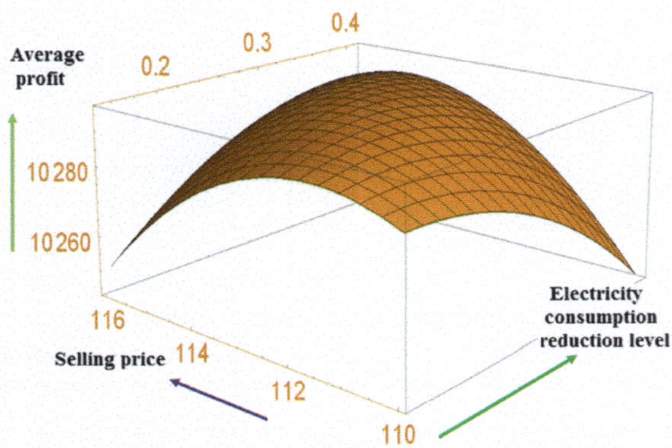

Table 2 Post optimality analysis w.r.t. different system parameters

Parameters	% of change of the parameters	% change in optimal values				
		$\pi(e_l, s_r, T_b)$	e_l	s_r	t_m	T_b
P_m	−10	+0.08	0	−0.04	+13.18	+1.83
	−05	+0.04	0	−0.02	+6.18	+0.84
	+05	−0.03	0	+0.02	−5.50	−0.74
	+10	−0.06	0	+0.03	−10.41	−1.40
α_0	−10	−24.53	0	−7.91	−8.51	+5.00
	−05	−12.69	0	−3.96	−4.24	+2.34
	+05	+13.53	0	+3.96	+4.23	−2.06
	+10	+27.90	0	+7.92	+8.46	−3.88
α_1	−10	−0.25	−98.76	−2.87	−0.09	+0.04
	−05	−0.19	−48.2	−1.46	−0.06	+0.03
	+05	+0.30	+46.09	+1.50	+0.09	−0.05
	+10	+0.71	+90.20	+3.04	+0.25	−0.12
α_2	−10	+18.62	+99.75	+12.24	+1.84	−1.24
	−05	+8.72	+48.45	+5.77	+0.92	−0.61
	+05	−7.74	−45.84	−5.19	−0.89	+0.59
	+10	−14.64	−89.39	−9.86	−1.78	+1.18
α_3	−10	+0.01	0	0	+0.18	+0.18
	−05	0	0	0	+0.09	+0.09
	+05	0	0	0	−0.09	−0.09
	+10	−0.01	0	0	−0.18	−0.18
c_p	−10	+5.97	+109.27	+1.45	+1.90	−0.95
	−05	+2.88	+53.21	+0.70	+0.92	−0.47
	+05	−2.71	−50.6	−0.67	−0.89	+0.47
	+10	−5.25	−98.84	−1.31	−0.18	+0.91
c_h	−10	+0.20	0	−0.02	+4.72	+4.69
	−05	+0.11	0	−0.01	+2.51	+2.51
	+05	−0.11	0	+0.01	−2.33	−2.32
	+10	−0.21	0	+0.02	−4.48	−4.45
A_m	−10	+0.23	0	−0.02	−5.11	−5.15
	−05	+0.11	0	−0.01	−2.52	−2.54
	+05	−0.11	0	+0.01	+2.46	+2.48
	+10	−0.22	0	+0.02	+4.86	+4.9

From Table 2, it is observed that

(i) $\pi(e_l, s_r, T_b)$ (average profit of the manufacturer) is equally sensitive directly and reversely w.r.t. α_0 (fixed demand rate) and α_2 (selling price sensitive demand parameter) respectively, whereas it is reversely less sensitive w.r.t. c_p (unit production cost). Also, it is insensitive w.r.t. c_h (holding cost), A_m (setup cost), P_m (production rate), α_1 (electric consumption reduction level sensitive demand parameter), and α_3 (time sensitive demand parameter).
(ii) the electric consumption reduction level (e_l) is highly sensitive directly w.r.t. α_1, whereas it is highly sensitive reversely w.r.t. α_2 and c_p. Moreover, it is insensitive w.r.t. α_0, α_3, c_h, A_m, and P_m.
(iii) s_r(selling price/items) is moderate sensitive directly and reversely w.r.t. α_0 and α_2 respectively, whereas it is less sensitive directly and reversely w.r.t. α_1 and c_p respectively. Also, it is insensitive w.r.t. α_3, c_h, A_m, and P_m.
(iv) the production period is moderate sensitive directly and reversely w.r.t. α_0 and P_m respectively, whereas it is less sensitive directly w.r.t. A_m. Also, less sensitive reversely w.r.t. c_h and α_2. Moreover, it is insensitive with respect to α_3, α_1, and c_p.
(v) the business period is less sensitive directly w.r.t. α_2 and A_m whereas it is less sensitive reversely w.r.t. α_0, c_h, P_m. Also, it is insensitive w.r.t. α_1 and α_3.

6 Conclusions

In the present work, the concept of sustainable development has been incorporated in a manufacturing system by considering electric consumption reduction level and selling price-dependent demand. On the other hand, customers prefer less electric consumption electric goods at the time of purchasing. In this connection, manufacturing companies invest an amount for new technology to reduce the electric consumption of each electronic goods and increase the customers' demand. Here, we have determined the optimal values of electric consumption reduction level and selling price per item by maximizing the manufacturer's average profit. From the post optimality analyses, the following conclusions are summarized as follows:

- Average profit is equally effective in a positive manner concerning fixed market demand, whereas it is equally effective in a negative sense w.r.t. the selling price per item. Furthermore, it is less effective in a negative sense w.r.t. the unit production cost.
- On the other hand, the selling price per unit item and the production period is moderately positive effective w.r.t. fixed market demand, whereas the electric consumption reduction level is insensitive w.r.t. fixed market demand.

For further investigation, one can extend this work considering backlogged shortages (fully/partially). Also, the proposed demand function can be applied in a supply chain model. Moreover, this model may be extended considering the imprecise parameters relating to demand rate, production and different inventory costs in fuzzy and interval environments.

Acknowledgements The first author thankful to UGC, India, for providing DSKPDF through The University of Burdwan(Research grant no. F.4-2/2006(BSR)/MA/18-19/0023). Also the second author would like to acknowledge the financial support provided by WBDST & BT, West Bengal, India for this research (Memo No: 429(Sanc.)/ST/P/S & T/16G-23/2018 dated 12/03/2019).

References

Abreu MF, Alves AC, Moreira F (2017) Lean-Green models for eco-efficient and sustainable production. Energy 137:846–853

Bhunia AK, Shaikh AA, Cárdenas-Barrón LE (2017) A partially integrated production-inventory model with interval valued inventory costs, variable demand and flexible reliability. Appl Soft Comput 55:491–502

Bhunia AK, Shaikh AA, Dhaka V, Pareek S, Cárdenas-Barrón LE (2018) An application of genetic algorithm and PSO in an inventory model for single deteriorating item with variable demand dependent on marketing strategy and displayed stock level. Sci Iran 25(3):1641–1655

Daryanto Y, Wee HM (2018) Sustainable economic production quantity models: an approach toward a cleaner production. J Adv Manag Sci 6(4):1–7

Kumar GM, Uthayakumar R (2019) Multi-item inventory model with variable backorder and price discount under trade credit policy in stochastic demand. Int J Prod Res 57(1):298–320

Hovelaque V, Bironneau L (2015) The carbon-constrained EOQ model with carbon emission dependent demand. Int J Prod Econ 164:285–291

Lu CJ, Lee TS, Gu M, Yang CT (2020) A multistage sustainable production inventory model with carbon emission reduction and price-dependent demand under Stackelberg game. Appl Sci 10(14):4878

Manna AK, Dey JK, Mondal SK (2019) Controlling GHG emission from industrial waste perusal of production inventory model with fuzzy pollution parameters. Int J Syst Sci Oper Logist 6(4):368–393

Manna AK, Das B, Tiwari S (2020a) Impact of carbon emission on imperfect production inventory system with advance payment base free transportation. RAIRO-Oper Res 54:1103–1117

Manna AK, Dey JK, Mondal SK (2020b) Effect of inspection errors on imperfect production inventory model with warranty and price discount dependent demand rate. RAIRO-Oper Res 54:1189–1213

Mishra U, Wu J-Z, Sarkar B (2020) A sustainable production-inventory model for a controllable carbon emissions rate under shortages. J Clean Prod 256:120268

Shah NH (2015) Manufacturer-retailer inventory model for deteriorating items with price-sensitive credit-linked demand under two-level trade credit financing and profit sharing contract. Cogent Eng 2(1):1012989

Shen Y, Shen K, Yang C (2019) A production inventory model for deteriorating items with collaborative preservation technology investment under carbon tax. Sustainability 11(18):5027

Tang Z, Liu X, Wang Y (2020) Integrated optimization of sustainable transportation and inventory with multiplayer dynamic game under carbon tax policy. Math Probl Eng 2020:1–16

Turkay M, Saraoglu O, Arslan MC (2016) Sustainability in supply chain management: aggregate planning from sustainability perspective. PloS One 11(1):e0147502

Wolf M-A, Chomkhamsri K (2015) From sustainable production to sustainable consumption. Life Cycle Manag 169–193

You P-S, Hsieh Y-C (2007) An EOQ model with stock and price sensitive demand. Math Comput Model 45(7–8):933–942

Zerang ES, Taleizadeh AA, Razmi J (2016) Analytical comparisons in a three-echelon closed-loop supply chain with price and marketing effort-dependent demand: game theory approaches. Environ Dev Sustain 20(1):451–478

Chapter 8
A Review on Cleaner Technology for Sustainable Industrial Practices: Global and Indian Scenario Genesis of Technologies to Control Pollution

X. Agnello J. Naveen, A. Arivoli, V. C. Malarmannan, and S. Boopathi

Abstract This paper aims to understand the international experience, most new policies relevant to industrial pollution control technology envisage by the environmental protection agency. This paper includes some countries' affairs, which have efficiently implemented cleaner technologies in developed nations like the United States of America, the United Kingdom, France, Canada, Norway, and Japan. This exercise may help India draw successful lessons to implement and support efficient driving forces and remove those barriers that impede adopting cleaner technology in large, medium, and small firms in India.

Keywords Cleaner technology · Co-processing · Co-generation · Process modification · Energy conservation

Abbreviations

UNEP	United Nation Environmental Protection
EPA	Environmental Protection Agency
EU	European Union
OECD	Organization for Economic Co-operation and Development

X. A. J. Naveen (✉)
Assistant Professor, Department of Environmental Science, Thanthai Roever Institute of Agriculture and Rural Development (TRIARD), Perambalur, Tamil Nadu, India

A. Arivoli
Assistant Professor, Department of Environmental Science, Government Arts College, Ariyalur, Tamil Nadu, India

V. C. Malarmannan
Professor, Department of Management, Nalanda School of Business, Tiruchirappalli, Tamil Nadu, India

S. Boopathi
Professor, Department of Economics, Bharathiar University, Coimbatore, Tamil Nadu, India

I. Ali et al. (eds.), *Computational Modelling in Industry 4.0*,
https://doi.org/10.1007/978-981-16-7723-6_8

CPCB	Central Pollution Control Board
MBI	Market based Investment
CC	Control and Command
PPP	Polluter pay principle
GHG	Green House gases
IPCC	Intergovernmental Panel on Climatic Change
UNFCCC	United Nations Framework for Convention of Climatic Change
SMEs	Small and Medium Enterprises
MoEFCC	Ministry of Environment Forest and Climatic Change
UNIDO	United Nations Industrial Development Organization
ISO	International Organization for Standardization
NPC	National Productivity Council

1 Introduction

Economic activities bring 'negative externalities' towards the environment while producing goods and services (Ravichandran 1989; Sathya and Ravichandran 2010). According to Porter and Linde, there is always a "Tradeoff" between cost analysis and environmental degradation (Porter and Linde 1995). Apart from setting economic policy for environmental protection such as Market Based Instruments (MBI), Control and Command (CC) (Frondel et al. 2007), and also Polluter Pays Principle (PPP) was initiated by the Organization of Economic Corporation and Development (OECD) in 1972 (Agnello et al. 2015). Standards for controlling industrial pollution from traditional methods to adopt Pigouvian taxes, technology adoption subsidies, and technology (Baumol and Oates 1971; Banzhaf 2020). Hence, in controlling industrial pollution, policymakers are not able to achieve practically (Kumar et al. 2014), because standards and permissible limits can be achieved through implementing technology tools for treatment; there is a variety of technology available from the production process itself handling can be chosen between End-of-Pipe technology solution or process modification (adaptation) or raw material usage (substitution) or Advance technology (Blackman and Bannister 1997). To control water, air, solid, hazardous waste, and energy to reduce emission Greenhouse Gas. There is a considerable concern for the Convention on Climatic Change by the Intergovernmental Panel on Climatic Change (IPCC) (Change IPOC 1990, 2014; Dincbas et al. 2021).

A hierarchy has been designed from the initial extraction of raw materials to the production line to their ultimate disposal to achieve the 'WINWIN' strategy in industrial pollution (Elkington 1994; Agnello et al. 2021). The conventional industrial operation method believed in designing technologies to produce goods and services and later decide on pollution control measures to create control technology (Wang et al. 2020). A Change in trend has facilitated a technological process that reduces output production without causing pollution to the environment. The leading industrial firms and research and development are looking for a new industrial sustainable mode based on waste minimization, energy-conserve, and reducing resource production (Aldieri and Vinci 2020). There is an essential strategy behind the use of 'Cleaner production technologies, which is a combination of sustained—technological approach to prevent pollution, conserve resources usage, minimize the solid waste generation and conserves energy in the production process itself' (Chan et al. 2018; Yin et al. 2020). Essentially, the study's objectives are twofold: to trace the background, the development, and the use of cleaner technologies at the global level. Moreover, to understand the driving and barrier forces for developing and using cleaner technologies in small, medium (SME), and large firms in India.

2 Materials and Methods

Based on the Secondary (Exploratory review-research) published sources and data like the concept, they were used to understand the genesis for cleaner production/technologies and growth over the period. This exercise helped the authors to narrow down the type of industrial units for case studies. Nevertheless, even the aforesaid secondary source of information served as a potential base for the research study underway to proceed further. As discerned from the published report, 'cleaner technologies' origin and growth are presented below.

3 Classification of Cleaner Production/Technologies

3.1 Processes Reduction–Modification

Cleaner production or technology is a concept based on a holistic environmental management system towards sustainable environment management on raw material resources used in the production process like consumption and energy conservation, less usage of toxic raw material, reducing harmful emission, waste minimization in the production process itself.

3.2 Eco-Friendly Products

The cleaner production or technology is an application process, which reduce the negative impact with a comprehensive life cycle assessment of the products, from material removal in the end life of process/product, life cycle assessment (LCA), and Total quality management (TQM) (UNEP 1994).

Table 1 shows the Road map and diffusion of new state-of-art technologies and regulation policies (Developed countries), EPA, EU-BAT, OECD, and UNEP to developing countries like India. It has included a hierarchy in an Indian agency pollution committee setting up (MoEFCC-CPCB-SPCB); this has a regularized diffusion/road map set-up system. These give an overview on a comparative study platform

Table 1 A review of comparative studies between the road map/diffusion of technological concepts (up-gradation with regulation) changes globally and in India towards industrial pollution control (cleaner technology)

S. no	Global technological changes have taken place to control industrial pollution	Official boards agencies	Year	CPCB year	Journal
1	Ignoring pollution	EPA/CPCB	1950s	–	(Afsah et al. 1998)
2	Dispersion and diffusion	EPA	1960s	–	(Hesketh 1996)
3	Clean air act	EPA/CPCB	1970s	1981	(Greenstone 2002; Waxman 1991)
4	Clean water act	EPA/CPCB	1972	1974	(Adler et al. 1993; Anju 2010)
5	Environmental regulation policy to countries	EPA/EU/OECD	1970–72	–	(Blackman et al. 2018)
6	Resources, conservation, and recovery act, (RCRA)	EPA/CPCB	1976	1990	(Lown 2002)
7	Control and command 3P (pollution prevention pay)	EPA, OECD/CPCB	1989	1992	(Ochsner et al. 1995)
8	Life cycle assessment (LCA), Cradle to Cradle (C2C)	EPA	1980	–	(Cabeza et al. 2014; Llorach-Massana et al. 2015)
9	Taxes, subsidies, tradable permit	OECD/CPCB	1996, 98	1977	(Requate and Unold 2001; Arguedas et al. 2010
10	Total quality management system (TQM)	EPA	1980	–	(Tarí 2005; Dahlgaard et al. 2008)
11	End of pipe/emission control technology	EPA/CPCB	1980	1980	(Zeng et al. 2020)
12	COINDS comprehensive industry document from EPA, E.U. set standards MINAS, GEMS minimal national standards	CPCB		1980	(Khoshoo 2008)
13	Environmental information system (ENVIS) database management system	CPCB	–	1983	(Singh 1999; Park et al. 2004)
14	Pollution prevention (P_2) waste minimization	EPA/CPCB	The 1990s	1990	(Ghassemi 2001; Bumble 2020)

(continued)

Table 1 (continued)

S. no	Global technological changes have taken place to control industrial pollution	Official boards agencies	Year	CPCB year	Journal
15	Indian policy statement cleaner technology in SMEs, waste minimization circle (WMC) and promotion of economic instrument for pollution technology, eco-mark in India	MoEFCC	–	1992	(Bhupendra and Sangle 2016; Satapathy 2017)
16	Corporate responsibility for environmental protection (CREP) polluting industry	CPCB	–	2003, 2016	(Sood and Arora 2006; Kumar and Shetty 2018)
17	DESIRE (demonstration in small industries for reducing waste) project, PRISMA, UPTECH, APCTT	UNEP	–	1993–94	(Prakash 1994)
18	Waste minimization (hazardous waste)	EU/CPCB	1991	2000	(Freeman 1992; Higgins 2018)
19	Cleaner production (GCPC) in India (national cleaner production centres)	UNIDO/CPCB	1994	1995	(Ashton and Shenoy 2015; Abhishek and Biswas 2018)
20	Environmental audit in polluting industries	CPCB	–	1995–96	(Mahwar et al. 1997)
21	Clean industrial product and process by biotechnology	OECD	1994	–	(Bull 2001)
22	Clean development mechanism (CDM) co-generation process/auxiliary energy conservation instruments	Kyoto/CPCB	1994	1995	(Glensk et al. 2015)
23	EU-BAT (best available technology) IPPC (integrated pollution prevention, control) air pollution control	EU/CPCB	1994	1994	(Helman and Parchomovsky 2011)

(continued)

Table 1 (continued)

S. no	Global technological changes have taken place to control industrial pollution	Official boards agencies	Year	CPCB year	Journal
24	Industrial ecology, creating (industrial estates)	OECD/CPCB	1970–80	1995	(Erkman 1997)
25	Life cycle energy assessment (LCEA)	EPA	1996		(Kofoworola and Gheewala 2009)
26	Guidelines for collecting and interpreting Technological innovation	OECD	1997	–	(OECD E.U. 1997)
27	Guidelines for green belt (carbon sequestration) air-polluting industry	EU/CPCB	1980	2000	(Rowe 2011)
28	EIA, EMP, and MEP	EPA, EU/CPCB	1990	1992	(Canter and Wood 1996; Glasson and Therive 2013)
29	Good housekeeping or best operating practices in industries	EPA/CPCB	1990	1994	(Becker 2001)
30	Green chemistry	EPA/OECD	1988–90	–	(Anastas and Warner 1998)
31	Information manual on pollution abatement and cleaner technology series (IMPACTS) industrial notification in cleaner technology	CPCB	–	1995, 2020	(Patil et al. 2015)
32	Hazardous waste management series (HWMS)	CPCB	–	1995	(Wang et al. 2009; Devi et al. 2018)
33	Common hazardous waste storage, treatment and disposal facilities (CHWSTDF), treatment, storage and disposal facilities (TSDF)	CPCB	–	1997	(Treatment 1989;Vidhya and Joseph 2010)
34	Programme objectives series (probes)	CPCB	–	1997	(Maji et al. 2020)
35	Resource recycling series (RERES)	CPCB			(Kaur et al. 2012)

(continued)

Table 1 (continued)

S. no	Global technological changes have taken place to control industrial pollution	Official boards agencies	Year	CPCB year	Journal
36	Green rating project (GRP)	CSE	–	1997	(Powers et al. 2011)
37	Biotechnological treatment of wastes	OECD/CPCB	1998	2001	(Macaskie 1991)
38	Environmental training units (ETU)	CPCB	–	2004	(Chand 2018)
39	Environmental surveillance squad (ESS), GIS mapping	CPCB	–	2010, 2011	(Singh 2018)
40	Rapid ecologically sustainable development on adopting cleaner technology	NPC	–	2011	
41	Energy conservation and technology	Company Act (1956)	–	1988, 2013	(Conservation 1988)
42	4R, 6R, 5R systems recycle, reduce, reuse and recharge (co-processing technology)	EPA/CPCB	1991	2014	(Jenssen et al. 2003)
43	Quality management system-ISO 9001	ISO	1987	–	(Priede 2012)
44	Environmental management system EMS-ISO 14001, Plan-Do-Check-Act (PDCA) cycle	ISO	1992	–	(Oliveira et al. 2016)
45	Green supply chain management (GSCM)	EMS/CPCB	1993	–	(Mathiyazhagan et al. 2013)
46	Hazardous waste generating industries	CPCB	–	2009	(Basu and Chakraborty 2016)
47	Energy management system (EnMS-ISO 50001)	ISO/CPCB	2011	2011	(Imes et al. 2013)
48	Comprehensive environmental pollution index (CEPI) Technical report	CPCB			(Sharma and Singh 2019)
49	ZLD (zero liquid discharge)	EPA/CPCB	1970	2015	(Shah et al. 2020)

(continued)

Table 1 (continued)

S. no	Global technological changes have taken place to control industrial pollution	Official boards agencies	Year	CPCB year	Journal
50	OCEMS-online continuous emission/effluent monitoring system	EPA/CPCB	1960	2015	(Kakarla et al. 2019)
51	Pollution monitoring-environmental surveillance system (ESS)	EPA/CPCB	1990, 2000	2015	(Chen et al. 2013)
52	Industrial energy efficiency (IEE) CP	UNIDO/CPCB	2017	2017	(Du et al. 2013)
5	Best available technique (health and safety) on air pollution	OECD	2018	–	(Milone et al. 2015)
54	Guidelines for taking up a demonstration of new/innovative technology for air-polluting control abatement	CPCB	–	2018	(CPCB 2018a, b, c)
55	Hazardous waste management has given 50 industrial guidelines on recycling, reusing, and co-processing in the red industry as an operating procedure	CPCB	–	2016–2019	(CPCB 2016)
56	Resources recycling series	CPCB	–	2020	(CPCB 2020a)
57	Environmental mapping and planning	CPCB	–	2020	(CPCB 2020b)
58	Groundwater quality series	CPCB	–	2020	(CPCB 2020c)
59	Hazardous waste management series	CPCB	–	2020	(CPCB 2020d)

of International with Indian regulation system, On policymaking amendment from the developed to developing countries to upgrade/update new technologies in every given aspect like concepts, permissible levels, Guidelines/guidance, research and development, institutional background, technical training, standard operating procedure (SOP), government support documentation facility like (MoU), cost-cutting framework analysis to control pollution.

3.3 Cleaner Production (CP)

The Cleaner production goal is to create a sustainable industrial pollution control, and awareness to promote a cleaner environment by technology/production was introduced in developing countries by UNEP in the 1990s. The 'Cleaner Production' operating group meeting was held in Copenhagen in the 1990s; two different sessions has taken into account in an initial project to promote in the international arena. As a result, the National Cleaner Production Centers (NCPC) were established at the national level initially in 20 developing countries (UNEP 1994) and 2014, they were set up in 58 countries, and later changed into Resource Efficient and Cleaner Production (RECP) to manage resources and generate multiple benefits. The main moot is to help small, medium, and large Industrial firms establish clean technology and create awareness towards clean technology to support Sustainable Development Goals that follow and expand Millennium Development Goals (Igere and Ekundayo 2020).

3.4 UNIDO-National Cleaner Production Centre (NCPC)

After the United Nations meeting on Environment and Development in Rio 1992, two for were discussed the United Nations Industrial Development Organization (UNIDO) and the United Nations Environment Programme (UNEP) both set preventive standards on environmental concepts to enhance in developing countries, for their successful completion, UNIDO and UNEP jointly launched a programme to establish National Cleaner Production Centers (NCPCs) in 58 countries with a moot to promote/analysis/review the sustainable production and consumption, 'Initiating from a simple concept to produce with fewer waste operations'. The concept has built-in an altering idea of (Pollution Prevention Pays) 3P programme (Shen 1995; Royston 2013) and to bring Environmental Sound Technologies (ESTs). The NCPC was established in 1994 for small and medium firms in industrial production. The first cleaner production concept has been adopted in European countries; the pioneers are Austria used from 1992 to 1994 to check based on trial and error depends on feasibility, viability, cost-cutting, and efficacy of technologies; the United Nations Industrial Development Organization has operated by National Cleaner Production Centre The same concept has established in Central America, South America, Africa, Asia, and Europe and this has been working in 58 countries were diffused on international scale. Recently UNEP and UNIDO have a portfolio of two concepts of industrial energy efficiency (IEE) (Cagno et al. 2013) and Global Clean-tech innovation index (GCII) Clean-tech Group (CTG) in the list (Xiang 2020). There are 10 Donor (Developed countries) and 54 Developing countries (Asam 2021).

3.5 European Union Directives

European Union directives have the Best available technology (BAT) and Best available control technology (BACT) Industrial Emissions Directive (Foster 2004).

3.6 Pollution Prevention (P_2)

According to the Pollution Prevention (P_2) Act of 1990, defines 'resource reduction, less use of raw materials, energy and water resources, conserve in the material manufacturing process' (Freeman et al. 1992; Bi et al. 2021).

The fundamental classification between an agency like Environmental pollution agency was pioneers promoting Pollution Prevention and Waste Minimization. Then European Union has developed the Best Available Technology; UNEP handles Cleaner Production, three strategies have made in the same concepts, but different procedure/standard operational procedures, handling method/institutional, background/funding/technologies/intellectual property rights with different organizational management set-up.

The studies have broadly classified into developed and developing countries. Studies show that the private corporate sector in developed countries has made significant progress towards CT Adaptation because of stringent government policy tools; evolution has significantly changed over the last two decades. Figure 1 shows that cleaner technologies have diffused worldwide, which indicate use in seven OECD countries like Canada, Japan, Norway, Hungary, Germany, and the USA to adapt and

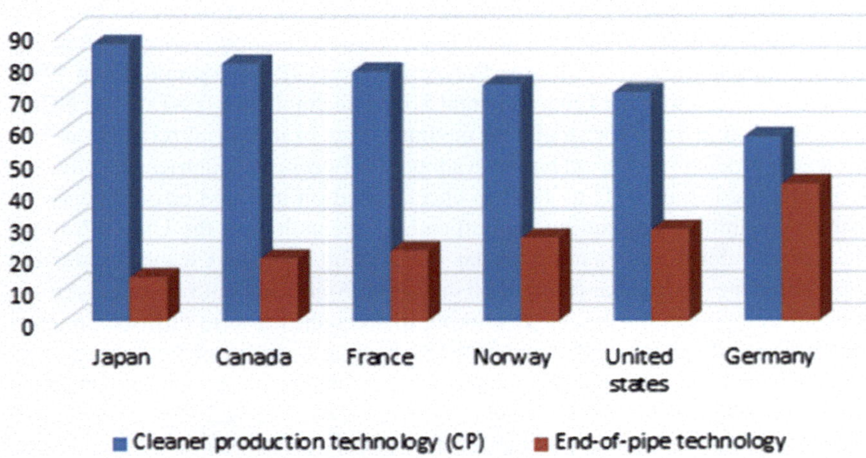

Fig. 1 The use of cleaner production technology and end-of-pipe technology in developed nation. *Source* Frondel et al. (2007)

implement cleaner production technology. Japan has 85.5% of its production through clean technologies. Germany uses 57.5% of its output through cleaner technology (Agnello et al. 2015). WRITE (Waste Reduction Innovative Technology Evaluation); the programme was conducted in the USA to promote cleaner technology waste minimization (W.M.), Total quality management (TQM) (Khanna et al. 2010), Resources, conservation and recovery Act, (RCRA) and life cycle Assessment (Dvarionienė et al. 2012). There are advanced concepts like Reducing, Rebuild, Remanufacturing, Consume Internally, Prolonging use, Returnable package, Waste incineration, Relocation has classified into reverse logistics are not used in this industry.

In Sweden, industries are using substitutes/add-on combining cleaner production, and End-of-Pipe technology has been a success. OECD countries have used cleaner technology (CT) for many years, and they have intellectual property rights and export goods to other developing countries. They have a massive market in cleaner technology (MoU) and research and development in countries like the USA, Japan, Germany, the U.K. (Frondel et al. 2007). The concept has improved further by adding economic (cost-cutting benefits) and Environmental sustainability (Ecological principles) and by promoting different types of terminology like Remanufacturing, Purification, Reverse logistics, Regeneration, Recovery, Environmental accounting, Eco-efficiency, Eco-design, waste minimization Integrated Pollution Prevention and Pollution Control (IPPC), Green Supply chain management, industrial ecosystem/symbiosis, Green chemistry, life cycle assessment, zero waste, Green engineering, Centre for Environmental Training and International Consulting (CENTRIC) (Peter and Rebeka 2007).

4 A Scenario of Cleaner Technology in Developing Countries

In 2015, 58 Cleaner production centres were operating in 56 countries (Luken et al. 2018). Among the less developed countries, India, China (Geng et al. 2010), Thailand (Visvanathan and Kumar 1999) are pioneers in implementing cleaner technologies in numerous industrial sectors. In these regions, existing policy initiatives is targeted towards pollution control management. In India, Cleaner Production was introduced in 1992 by the Ministry of Environment and forest's stated policy to decrease pollution. The concept of Cleaner Technology has performed in many African countries like the Republic of Tunisia, Zambia, Zimbabwe, Ethiopia, etc. (Sakr and Sena 2017; Chen et al. 2021). In the Middle East and Iran, cleaner production (Ghaiznoory 2005; Sakr and Sena 2017) and South American countries show higher economic and monetary development degrees to decrease environmental degradation. Scientist suggests money alleviation and openness are essential factors for CO_2 reduction. Adopting policies directed to money, exposure, and comfort to draw in higher research and development levels related to foreign direct investment may cut back the environmental degradation in countries. Also, the hardiness check through people's inclusion and Japan does not alter our main findings. This study measures the G7 and BRICS group's improvements, showing low-carbon cost-cutting based on the analysis of energy efficiency, benefit/profit, and environmental Kuznets curves. The G7 group has a significant room for improvement in carbonization value than BRICS group added before 2005 in developing countries sighing in (MoU) for the advancement of research and development in upgrading the technology with developed countries giving helping organizations like OECD, UNIDO, EPA, and European Union (Santra 2017). In developing countries like China, India, Brazil, some African and Middle East countries are encouraging policy on adopting cleaner technology in their Industrial firms.

4.1 Road Map and Diffusion of Cleaner Technology in India

The cleaner technology concepts gave importance in 1992, introducing (Indian policy statement) to abate pollution (Peattie and Charter 1992) and promote cleaner technology in small, medium, and large scale industries. "Towards initiating environmental sustainability by cost-cutting aspects in developing a proactive process like planning, preventive measures to decrease pollution, and promotion of new technological instruments to reduce pollutants from the production process itself and Sect. 4 regulation emphasized to implement integrating pollution abatement process on infrastructure set-up starting from industrial ecology towards Estate for small scale industrial (SSI), units in rural areas assisting in the adoption of cleaner technologies. Under policy Sect. 3, a comprehensive approach has proposed (MoEFCC) stated 'Clean technology cell' that will only focus on holistic environmental sustainability (Hughes and Meckling 2018). There are 88 critical polluting industrial areas in a cluster (Satyanarayana et al. 2012; Joshi et al. 2015; Dorligjav and Gharai 2021). The Supreme Court has announced that implementing clean technology is mandatory (SMEs).

By installing ETP technology, it is mandatory to check physico-chemical, biological, tertiary treatments, advanced treatments like Ion exchange, Nanofiltration, Ultrafiltration, Microfilter, and Reverse osmosis (R.O.) (Rajakumar and Nagan 2006). As a result, the effluent water is recycled and recovered by advanced technology. Therefore, fresh water consumption is reduced by 75–85%. These cleaner practices have been made mandatory in SME clusters like Textile (wet) Processing, Tanneries, Pulp, and paper industries. As a result, the effluent has disposed into the Ganga river basin without recycling water and pollution prevention (Ranade and Bhandar 2014). Now, in India used in the Cleaner coal sector like integrated gasification with the combined-cycle operation (IGCC). In Indian hosiery clusters, cleaner production technology has been used in resource conservation on energy and water-saving (Narayanasamy and Scott 2001); Wastewater has recycled using a Zero liquid discharge process (ZLD) in the tannery industry (Shah et al. 2020) and foundry industry cleaner production has used to optimize energy (Mukherjee 2011).

In India, small and medium scale industries contribute half of its industrial output and nearly 40% of its exports (Agarwal 2009). Still, only a fraction of the 3 million small-medium scale industrial units adhere to environmental norms because of the heavy water, air, and solid waste polluting industry, so the government had ordered to relocate the industries. This industry is responsible for approximately 65% of the total pollution load (Rajaram and Das 2008). Five significant barriers in adopting "Cleaner technology process in Indian manufacturing industries are classified as given:

1. Need of financial assistance, technical skills, research and development, commitment from top-level management, reliable information, and up-gradation technology (Agnello et al. 2015) UNEP and UNIDO have successfully implemented cleaner production (CP) options.

A CT scheme to assist the Small and Medium Scale industry in adopting cleaner production practices and reducing waste generation has also been in place. 118 Waste minimization Circles have established in forty-one sectors covering 600 SSI. The Supreme Court has filed a case against the oil industry towards refining/recycling used oil waste through cleaner technology. All industrial operations must engage the rule number 21 of the H.W. 1989 amendments that in recycling refining of the waste/used oil to use only by providing cleaner technologies compliances within six months from the date of publication amendments and this stated in standard operating procedure given by CPCB (ENVIS Centre 2003) such as Common Hazardous Waste Storage, Treatment and Disposal Facilities (CHWSTDF), Treatment and Disposal Facility (TSDF), 'Environmentally sound technologies' (EST), and hazardous waste management. A list of hazardous waste management (HWM) standard operating procedures has by the Hazardous waste management series (HWMS), Common Hazardous Waste Storage, Treatment and Disposal Facilities (CHWSTDF), Information manual on pollution abatement and cleaner technology series (IMPACTS), Program objectives series (PROBES) are used (Williams 1992; Mahwar et al. 1997; Sudhakar and Rao 2008; Khanna et al. 2010; CPCB 2010a, b). The scheme promotes clean technology division, cell development, and clean technology and waste minimization strategies developed by Ministry of Environmental framework for climatic change (MoEFCC). This regulation applies to small, medium, and large firms (CPCB 2016). Guidelines for demonstrating New and Innovative technology for air-polluting control Abatement Hazardous waste management has given 50 industrial guidelines on recycling, reusing, and co-processing in the red industry has an operating procedure (CPCB 2009). The online monitoring system is connected (CPCB 2018a, b, c).

4.2 The Scenario of Cleaner Technology Adoption in Large Scale Industry in India

The guidelines towards norms given by Corporate Responsibility on Environment Protection by central pollution control board for 17 "Ultra Red" category industries for (large scale firms and most polluting enterprises), because after the amendment of globalization policy in 1990, there are Multinational National Companies (MNC) investing foreign development funding in manufacturing industries came to build industry more in India like large and medium-size, small towards services orientation for this strict regulation policy which has made mandatory requirement by Environment Protection Agency, Best Available Technology, United Nations Organization and European Union environment protection, which influenced the Indian market. The pollution control board has also given guidelines/guidance for technology with unique features like waste is recycled by a method called Co-processing in (Cement, Power, Steel) Industry (Saha and Karstensen 2019). Setting of plants like Co-generation captive energy (Renewable energy) using windmill farms and solar panel, and 'Green belt' under green development framing improvise air quality towards corporate social responsibility by bioremediation concept (Tak and Kakde 2020), Concepts like 5R system, green chemistry (Lancaster 2020), and less raw material, utilization of energy, Process modification, and submitting the order of consent in the conservation of energy, technology upgrading, Companies Act, List of Industries which should furnish information in Form A (Red category industry/most polluting industry) has given primary perspective like India prepared Eco-Mark criteria for 14 classified products.

4.3 The Scenario of Water Pollution Industries in India

Corporate responsibility for environmental protection (CREP) has given specific norms to industries that use high-water intensive with raw material usage and pollution to the environment. The water pollution industries have classified into Distillery, Sugar, Textiles, Tannery Thermal power plants, Paper, and Pulp and Dyes dye intermediates. The pollution board committee has given a list of advanced effluent treatment technology that are Membrane filtration (M.F.) (Scott 1995), Reverse osmosis (R.O.) (Manyuchi and Ketiwa 2013), Nanofiltration (N.F.) (Babu et al. 2007; Guo et al. 2020), Ultrafiltration (U.F.) (Cheryan 1986; Liu et al. 2020). A Multigrade filter (MGF) (Chaitanyakumar et al. 2011) is used in the wastewater treatment process to achieve zero liquid discharge (CPCB 2015). Effluent treatment technologies have the tertiary technology (Ray and Ghangrekar 2019; Liu and Lipták 2020) like Bio-oxidation process, Advanced-Oxidation process (Culp et al. 1978; Diya'uddeen et al. 2011; Appelhaus et al. 2020), Activated carbon adsorption, Pressure sand filter, Activated Carbon filter, Tertiary chemical treatment, Sand filtration, Micro-filtration, Agitator Thin Film Drier (ATFD), Advanced/absorption (Cheremisinoff

2001). Advanced technologies like Bio-filter, Brine system, Bio-oxidation (Giannakis et al. 2020), Fixed bed biofilm reactor (Leyva-Díaz et al. 2020), the Multiple evaporator effect (Esfahani et al. 2014), Mechanical vapor recompression (Zhou et al. 2014; Si et al. 2020), Carbon absorption Electro-coagulation, Electrodylsis, Multi-stage flash, Forward Osmosis (Wang et al. 2007), aerated lagoons, Bioreactor (Hua et al. 2015) Membrane bio-reactor, Upflow anaerobic sludge blanket digestion (Vassalle et al. 2020) moving bed biofilm reactor, an advance Fenton regent is used for effluent treatment methods.

4.4 The Scenario of High Energy Consumption and Air Pollution Industry in India

The Pollution Control Committee (PCB) has identified ten different manufacturing industries (large and medium scale) as significantly polluting. The list includes highly air-polluting industries such as integrated iron and steel, thermal power plants, copper/zinc/aluminium smelters, cement, oil refineries, petrochemicals, pesticides, and fertilizer. There are about 43 critically air polluted industrial areas in India. All these factors are related to high-energy consumption levels that lead to environmental problems like air pollution industry are classified as an Aluminum smelter, copper smelter, iron and steel, cement, paper and pulp, and fertilizer industry (Dutta and Mukherjee 2010; Sengupta et al. 2019; Napp et al. 2014; Dasgupta and Roy 2015).

4.5 Air Pollution Control Technology

Most of the large and medium scale industries are energy-intensive and consume millions of tonnes of solid, liquid, gaseous fuels, debris and emit a considerable quantity of (SO_2, Oxides of nitrogen-NO_X, CO (Carbon monoxide), HC (Hydrocarbon), Respirable suspended particulate matter), Chlorine Cl_2, Ammonia (NH_3), Pb (Lead), Particulate matter (PM_{10}), Persistent Organic Pollutants (POP), Persistent Bio Accumulative Toxic Chemical (PBTs), Volatile organic compounds (VOC), Mercury (Hg) Chromium (Cr), (Tong et al. 2020), Polycyclic Aromatic Hydrocarbon (PAH), ammonia, chlorofluorocarbon (CFC), Persistent free radicals, radioactive pollutants, ground-level ozone (O_3), (Wang 2020) for energy production. Radioactive pollutants Peroxyacetyl nitrate ($C_2H_3NO_5$) (Theodore and Buonicore 1976; Bretschneider and Kurfurst 1987; Hettige et al. 1995; Wani et al. 1997; Tamhane 2008; Vahlsing and Smith 2012; Giridhar and Neeraja 2020; Zhang et al. 2020). Effective control technology includes advanced air pollution treatment technology like use of dust control systems like cyclones, multi cyclones, Electrostatic precipitators, Double ESP, scrubber, Baghouse, baffle spray scrubber, Particulate scrubber, Calayatic converter (To control VOC), Selective non-catalytic, Electro venturi scrubber,

NO_x scrubber, Flue gas desulphurization, spray tower (Cooper and Alley 2010; Ahn and Yoon 2020) Absorption, Dry and Wet scrubber and bag filter and water spray for dust suppression (Schnelle et al. 2015; Gupta 2020). National Ambient Air Quality Standards is a national air quality monitoring agency set by the pollution control committee in India acts under the regulation policy given by the environmental protection agency, and best available technology; such as green belt management using Phytoremediation through Biotechnology and the online emission monitoring system is the new initiatives to control air pollution (Crandall 1983; Rojo et al. 2010; Mudd 2012; Shareefdeen et al. 2005; Idrees and Zheng 2020; Lee et al. 2020; West 2020).

4.6 Documentation and Reporting

Now each large and medium scale industry has to monitor every aspect to save nature, and this is made mandatory by accountability from the firm's side with strict rules. These are classified into an Environment statement report (ESR) for industrial pollution reporting three months ones, Pre and post Environmental Impact assessment (EIA) for post-planning, Environment management system (EMS). An energy management system (EnMS) is an audit towards sustainability (Gargeya and Gangishetti 2013).

5 Findings and Remarks

Cleaner technology envisages qualitative and quantitative pollution-free production levels. It's essential to implement advanced technology for clean, production-technology systems like reverse environmental logistics, Green chemistry/marketing, carbon reduction projects, sustainable engineering, information technology, green transportation, and industrial symbiosis for a sustainable environment in every industry grassroots of pollution control technology. Cleaner technology production depends on most developed countries like Europe, the U.S., Japan, Canada for import, manufacturing rights, research and development, policy set-up to their terms and conditions like EPA ad E.U., which shows their dominance in manufacturing CT.

The Operational and Maintenance is very high due to tax, including export duty taxes for developing countries. Due to this, lacking in Top-level management, institutional background, R and D, finance, maintaining records, and government regulatory policy on setting limits are a pressing issue for small and medium industries. Large-scale industries can manage due to higher production output. It has been a mandatory objective that includes audits like QMS, EMS, EnMS, which brings the cleaner production technology used. But still, in developing countries, using closed-loop systems for water pollution control system, customizing older and advanced technology in a zero liquid discharge in a 70%. Closed-loop system or Add-on (system modification) in the energy conservation system, supply chain. The air pollution control system operates since the 90s using an older technology like gas flue, scrubbers, etc. Due to using cost-cutting technology, Cleaner technology cost is of higher health and safety. Less training, behaviour towards CT is an essential factor to be considered. Because this is new, there should be adaptive guidelines to the operators and frequency training should be given; developing countries should be given the manufacturing status of cleaner technology for production, and more focus should be given to in-house staff, information on CT, Standard operating procedure (SOP), Industry Associations and Research and Development Organizations. Finally, this paper tries to emphasize the reviewed paper's role in identifying gaps between the developed and developing and the importance of adapting CT for the social benefit of future generations and research.

6 Future Scope

Cleaner technology and production have an advanced inbuilt treatment and automated process of booting with a different type of domains (Software) for calibrating the numbers in trial and error modification, Physico-chemical, water, air, noise, soil, geographical information system, parameter using sensor and online database management system (pollution) using structural computing unit. Zhang et al. 2021; Cui et al. 2021; Li et al. 2021, da Silva et al. 2021; Amjad et al. 2021; Rasheed et al. 2021; Wang et al. 2021; Liu 2021).

7 Conclusion

In recent years, cleaner technology has become increasingly significant as focused by researchers, policymakers, and large, medium, and small industries. It proceeds to address the issues like economic, scientific and technological, legal, and behavioural dimensions. Studies concerning financial aspects suggest the need for an economic evaluation of Cleaner Technology by way of regulatory policy and other decision-making. The other side explains scientific and technological developments like an add-on, inbuilt, process modification, closed-loop system, green tech, co-processing, and co-generation availability of different technologies in a cross-section of industries. Coming to legal dimensions, a comparative picture of the prevailing scenario in the developed and developing countries' adaptation needs to be understood. Also, it drives the need for effective pollution control regulations. Studies on social (people) and employer behaviour are essential to focus on the market to better society's awareness. Providing better information and networking system gives an impetus to the driving forces and removes the barriers of cleaner technology. Provision of institutional background, Government support and technological support, self-reliance on manufacturing material of cleaner technology, research, and development, finances for the development of Cleaner Technology adoption. However, this study shows the efficiency of cleaner technology, which has brought a significant change to minimize industrial pollution, leading to a 'WIN–WIN' situation.

Acknowledgements The author sincerely thanks for the help rendered by Dr M. Ravichandran, Professor (Retired), Department of Environmental Management, Bharathidasan University, Tiruchirappalli, for selecting this topic and editing the final manuscript for this review article.

References

Afsah S (1998) Impact of financial crisis on industrial growth and environmental performance in Indonesia. World Bank, Washington, DC

Abhishek K, Biswas A (2018) Practices in sustainable industrial infrastructure and technology: a case study of Gujarat, India. IACSIT Int J Eng Technol 7(4):4164–4169

Adler RW, Landman JC, Cameron DM (1993) The clean water act 20 years later. Island Press, Washington, D.C.

Agarwal B (2009) Environmental issues in India: a reader. Pearson Education India, London

Agnello X, Naveen J, Ravichandran M, Balamurugan J (2015) Clean technology and its efficacy: strategies of environmental management. J Environ Soc Sci 2(2):1–10

Ahn SJ, Yoon HY (2020) 'Green chasm' in clean-tech for air pollution: patent evidence of a long innovation cycle and a technological level gap. J Clean Prod 272:122726

Aldieri L, Vinci CP (2020) Climate change and knowledge spillovers for cleaner production: new insights. J Clean Prod 271:122729

Alreck PL, Alreck PL, Settle RB, Robert S (1995) The survey research handbook. McGraw-Hill/Irwin, New York

Amjad MS, Rafique MZ, Khan MA (2021) Leveraging optimized and cleaner production through industry 4.0. Sustain Prod Consum 26:859–871

Anastas PT, Warner JC (1998) Green chemistry. Frontiers 640

Appelhaus D, Jasch K, Jahnke S, Bin HHK, Tegethoff W, Köhler J, Scholl S (2020) A new approach to simulate the fluid dynamics in a wiped film evaporator using Modelica. Chem Eng Res Des 161:115–124

Arguedas C, Camacho E, Zofío JL (2010) Environmental policy instruments: technology adoption incentives with imperfect compliance. Environ Resource Econ 47(2):261–274

Ashton W, Shenoy M (2015) Industrial ecology in India: converging traditional practice and modern environmental protection. In: International perspectives on industrial ecology. Edward Elgar Publishing, Cheltenham

Azam M, Liu L, Ahmad N (2021) Impact of institutional quality on environment and energy consumption: evidence from developing world. Environ Dev Sustain 23(2):1646–1667

Babu BR, Parande AK, Raghu S, Kumar TP (2007) Cotton textile processing: waste generation and effluent treatment. J Cotton Sci

Banzhaf HS (2020) A history of pricing pollution (or, why Pigouvian taxes are not necessarily Pigouvian), No. w27683. National Bureau of Economic Research

Barrera-Díaz C, Canizares P, Fernández FJ, Natividad R, Rodrigo MA (2014) Electrochemical advanced oxidation processes: an overview of the current applications to actual industrial effluents. J Mex Chem Soc 58(3):256–275

Basu P, Chakraborty J (2016) Environmental justice implications of industrial hazardous waste generation in India: a national scale analysis. Environ Res Lett 11(12):125001

Baumol WJ, Oates WE (1971) The use of standards and prices for protection of the environment. In: The economics of environment. Palgrave Macmillan, London, pp. 53–65

Becker JE (2001) Implementing 5S: to promote safety and housekeeping. Prof Saf 46(8):29
Bhupendra KV, Sangle S (2016) Strategy to derive benefits of radical cleaner production, products and technologies: a study of Indian firms. J Clean Prod 126:236–247
Bi X, Mullally C, Gaonkar S (2021) Does peer adoption increase the diffusion of pollution prevention practices? Land Econ 97(1):224–245
Blackman A, Bannister GJ (1997) Pollution control in the informal sector: The Ciudad Juárez brickmakers' project. Nat Resourc J 37:829
Blackman A, Li Z, Liu AA (2018) Efficacy of command-and-control and market-based environmental regulation in developing countries. Annu Rev Resour Econ 10:381–404
Bretschneider B and Kurfurst J (1987) Air pollution control technology
Bull AT (2001) Biotechnology for industrial sustainability. Korean J Chem Eng 18(2):137
Bumble S (2020) Computer simulated plant design for waste minimization/pollution prevention. CRC Press, Boca Raton
Burritt RL, Schaltegger S, Ferreira A, Moulang C, Hendro B (2010) Environmental management accounting and innovation: an exploratory analysis. Account Audit Account J
Cabeza LF, Rincón L, Vilariño V, Pérez G, Castell A (2014) Life cycle assessment (LCA) and life cycle energy analysis (LCEA) of buildings and the building sector: a review. Renew Sustain Energy Rev 29:394–416
Cagno E, Worrell E, Trianni A, Pugliese G (2013) A novel approach for barriers to industrial energy efficiency. Renew Sustain Energy Rev 19:290–308
Canter LW, Wood C (1996). Environmental impact assessment
Central Pollution Control Board (2009) National inventory of hazardous wastes generating industries and hazardous waste management in India, Hazardous Waste Management Division Delhi, pp 1–89
Central Pollution Control Board (1996) Guidelines for environmental management in industrial estates, pp 1–33
Central Pollution Control Board (2000) Guidelines for development of green belts. Programme Objective Series, PROBES/75/1999–2000, ENVIS
Central Pollution Control Board (2009) Criteria for comprehensive environmental assessment industrial clusters, series EIAS/4/2009–10, December 2009. http://cpcb.nic.in/divisions-of-head-office/ess/New-Item_151_Final_Book1.pdf2. CPCB, 2016. Final Document on Revised CEPI Version—2016, No. B-29012/ESS(CPA)/2015–16/April 26
Central Pollution Control Board (2010a) Pollution control acts, rules and notification, pp 1–1304. https://www.cpcb.nic.in/Pollution-Control-Law.pdf
Central Pollution Control Board (2010b) Protocol for performance evaluation and monitoring of the common hazardous waste treatment storage and disposal facilities including common hazardous waste incinerators. Hazardous Waste Management Series, pp 1–51
Central Pollution Control Board (2015) Guidelines on techno-economic feasibility of implementation of zero liquid discharge (ZLD) for water-polluting industries, p 14
Central Pollution Control Board (2016) Annual report 2015–16. http://cpcb.nic.in/annual-report
Central Pollution Control Board (2018a) Environmental training unit, (Ministry of Environment, Forests and Climate Change), Central pollution control board training calendar 2017–2018, pp 1–7
Central Pollution Control Board (2018b) Environmental training unit, (Ministry of Environment, Forests and Climate Change), Central pollution control board training calendar 2017–2018, pp 1–7
Central Pollution Control Board (2018c) Guidelines for continuous emission monitoring systems OCEMS. https://cpcb.nic.in/upload/thrust-area/Final_OCEMS_Guidelines_14-June-2018.pdf
Centre ENVIS (2016) Air pollution on control of pollution (water, air, and noise). Energy demand in manufacturing industries in India. Energy Policy 83:1–13
CERC (2008) Promotion of co-generation and generation of electricity from renewable sources of energy. Discussion paper. www.cercind.gov.in/09012008/Comments-invited-on16-05-2008.pdf

Chaitanyakumar D, Unnisa SA, Rao B, Kumar GV (2011) Efficiency assessment of combined treatment technologies: a case study of Charminar brewery wastewater treatment plant. Indian J Fundam Appl Life Sci 1(2):138–145
Chan HL, Shen B, Cai Y (2018) Quick response strategy with cleaner technology in a supply chain: coordination and win-win situation analysis. Int J Prod Res 56(10):3397–3408
Chand P (2018) Environmental protection and regulations in India: role of the central pollution control board. Indian J Public Adm 64(4):645–663
Change IPOC (1990) Climate change: The IPCC scientific assessment. Mass, Cambridge
Change IPOC (2014) IPCC. Climate change
Chen PJ, Chen CH, Chen YC, Wang JH, Liao TS, Hwang CH (2013) Universal environmental surveillance system with instrument cloud technology. In: IEEE international instrumentation and measurement technology conference, pp 845–849
Chen M, Sinha A, Hu K, Shah MI (2021) Impact of technological innovation on energy efficiency in industry 4.0 era: Moderation of shadow economy in sustainable development. Technol Forecast Soc Change 164:120521
Cheremisinoff NP (2001) Handbook of water and wastewater treatment technologies. Butterworth-Heinemann, Oxford
Cheryan M (1986) Ultrafiltration handbook. Technomic Publishing Co. Inc., Lancaster
Chikkatur AP, Sagar AD (2007) Cleaner power in India: towards a clean-coal-technology roadmap. Belfer center for science and international affairs discussion paper, vol 6. pp 1–261
Comprehensive Environmental Protection Index (2016). http://www.cpcb.nic.in/upload/Latest/Latest_120_Directions_on_Revised_CEPI.pdf
Conservation OE (1988) Technology absorption, foreign exchange earnings and outgo. Details of conservation of energy, technology absorption, foreign exchange earnings and outgo: in accordance with the Companies (disclosure of particulars in the report of board of directors) rules
Cooper CD, Alley FC (2010) Air pollution control: a design approach. Waveland Press, Long Grove
CPCB (2020a) Resources recycling series technical publication. https://cpcb.nic.in/
CPCB (2020b) Environmental mapping and planning publication. https://cpcb.nic.in/
CPCB (2020c) Groundwater quality series publication. https://cpcb.nic.in/
CPCB (2020d) Hazardous waste management publication. https://cpcb.nic.in/
Crandall RW (1983) Controlling industrial pollution: the economics and politics of clean air
Cui, Z, Tian W, Zhang H, Guo Q (2021) Multi-scale modeling and control of chemical looping gasification coupled coal pyrolysis system for cleaner production of synthesis gas. J Clean Prod 126903
Culp RL, Wesner GM, Culp GL (1978) Handbook of advanced wastewater treatment, 2nd edn. Van Nostrand Reinhold Co. Ltd, New York
da Silva PC, de Oliveira Neto GC, Correia JMF, Tucci HNP (2021) Evaluation of the economic, environmental and operational performance of the adoption of cleaner production: survey in large textile industries. J Clean Prod 278:123855
Dahlgaard JJ, Khanji GK, Kristensen K (2008) Fundamentals of total quality management. Routledge, London
Dasgupta S, Roy J (2015) Understanding technological progress and input price as drivers of energy demand in manufacturing industries in India. Energy Policy 3:1–13
Devi KS, Sujana O, Singh TC (2018) Hazardous waste management in India—a review. Int J Creat Res Tho 6:1547
Dincbas T, Ergeneli A, Yigitbasioglu H (2021) Clean technology adoption in the context of climate change: Application in the mineral products industry. Technol Soc 64:101478
Diya'uddeen BH, Daud WM, Aziz AA (2011) Treatment technologies for petroleum refinery effluents: A review. Process Saf Environ Prot 89(2):95–105
Dorligjav S, Gharai B (2021) Satellite-based air pollution potential climatology over India. Water Air Soil Pollut 232(9):1–19

Du H, Wei L, Brown MA, Wang Y, Shi Z (2013) A bibliometric analysis of recent energy efficiency literature: an expanding and shifting focus. Energ Effi 6(1):177–190
Dutta M, Mukherjee S (2010) An outlook into energy consumption in large scale industries in India: the cases of steel, aluminium and cement. Energy Policy 38(11):7286–7298
Dvarionienė J, Zobėlaitė-Noreikienė G, Kruopienė J, Stasiškienė Ž (2012) Application of the life cycle assessment method for pollution prevention in Klaipėda Sea Port, Lithuania. J Coast Res 29(5):1083–1091
Elkington J (1994) Towards the sustainable corporation: Win-win-win business strategies for sustainable development. Calif Manag Rev 36(2):90–100
Erkman S (1997) Industrial ecology: a historical view. J Clean Prod 5(1–2):1–10
Esfahani IJ, Kang YT, Yoo C (2014) A high efficient combined multi-effect evaporation–absorption heat pump and vapour-compression refrigeration part 1: energy and economic modelling and analysis. Energy 75:312–326
Foster JAD (2004) EPA oversight in determining best available control technology: The supreme court determines the proper scope of enforcement. Mo L Rev 69:1157
Freeman H, Harten T, Springer J, Randall P, Curran MA, Stone K (1992) industrial pollution prevention, a critical review. J Air Waste Manag Assoc 42(5):618–656
Frondel M, Horbach J, Rennings K (2007) End of pipe or cleaner production? An empirical comparison of environmental innovation decisions across OECD countries. Bus Strateg Environ 16:571–584
Gargeya MA, Gangishetti S (2013) Energy audit in industrial drives. Int J Eng Res Technol (IJERT) 2(3):1–5. ISSN: 2278-0181
Geng Y, Xienbi W, Hiengxia ZQZ (2010) Regional initiatives on promoting cleaner production in China: a case of Liaoning. J Clean Prod 18(15):1502–1508
Ghaiznoory S (2005) Cleaner production in Iran: necessities and priorities. J Clean Prod 13(8):755–762
Ghassemi A (ed) (2001) Handbook of pollution control and waste minimization. CRC Press, Boca Raton
Giannakis S, Lin KYA, Ghanbari F (2020) A review of the recent advances in industrial waste waters' treatment by sulfate radical-based Advanced Oxidation Processes (SR-AOPs). Chem Eng J 127083
Giridhar P, Neeraja SRK (2020) Air pollution a major health hazard in future. Asian J Pharm Res Dev 8(3):118–121
Glasson J, Therive R (2013) Introduction to environmental impact assessment. Routledge, London
Glensk B, Rosen C, Schiavo RB, Rabiee S, Madlener R, De Doncker R (2015) Economic and technical evaluation of enhancing the flexibility of conventional power plants. ON Energy Research Center Series 7(3)
Greenstone M (2002) The impacts of environmental regulations on industrial activity: evidence from the 1970 and 1977 clean air act amendments and the census of manufactures. J Polit Econ 110(6):1175–1219
Gunawan M, Asyahira R, Sidjabat FM (2020) Environmental management system implementation in MSMEs: a literature review. J Serambi Eng 5(2)
Guo D, Xiao Y, Li T, Zhou Q, Shen L, Li R, Lin H (2020) Fabrication of high-performance composite nanofiltration membranes for dye wastewater treatment: mussel-inspired layer-by-layer self-assembly. J Colloid Interface Sci 560:273–283
Gupta R (2020) Prevention and control of air pollution: legislative and administrative framework in India. Asian J Multidimens Res 9(2):56–62
Helman L, Parchomovsky G (2011) The best available technology standard. Colum L Rev 111:1194
Hesketh HD (1996). Air pollution control: traditional hazardous pollutants. CRC Press, Boca Raton
Hettige H, Martin P, Singh M, Wheeler D (1995) The industrial pollution projection system. World Bank policy research working paper, p 1431
Higgins TE (2018) Hazardous waste minimization handbook. CRC Press, Boca Raton

Howes R, Skea J, Whelan B (1997) Clean and competitive? Motivating environmental performance in industry. ISBN: 978-1-85383-490-5:1-188
Hua LC, Huang C, Su YC, Chen PC (2015) Effects of electro-coagulation on fouling mitigation and sludge characteristics in a coagulation-assisted membrane bioreactor. J Membr Sci 495:29–36
Hughes L, Meckling J (2018) Policy competition in clean technology: Scaling up or innovating up? Bus Polit 20(4):588–614
Idrees Z, Zheng L (2020) Low-cost air pollution monitoring systems: a review of protocols and enabling technologies. J Ind Inf Integr 17:100123
Igere BE, Ekundayo TC (2020) Global mapping of cholera Vibrio and outbreaks in the Pre-Millennium Development Goals (MDG)/Sustainable Development Goals (SDG) and MDGs/SDGs of 1990–2019. Microb Pathog 149:104319
Imes KR, Hollister J (2013) US Patent 8,509,954. US Patent and Trademark Office, Washington, DC
Jenssen PD, Heyerdahl PH, Warner WS, Greatorex JM (2003) Local recycling of wastewater and wet organic waste—a step towards the zero-emission community. In: Proceedings of the 8th international conference on environmental science and technology, pp 8–10
Joshi N, Joshi A, Bist B (2015) Air pollution tolerance index of some trees species from the industrial area of Tarapur. Int J Life Sci Sci Res 2(2)
Kakarla A, Munagala VSKR, Qureshi A, Thatikonda S, De S, Ishizaka T, Jana S (2019) Comprehensive air quality management system for rapidly growing cities in developing countries. In: Global humanitarian technology conference, pp 1–7
Kaur R, Wani, SP, Singh AK, Lal K (2012) Wastewater production, treatment and use in India. In: National report presented at the 2nd regional workshop on safe use of wastewater in agriculture
Khanna H, Laroiya S, Sharma D (2010) Integrated management systems in Indian manufacturing organizations. TQM J 22(6):670–686
Khoshoo TN (2008) Environmental concerns and strategics. APH Publishing, New Delhi
Klassen RD, Whybark DC (1999) The impact of environmental technologies on manufacturing performance. Acad Manag J 42(6):599–615
Kofoworola OF, Gheewala SH (2009) Life cycle energy assessment of a typical office building in Thailand. Energy Build 41(10):1076–1083
Kumar S, Shetty S (2018) Corporate participation in voluntary environmental programs in India: determinants and deterrence. Ecol Econ 147:1–10
Kumar A, Jain V, Kumar S (2014) A comprehensive environment friendly approach for supplier selection. Omega 42(1):109–123
Lancaster M (2020) Green chemistry: an introductory text. Royal Society of Chemistry, London
Lăzăroiu G, Ionescu L, Andronie M, Dijmărescu I (2020) Sustainability management and performance in the urban corporate economy: a systematic literature review. Sustainability 12(18):17705
Lee BXY, Hadibarata T, Yuniarto A (2020) Phytoremediation mechanisms in air pollution control: a review. Water Air Soil Pollut 231(8):1–13
Leyva-Díaz JC, Monteoliva-García A, Martín-Pascual J, Munio MM, García-Mesa J, Poyatos JM (2020) Moving bed biofilm reactor as an alternative wastewater treatment process for nutrient removal and recovery in the circular economy model. Bioresour Technol 299:122631
Li X, Yi X, Liu Z, Liu H, Chen T, Niu G, Ying G (2021) Application of novel hybrid deep leaning model for cleaner production in a paper industrial wastewater treatment system. J Clean Prod 294:126343
Liu X (2021) The online monitoring method of non-point source pollution of water resources in river scenic spots. Arab J Geosci 14(7):1–10
Liu DH, Lipták BG (eds) (2020) Wastewater treatment. CRC Press, Boca Raton
Liu B, Zhu T, Liu W, Zhou R, Zhou S, Wu R, Van der Bruggen B (2020) Ultrafiltration pre-oxidation by boron-doped diamond anode for algae-laden water treatment: membrane fouling mitigation, interface characteristics and cake layer organic release. Water Res 187:116435

Llorach-Massana P, Farreny R, Oliver-Sola J (2015) Are cradle to cradle certified products environmentally preferable? Analysis from an LCA approach. J Clean Prod 93:243–250
Lown JJ (2002) Eco-industrial development and the resources conservation and recovery act: examining the barrier presumption. BC Envtl Aff L Rev 30:275
Luken A, Van Berkel R, Leuenberger H, Schwager P (2016) A 20-year retrospective of the National Cleaner Production Centres programme. J Clean Prod 112:1165–1174
Macaskie LE (1991) The application of biotechnology to the treatment of wastes produced from the nuclear fuel cycle: biodegradation and bioaccumulation as a means of treating radio nuclide-containing streams. Crit Rev Biotechnol 11(1):41–112
Mahwar RS, Verma NK, Chakrabarti SP, Biswas DK (1997) Environmental auditing programme in India. Sci Total Environ 204(1):11–26
Maji S, Ahmed S, Ghosh S, Garg SK, Sheikh T (2020) Role of Particulate matter on air quality assessment of Delhi. In: Smart cities—opportunities and challenges, pp 63–78
Manyuchi MM, Ketiwa E (2013) Distillery effluent treatment using membrane bioreactor technology utilising pseudomonas fluorescens
Mathiyazhagan K, Govindan K, Noorul Haq A, Geng Y (2013) An ISM approach for the barrier analysis in implementing green supply chain management. J Clean Prod 47:283–297
Milone D, Peri G, Pitruzzella S, Rizzo G (2015) Are the best available technologies the only viable for energy interventions in historical buildings? Energy Build 95:39–46
Mudd JB (ed) (2012) Responses of plants to air pollution. Elsevier, Amsterdam
Mukherjee DP (2011) Barriers towards cleaner production for optimizing energy use and pollution control for foundry sector in Howrah, India. Clean Technol Environ Policy 13(1):111–123
Napp TA, Gambhir A, Hills TP, Florin N, Fennell PS (2014) A review of the technologies, economics and policy instruments for decarbonizing energy-intensive manufacturing industries. Renew Sustain Energy Rev 30:616–640
Narayanasamy V, Scott JA (2001) Lessons from cleaner production experiences in Indian hosiery clusters. J Clean Prod 9(4):325–340
Naveen AJ, Boopathi S, Arivoli A (2021) Case experience on WIN WIN approach towards industrial pollution control-economic and environmental efficacy of cleaner technology in select cement industry in Tamil Nadu. In: Bioremediation and green technologies. Springer, Cham, pp 209–234
Ochsner M, Chess C, Greenberg M (1995) Pollution prevention at the 3M corporation: case study insights into organizational incentives, resources, and strategies. Waste Manag 15(8):663–672
Oliveira JA, Oliveira OJ, Ometto AR, Ferraudo AS, Salgado MH (2016) Environmental management system ISO 14001 factors for promoting the adoption of cleaner production practices. J Clean Prod 133:1384–1394
Organization for Economic Co-operation and Development: Statistical Office of the European Communities (1997) Oslo manual: proposed guidelines for collecting and interpreting technological innovation data. Organization for Economic Co-operation and Development
Ou Y, West JJ, Smith SJ, Nolte CG, Loughlin DH (2020) Air pollution control strategies directly limiting national health damages in the U.S. Nat Commun 11(1):1–11
Park SM, Rhee CG, Park JB, Lee HJ, Kim CL (2004) Database modeling and environmental information. J Korean Nucl Soc 36(3):263–275
Patil P, Ghosh O, Jeswani H (2015) Sustainable industrial waste water management for industrial development cooperation in Thane, Maharashtra
Peattie K, Charter M (1992) Green marketing. In: The marketing book, p 726
Peter G, Rebeka L (2007) Review of sustainability terms and definitions. J Clean Prod 15:1875–1885
Porter ME, van der Linde C (1995) Green and competitive: ending the stalemate. Harv Bus Rev 73(5):120–137
Powers N, Blackman A, Lyon TP, Narain U (2011) Does disclosure reduce pollution? Evidence from India's green rating project. Environ Resour Econ 50(1):131–155
Prakash Chandak S (1994) DESIRE, demonstration in small industries for reducing waste. Ind Environ 41–5

Priede J (2012) Implementation of quality management system ISO 9001 in the world and its strategic necessity. Procedia Soc Behav Sci 58:1466–1475
Rajakumar AS, Nagan S (2006) Study on the quality of effluent discharged by the Tiruppur textile dyeing units and its impact on river Noyyal, Tamil Nadu (India). J Environ Sci Eng 52(4):333–342
Rajaram T, Das A (2008) Water pollution by industrial effluents in India: Discharge scenarios and case for participatory ecosystem specific local regulation. Futures 40(1):56–69
Ranade VV, Bhandari VM (2014) Industrial wastewater treatment
Ranade VV, Bhandari VM (2014) Industrial wastewater treatment, recycling and reuse. Butterworth-Heinemann, Oxford
Rasheed R, Javed H, Rizwan A, Sharif F, Yasar A, Tabinda AB, Su Y (2021) Life cycle assessment of a cleaner supercritical coal-fired power plant. J Clean Prod 279:123869
Ravichandran M (1989) Economics of pollution prevention and control: A study of industrial pollution in Karnataka
Ray SG, Ghangrekar MM (2019) A comprehensive review on the treatment of high-strength distillery wastewater in advanced physico-chemical and biological degradation pathways. Int J Environ Sci Technol 16(1):527–546
Reddy BS, Ray BK (2010) Decomposition of energy consumption and energy intensity in Indian manufacturing industries. Energy Sustain Dev 14(1):35–47
Reed MS (2008) Stakeholder participation environmental management: a literature review. Biol Cons 141(10):2417–2431
Requate T, Unold W (2001) On the incentives created by policy instruments to adopt advanced abatement technology if firms are asymmetric. J Inst Theor Econ (JITE)/Zeitschriftfür die gesamte Staatswissens chaft 536–554
Rojo N, Gallastegi G, Barona A, Gurtubay L, Ibarra-Berastegi G, Elías A (2010) Biotechnology as an alternative for carbon disulfide treatment in air pollution control. Environ Rev 18:321–332
Rourka DO, Connelly L, Koshland CP (1996) Industrial ecology: a critical review. Int J Environ Pollut 6:89–112
Rowe DB (2011) Green roofs as a means of pollution abatement. Environ Pollut 159(8–9):2100–2110
Rowe DR, Abdel-Magid IM (1995) Handbook of wastewater reclamation and reuse. CRC Press, Boca Raton
Royston MG (2013) Pollution prevention pays. Elsevier
Saha PK, Karstensen KH (2019) Co-processing of Alternative Fuels and Resources in Indian Cement Industry—Baseline and Potential. In: Waste valorisation and recycling. Springer, Singapore, pp 19–29
Sakr D, Sena AA (2017) Cleaner production status in the Middle East and North Africa region with special focus on Egypt. J Clean Prod 141(10):1074–1086
Santra S (2017) The effect of technological innovation on production-based energy and CO_2 emission productivity: evidence from BRICS countries. Afr J Sci Technol Innov Dev 9(5):503–512
Sathya J, Ravichandran M (2010) Environmental management. An economic approach to pollution control in Sago Industry. Abhijeet Publications
Satapathy SK, Sangle S, Unnikrishnan S (2017) Role of government towards adoption of cleaner technologies for climate proactivity: A survey-based empirical study of Indian firms. World J Sci Technol Sustain Dev
Satyanarayana T, Johri BN, Prakash A (eds) (2012) Microorganisms in environmental management: microbes and environment. Springer Science & Business Media
Schnelle KB Jr, Dunn RF, Ternes ME (2015) Air pollution control technology handbook. CRC Press, Boca Raton
Scott K (1995) Handbook of industrial membranes. Elsevier, Amsterdam
Sengupta PP, Sinha M, Dutta UP (2019) Economic and environmental performances in manufacturing industries: a comparative study. Period Eng Nat Sci 7(1):99–108
Shah AV, Varjani S, Srivastava VK, Bhatnagar A (2020) Zero liquid discharge (ZLD) as sustainable technology—challenges and perspectives. Indian J Exp Biol (IJEB) 58(08):508–514

Shareefdeen Z, Herner B, Singh A (2005) Biotechnology for air pollution control—an overview. In: Biotechnology for odour and air pollution control, pp 3–15
Sharma M, Singh G (2019) Critically polluted area analysis using the comprehensive environmental pollution index
Shen TT (1995) Industrial pollution prevention. In: Industrial pollution prevention. Springer, Berlin, Heidelberg, pp 15–35
Si Z, Han D, Gu J, Song Y, Liu Y (2020). Exergy analysis of a vacuum membrane distillation system integrated with mechanical vapour recompression for sulfuric acid waste treatment. Appl Therm Eng 178:115516
Singh RP, Kolok AS and Bartelt-Hunt SL (eds) (2019) Water conservation, recycling and reuse: issues and challenges. Springer, Singapore
Singh H (1999) Environmental information system in India. DESIDOC J Libr Inf Technol 19(2)
Singh A (2018) Flexible waste management practices in service sector: a case study. In: Global value chains, flexibility and sustainability, pp 301–318
Sood A, Arora B (2006) The political economy of corporate responsibility in India. UNRISD, Geneva
Sudhakar Rao M, Monto M, Ravindranath NH (2008) Advances in water quality and management. ISBN: 978-981-05-9012-3:1-331
Tak AA, Kakde UB (2020) Evaluation of air pollution tolerance and performance index of plants growing in industrial areas. Int J Ecol Environ Sci 2(1):1–5
Tarí, JJ (2005) Components of successful total quality management. The TQM Magazine
Theodore L, Buonicore A (1976) Industrial air pollution control equipment for particulates
Tong Y, Yue T, Gao J, Wang K, Wang C, Zuo P, Liang Q (2020) Partitioning and emission characteristics of Hg, Cr, Pb, and as among air pollution control devices in Chinese coal-fired industrial boilers. Energy Fuels 34(6):7067–7075
Treatment HW (1989) Storage and disposal facilities (TSDF)—air emission models. EPA-450/3-87-026 review draft
Turkar SS, Bharti DB, Gaikwad GS (2011) Various methods involved in wastewater treatment to control water pollution. J Chem Pharm Res 3(2):58–65
UNEP (1994) Industry and environment. Report on cleaner production, vol 17, pp 4–22
UNEP (2004) The use of economic instruments in environmental policy: opportunities and challenges
UNEP (2013) Cost of inaction on the sound management of chemicals. http://www.unep.org/chemicals-and-waste/Portals/9/Mainstreaming/Cost-Of-Inaction/Report-Cost-of-Inaction-Feb2013.pdf
UNFCC (2014) GHG data from UNFCC. United Nations Framework Convention on Climate Change. New York City, U.S. http://unfccc.int/ghg_data/ghg_data_unfccc/items/4146
United Nation Environment programme on Industry and Environment (1994) Government strategies and polices for cleaner production, pp 1–35
Vahlsing C, Smith KR (2012) Global review of national ambient air quality standards for PM_{10} and SO_2 (24 h). Air Qual Atmos Health 5(4):393–399
Vassalle L, Díez-Montero R, Machado ATR, Moreira C, Ferrer I, Mota CR, Passos F (2020) Upflow anaerobic sludge blanket in microalgae-based sewage treatment: co-digestion for improving biogas production. Bioresour Technol 300:122677
Vidhya Lakshmi D, Joseph K (2010) Physico-chemical characteristics of leachate from a common hazardous waste disposal facility in South India. Int J Environ Technol Manage 13(3–4):348–361
Visvanathan C, Kumar S (1999) Issues for better implementation of cleaner production III Asian small and medium industries. J Clean Prod 7:27–34
Wang X, Zhou D (2021) Spatial agglomeration and driving factors of environmental pollution: a spatial analysis. J Clean Prod 279:123839
Wang LK, Hung YT, Shammas NK (eds) (2007) Advanced physico-chemical treatment technologies. Humana Press, Totowa, NJ, pp 57–83
Wang LK, Hung YT, Shammas NK (eds) (2009) Handbook of advanced industrial and hazardous waste treatment. CRC press, Boca Raton

Wang Z, Wang X, Cong S, Geng F, Zhao Z (2020) Fusing electrochromic technology with other advanced technologies: a new roadmap for future development. Mater Sci Eng 140:100524

Wang Z (2020) Air quality analysis of nitrogen oxides and relationships with ozone pollution. Doctoral dissertation

Wani AH, Branion RM, Lau AK (1997) Biofiltration: a promising and cost-effective control technology for Odors, VOCs and air toxics. J Environ Sci Health Part A 32(7):2027–2055

Waxman HA (1991) An overview of the clean air act amendments of 1990. Envtl L 21:1721

Wheeler D, Afsah S (1996) Controlling industrial pollution: a new paradigm. The World Bank, Washington, D.C.

Williams TA (1992) Pollution and hazardous waste on Indian lands: do federal laws apply and who may enforce them? Am Indian L Rev 269–290. https://digitalcommons.law.ou.edu/ailr/vol17/iss1/11

Wong CY, Wong CW, Boon-Itt S (2015) Integrating environmental management into supply chains: a systematic literature review and theoretical framework. Int J Phys Distrib Logist Manag 45(1/2):43–68

Xiang JY (2020) Cleantech innovations by developing countries. BU Int'l LJ 38:183

Yin S, Shao Y, Wu A, Wang H, Liu X, Wang Y (2020) A systematic review of paste technology in metal mines for cleaner production in China. J Clean Prod 247:119590

Zeng B, Zhu L, Yao X (2020) Policy choice for end-of-pipe abatement technology adoption under technological uncertainty. Econ Model 87:121–130

Zhang T, Li G, Yu Y, Ji Y, An T (2020) Atmospheric diffusion profiles and health risks of typical VOC: numerical modelling study. J Clean Prod 275:122982

Zhang CM, Shi Y, Jiang LH, Hu YY, Li Q, Li HQ (2021) Analysis of lead pollution control in anode slime micromorphology evolution induced by Mn^{2+} ions for cleaner production of zinc electrolysis. J Clean Prod 126700

Zhou Y, Shi C, Dong G (2014) Analysis of a mechanical vapour recompression wastewater distillation system. Desalination 353:91–97

Chapter 9
Big Data Applications in Smart Sustainable Energy Systems and E-Mobility: Review and Case Study

Tayyibah Khanam, Mohammad Saad Alam, Sanchari Deb, and Yasser Rafat

Abstract Since sustainability-related concerns need to be answered beforehand, a vast majority of research in the recent past has been focussing on the potential utilization of data-driven approaches to move a step ahead in achieving sustainable development goals. While big data in itself is not technological advancement, practices such as analysis of data sets and implementation of Machine Learning algorithms on data sets can give deeper insights and understanding to aid the decision-making process of organizations. With the increasing amount of data in almost every sector, it is now even possible to get accurate predictions for practical purposes. Hence, big data sciences can be viewed as a paradigm shift in achieving sustainable development goals. First-hand, this chapter gives necessary background about sustainable development, renewable energy, big data, and E-mobility. Since smart cities are expected to integrate several important components of human–nature interactions in the near future, we perform a review of the existing literature to examine the current research directions in the role of big data tools and technologies in different components of smart cities. Additionally, pieces of literature on the applications of big data in electric vehicles are closely reviewed for better understanding & problem formulation. Lastly, this chapter also presents a case study based on traffic congestion data to analyse and understand the charging behaviour of users across the city of Mumbai and thus draw relevant conclusions regarding the Charging Pile Network of the city.

Keywords Big data · Charging infrastructure · Renewable energy · Smart cities · Sustainability

---

T. Khanam (✉) · M. S. Alam
Department of Electrical Engineering, Aligarh Muslim University, Aligarh, India
e-mail: tayyibahkhanam@zhcet.ac.in

M. S. Alam
e-mail: saad.alam@zhcet.ac.in

S. Deb
Electric Power Train Group, VTT Technical Research Centre, Espoo, Finland

Y. Rafat
Department of Mechanical Engineering, Aligarh Muslim University, Aligarh, India
e-mail: yasser.rafat@zhcet.ac.in

I. Ali et al. (eds.), *Computational Modelling in Industry 4.0*,
https://doi.org/10.1007/978-981-16-7723-6_9

1 Introduction

The requirement for sustainability emerges from the right to extremely essential needs like food, water, energy, sanitation, and work to everybody. Absence of these essential needs brings about poverty. Till date, almost half of the world's total populace, i.e. 3 billion people live in poverty and about 297,000 children under five kick the bucket every year because of destitution (11 Facts About Global Poverty 2021). Apart from poverty issues, another significant danger to life on earth is air contamination due to the consumption of fossil fuels. These fossil fuels discharge GHGs (green-house gases) and other harmful gases which are an immediate danger to an unnatural weather change (increasing of the world's temperature) and deteriorate the health of living creatures. One can only imagine the interconnections between the different threats to sustainability, being the cause and effects of each other.

The origin of sustainability dates back to the Brundtland report (1987) (Kuhlman and Farrington 2010). The main concern highlighted in this report was regarding the arising tensions between the human desires of a superior lifestyle and the restrictions put by nature. Thus, over the period, the idea of sustainability has been re-deciphered as enveloping three aspects specifically—social, economic, and environmental. Each of these three aspects needs to be examined on different but suitable parameters by professionals of different fields depending upon their areas of direct impact as the basic grounds for such a classification of sustainability.

While socio-economic sustainability focuses on several factors of social equality, citizen welfare, economic growth, and expenditures, the goal of environmental sustainability is to moderate the consumption of natural non-renewable energy sources and to create alternate sources of energy while diminishing contamination and damage to the earth. Thus, the intention behind achieving environmental sustainability is to address the issues of the present without compromising the limits of individuals to address their issues in the future. In light of this affair, it is reasonable to state that Renewable Energy Sources (RES) are viable direct arrangements for sustainable development as they are clean and non-exhaustible sources of energy. The next major concern regarding environmental sustainability is the lack of adequate technologies to capture and effectively utilize energy from renewable sources due to knowledge gaps. Such an issue is quite common in Less Economically Developed Countries (LEDCs) as in the case of Nigeria. Even though Nigeria has a huge potential for the growth of solar power due to high values of annual incident solar radiations, solar radiations are being wasted due to the lack of local investments in solar modules, PV arrays, and well configured infrastructural architecture. It is reported that solar modules covering 1% of Nigeria's landmass can generate up to 627,260 MW of electricity yearly as compared to the current generation of 6803 MW (Ndukwu et al. 2020).

Thus, nations need to focus on overcoming the connected web of challenges related to all three aspects of sustainability. In this regard, several data-driven emerging trends such as Artificial Intelligence (AI), Data Sciences, Internet of Things (IoT) integrated have revolutionized several industries and organizations and have a much larger potential in the future. Thus, data is seen as a resource powering up the informational economy.

From the relevance of this chapter, we will mainly be focusing on research concerning the applications of big data-driven technologies in achieving sustainable development goals.

But what exactly is big data? Big data as the name suggests refers to huge collections of data, i.e. can be anything varying from records, facts, statistics, audios, videos, photos, figures, etc. Having said that, big data is worthless on its own, unless we associate it with added values. We can explain this dubious property of big data with an analogy similar to that of currency notes. In their true sense, banknotes and coins do not have any physical significance than just being pieces of paper and metal chunks. However, humans have added value to the banknotes by authorizing them from legal authorities such as banks. Similarly, big data sets need to be processed and then examined for deeper insights relative to the field of impact.

What is Big data analytics? Big data analytics is effectively breaking down the data collections to information, using the information to increase the existing knowledge and finally attaining wisdom from knowledge (Tien 2013) to make more intelligent business moves for an efficient future. Hence, the idea of Big Data analytics is viewed as a development of Business Intelligence (Big Data vs. Business Intelligence vs. Data Mining|The Differences 2021), since it incorporates new sorts of data that were already too complex to even think about handling them with usual procedures (Sanila et al. 2017). Cost assumes a major role in demonstrating the significance of big data where organizations tend to achieve ideal costs for products to expand their benefit as well as to ease the cost-related concerns of the customers.

The motivation behind this chapter is to review and highlight the role of big data in accomplishing sustainable development goals through various sectors of a smart city. A smart city is essentially expected to integrate all important sectors of human interference having an impact on sustainability. Since big data is one of the emerging technologies since the past few years, this chapter also contributes to the existing research by reviewing previous research papers that integrate various other emerging technologies such as Machine Learning (ML)/Deep Learning, Internet of Things (IoT), etc. with big data analytics to present a holistic view of the current research in sustainable development. We will also emphasize on how professionals can leverage the data collected, recorded, and generated on large scales using big data techniques to move closer in achieving sustainable development goals. Close attention is also paid to the prevailing challenges and their possible solutions.

The rest of the chapter is divided as follows—Sect. 2 provides essential background about sustainability, renewable energy, big data analytics, and E-mobility. Following up with the background, a comprehensive review of the existing literature is done for the same in Sect. 3. Section 4, 5 and 6 summarize the literature review, discuss the limitations of existing literature, and explore the scope of big data analytics in sustainable development and e-mobility, respectively. Section 7 presents a case study based on electric vehicle (EV) charging data analytics performed for the city of Mumbai, India to identify issues in the existing charging infrastructure and prepare for its future challenges. Finally, Sect. 8 summarizes the objectives of this chapter.

2 Background

2.1 Scenario of Global Sustainability

Global Warming has been one of the most severe consequences of air pollution due to green-house gases. Recently, one of the most environment triggering news that surfaced across the globe indeed concerned the topic of Global Warming. Scientists from the UK have concluded that the Earth has lost about 28 trillion tons of ice in 30 years as a result of global warming (Earth has lost 28 trillion tonnes of ice in less than 30 years 2021). This comes after similar harsh remainders about severe glacier melting in Greenland were put up in several reports in the first half of 2020 (Snowden 2021). A relatively new threat to sustainability is the increased number of wildfires in several parts of the world including California in the USA, Uttarakhand in India and Siberian regions. The European Union hence estimated that the Arctic wildfires alone contributed to about 35% more carbon dioxide generation up till 24 August 2020, than the whole year of 2019 (Siberia heatwave: why the Arctic is warming so much faster than the rest of the world 2021). Increasing records of melting of glaciers and wildfires suggest to us that the temperature of the earth is ever increasing and hence a major threat to global sustainability.

Likewise, COVID-19 times have worsened the situation and made it even more challenging to achieve the 2030 agenda for sustainable development put up by the United Nations. Some positive impacts have also been noticed, such as reduced planned travels (which resulted in less vehicular GHG emissions and hence smaller steps towards better air quality), fewer traffic congestions (which resulted in less travel time for emergencies), increased hygiene levels in several households (which resulted in the prevention of various diseases, not just COVID-19), etc. However, negative impacts have affected the economy and the environment much more severely. More plastic and related wastes are generated due to the huge demand for PPE (personal protective equipment) kits, gloves, and masks. Usually, these materials are also discarded after one-time use and hence the amount of waste generated increased considerably. As per the Chinese Ministry of Ecology and Environment, medical centres in Wuhan delivered an excess of 240 tons of waste every day at the height of the pandemic, in contrast with the 40 tons of waste during normal occasions (The plastic pandemic is only getting worse during COVID-19 2021). Six months since the outbreak now, and the world already has 105 million COVID-19 cases with 25 million of them being active as on 5 February 2021 (COVID Live Update: 171,029,617 Cases and 3,556,677 Deaths from the Coronavirus—Worldometer 2021). With such figures, it is expected that the per day waste generation has crossed records by now. Alongside, a great recession is being forecast, bringing the world's economies down to their knees, especially the Less Economically Developed Countries (LEDCs). The loss of jobs and the temporary shut-down of various offices has resulted in the reduction of annual GDP of several developing countries including India. The World Bank had already estimated about 3.1% loss in the Indian GDP for the fiscal year 2020–2021 (Data|An estimated 12.2 crore Indians lost their jobs during the coronavirus lockdown in April: CMIE 2021). Making the situation worse, the official data released by the National Statistics Office (NSO), India on August 31, 2020, revealed the first economic contraction of India in 4 decades, as much as 23.9% decrease in the April-June quarter of 2020 as compared to the same quarter of 2019 (Team 2021).

Figure 1 represents the Global Sustainable Competitiveness Index (GSCI) for the year of 2019 which measures the seriousness of nations in terms of sustainability in a coordinated way. It is determined dependent on 116 quantifiable, quantitative markers acquired from many reliable sources, for example, the World Bank, the IMF, and diverse UN organizations (The Global Sustainable Competitiveness Index 2021). Sweden tops the list with a score of 60.6. Finland and Iceland bag the 2nd and 3rd positions with scores of 59.5 and 57.3. The United Kingdom is on the 17th rank and one of the countries to cross the 50 mark with a score of 52.8. USA (34), China (37), & Nepal (64) follow up with scores of 49.1, 48.5, and 45.6, respectively. India comes on the 130th rank with a score of 39.5. India's neighbour Pakistan secures the 145th positions with a score of 38.3 whereas the last country in the rank-list is the Bahamas with a score of 30.5. Figures suggest, there is a long way to go!

Thus, due to the largely disruptive impact of COVID-19, the world is in an alarming position. More judicious utilization of resources is necessary to bring back the normal. Developing countries especially need to manage finances and resources

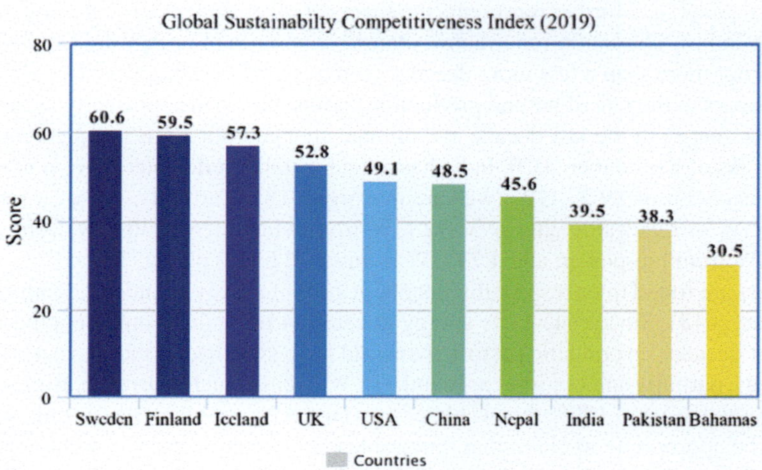

Fig. 1 The GSCI indexing scores for major countries of the world in the year 2019

whereas the developed countries need to severe down the waste generation and GHGs emissions.

The threats to sustainability are larger for India. India is the world's second-largest country by population and among the most densely populated countries. The per capita income of India remains among the lowest in the world even though India is the world's largest economy. Thus, the results of 'un-sustainability' not only affect the demands of the future, but also the public health of the present. It also puts enormous pressure on carrying capacity of the natural resource system as well as the inability of the environment to absorb the waste generated.

As a conscious aspirant, India has played a lead role in tackling climate change and putting forward policies that align with sustainable development goals of 2015. That being the case, endeavours have been made to include all partners—from people in general and private divisions—in the framing process of efficient policies, strategies to be implemented, and helpful practices. The international report on sustainable living—'Greendex' measures how consumers all over the world are reacting to sustainability concerns (Greendex Ranks 18 Countries Based On Sustainability 2021). Indian consumers occupy the second spot on this index, particularly receiving high grades in the sectors of housing, transport, and food choices (INDIAN CONSUMERS: Second place, n.d.).

2.2 *Green Renewable Energy*

Around 75% of electric generation systems in India use fossil fuels (Growth of Electricity Sector in India 1947–2020 2020) (coal, diesel and natural gas) as their major source and almost all vehicles run on crude oil derivatives such as petrol and diesel or on fossil gas (Fig. 2 further represents the shares of other sources of electricity generation in India). These non-renewable exhaustible sources of energy are estimated to last for not more than a few more decades (if the rate of consumption doesn't lower with respect to the rate of natural production). Thus, due to the constantly increasing need for energy in the last decade and climate change issues, RES implementation is vital. As of September 2020, India has a total installed grid-interactive renewable power capacity of 89.23 GW (web.archive.org 2018). Currently, out of a total of 89.23 GW of renewable energy 38.12 GW from wind energy, 36 GW from solar, 10.1 GW from bio-power, and 4.73 GW from small hydro plants.

Cities are found to be responsible for about 70% of energy-related CO2 emissions (Science 2018). Since renewable energy generation has gained much popularity in the past decade, several cities across the world have switched to these clean sources of energy, partially and in some cases entirely. With cities such as Brasíla, Burlington, Sydney switching to 100% renewable energy (www.power-technology.com, n.d.), it would be appropriate to claim that renewable energy can be used as a major power source in several leading cities and there is an immense potential in cities to lead on building a sustainable economy. India is taking baby steps in the same direction, Jaipur and Bengaluru being the first Indian cities in world's top 10 renewable smart

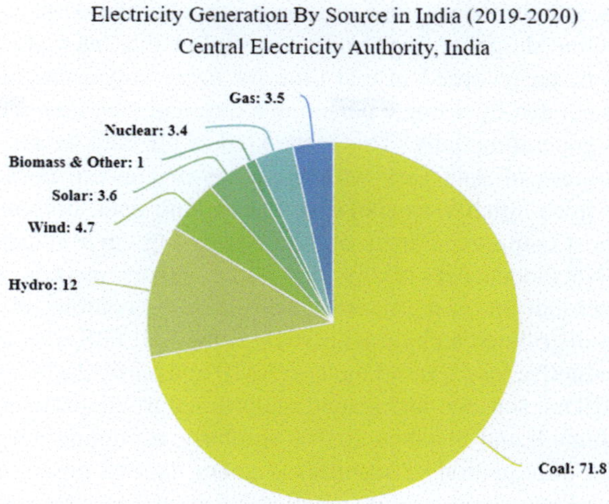

Fig. 2 Sources of electricity generation in India with their respective shares in 2019–2020

cities list bag the fourth and the sixth positions, respectively (www.power-technology.com, n.d.). India is also expected to be the largest contributor to the renewables by 2021, with the country's annual additions doubling from 2020 (EDF Energy 2015). Recently, Asia's largest solar power plant was inaugurated by Prime Minister in Madhya Pradesh's Rewa with a capacity of 750 megawatts. With the deployment of this power plant, it is estimated that the reduction in carbon dioxide emissions will reduce by 15 lakh tonnes annually (The Indian Express 2020).

2.3 Big Data

Data is being generated at a fast rate; almost more than 5 Exabyte per day increase due to use of emails, audio, video and data streams, health information, various queries, social networks, scientific data, and on-going mobile phone applications data (Sanila et al. 2017). It involves highly delicate multiple datasets owing to different aspects such as personal data, organizational data, environmental data, data of transport systems, and economic data. Thus, the applications of big data analytics in several aspects of day-to-day life as well as in several industrial, organizational, and business aspects is where humans benefit from the big data.

Large data production in the form of government records and surveys can be termed as universal data sets since they provide information related to both—the city and its citizens on a whole. However, currently one of the most popular and common data sources is the internet. Internet websites are not only based on data but also record data in the form of user activities (Einav and Levin 2014) and hence largely contribute to big data sets. Almost, 59% of the world's population is actively using the internet on an everyday basis (Johnson 2021). From personal data such as search history, text messages, emails, photos to relatively public data such as online shopping, registrations, social media, interactive websites, etc. each activity is being recorded in real time for every single user. Imagine, the amount of data generated by a single person in a day, and then imagine how much the whole world is generating daily! This is the kind of big data we are talking about. Other specific sources of data vary with the areas of concern because internet activity records majorly qualify as social and economic data, not environmental data. While the most common sources of data addressing environmental concerns include photos, SNS (social networking sites) data, survey records, and machine data (which is then put up on the internet websites), the continuous technological improvements in distributed technologies such as the IoT, GIS (Geographic Information System), inductive loops, smart meters (AMI) and particularly in intelligent electronic devices (IEDs), sensors, and actuators, have led to the production of huge amounts of data (Junaidi and Shaaban 2018). Similarly, economic data sources have evolved. With the introduction of scanners, scanner data seems to have revolutionized sales and marketing by providing detailed information such as quantity, type, prices, store ID, etc. about the goods and products sold. Scanner data and electronic inventory records are the current major sources of economic sales data (Einav and Levin 2014). Big data applications have a large role in improving environmental sustainability in the realm of smart cities and their activities. They have the potential to serve a variety of domains of smart cities such as power grids and/or smart grids, transport, traffic, urban design and planning, infrastructure monitoring and management, etc.

Although, data storage is one of the major challenges faced in big data analytics, relatively big organizations have achieved convenient data storage facilities for current as well as future use through constructive planning. For instance, through the GitHub Archive programme, several terabytes of data comprising of mainly open-source software were safely stored for use by future generations in a vault in the Arctic for as much as 1000 years (Months and Weeks 2020). Also, the demand for stream processing is increasing since most of the data generated in the modern world is real time (Vanani and Majidian 2019). When big datasets are combined, cases might be distinguished by the integration of data which then provides countless opportunities to apply reasoning in order to think so as to grow our natural organizational structure (Wu and Chen 2017).

Big data characteristics allow the utilization of big data methods to extract value from the business. The 5 V's are Volume, Velocity, Variety, Veracity, and Value (Amaro and Pina 2017). Figure 3 represents the basic underlying knowledge behind the 5 V's of big data.

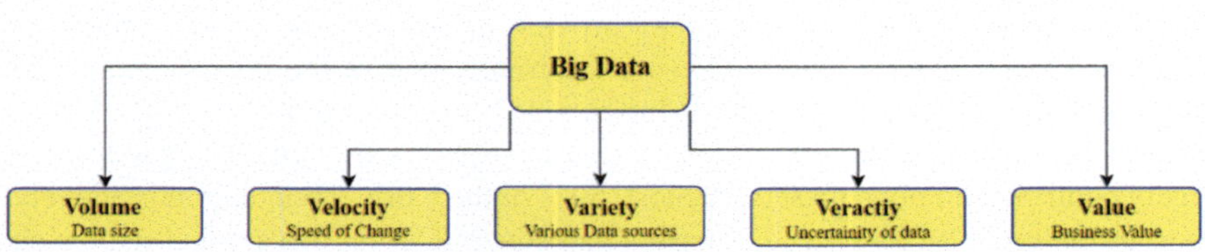

Fig. 3 The 5 V's of big data with basic underlying knowledge

Usually, when dealing with a limited amount of data with less variability as compared to big data, the MySQL database is used. It is an open-source relational database management system (RDBMS) that collects, cleans, analyses, and visualizes the data using Structured Query language (SQL). However, for big data sets possessing the 5 V's, the operating performance for transferring the data to RDBMS is greatly compromised. The current industry standards for decision support system are based on Data warehousing (Nitesh Jain 2015). Data warehousing is simply leveraging the big data produced in repositories, websites, weblogs, market trends, transition logs of customers, etc. by industrial data analysts. Similar to the drawbacks faced by the traditional RDBMS in querying through big data, the algorithms of data warehousing supported only small data sets where increasing the sample size of data sets would take a high turnaround time.

Hence, an open-source framework capable of handling big data sets at quicker throughputs were needed. This led to the evolution of several big data handling frameworks, databases, data warehouses, etc. such as Apache Hadoop, Apache Cassandra, Apache Spark, Google Big query, MongoDB, Xplenty, and a few more. Table 1 derived from (Apache.org 2019; Softwaretestinghelp.com 2019) gives a basic overview of a few commonly used big data tools and technologies and the languages supported by them.

Hadoop is known as one of the first and the most efficient big data software frameworks to date. The basic idea behind Hadoop and several others is using a cluster of computers instead of increasing the computation power of one computer. Hadoop is based on the idea of distributed computing where distributed frameworks use a cluster of computers for enormous amounts of constantly increasing data. Bill Gates thus popularized the idea of business at the speed of thought enabled by Hadoop. (Nitesh Jain 2015).

The basic two advantages of clustering are:

- Clusters are commodity hardware, cheaper than one expensive server (studies have shown that Hadoop clusters can give 10 times more throughput at 1/10th of the cost in certain situations).
- The license fee of traditional RDBMS is expensive, but Hadoop's software license is free and open source.

Hadoop core provides a layer between the user and cluster of machines under it. It comprises of three core components. The Hadoop distributed file system (HDFS), the programming framework or Map-reduce and Yet Another Resource Negotiator (YARN). Through Map-reduce, the code is moved to the data, unlike of what happened previously, where data was sent to the code. YARN assigns computational resources for execution of an application (Datadog 2016).

What happens inside Hadoop is an interesting question to answer. The programming or the MapReduce framework enables users to have full control on the processing of input datasets (Dittrich and Quiané-Ruiz 2012). A file when given as input, is broken down into smaller portions known as blocks. The blocks are then replicated and distributed over a cluster of computers. It is Hadoop that manages the

Table 1 Big data tools and technologies

Name	Owner	Type and description	Language
Hadoop	Apache Software Foundation	Java-based framework which runs applications on clusters of commodity hardware • Basic Components—Hadoop Distributed File System (HDFS), YARN, Hadoop MapReduce & Hadoop Common • Additional Software packages from Apache capable of running alongside Hadoop- Hive, Pig, HBase, Spark etc	Written in Java. Java is also preferred for programming however other languages such as Python, Scala, and SQL are supported depending on the type of additional software packages
Cassandra	Apache Software Foundation	NoSQL database management system which handles data across many commodity hardware	Written in Java, employs Cassandra Query Language (CQL)
Kafka	Apache Software Foundation	Stream processing software platform	Supported languages—Scala and Java
Big Query	Google	Serverless data warehouse for scalable analysis	Supports ANSI SQL
MongoDB	MongoDB Inc	Document oriented database program	Written in C++, Go, JavaScript, Python. Supports all popular programming languages
Xplenty	Xenon partners	Scalable cloud platform to integrate and process data from various sources	Supports SQL, Python and R

distribution, division, and management of the blocks, hence easing the task for users. Overall, programmers just need to write the Map and Reduce codes and the rest of the work is done by Hadoop distributed file system (Udemy, n.d.).

2.4 The Analytics Procedure

By now, it is fair to comprehend that the basis of any organization is formed by data analytics and all big data applications deal with a similar procedure of data acquisition, cleaning, processing, exploration, in-depth analysis and finally communicating the results for smart decision-making. In this section, we attempt to give the reader a general overview of the stages of a data processing procedure which then forms the foundation for implementation of various ML algorithms.

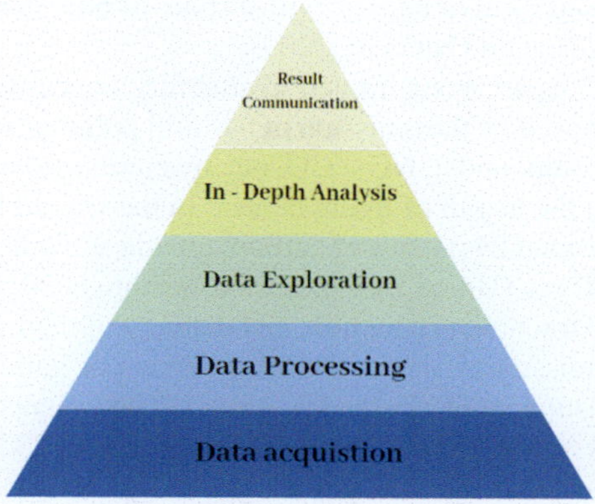

Fig. 4 The pyramid represents steps involved in a data analytics procedure in an orderly manner

The data acquisition process involves the collection of raw data from all possible data sources and acquires the data in usable formats such as.csv,.json,.xml, etc. The stage of data processing usually needs the analyst to examine and understand each column. Data processing involves two stages where the first stage also known as data pre-processing aims to prepare data for analysing by defining a common standard (Amaro and Pina 2017) or by using the concept of 'virtual objects' (Linder et al. 2017) to solve compatibility issues. Further, in the second and final stage of processing, it is important to filter out/replace errors and missing values to refine the database. The data exploration stage incorporates splitting, grouping, and plotting the data in different forms of visualization to identify patterns, trends, and correlations. At this stage, the data is ready for an in-depth analysis using predictive models based on big data techniques such as ML/Deep learning for several purposes such as clustering, pattern recognition, regression, classification, feature selection, or time series forecasting, etc. Finally, from visual observation of graphs, charts, and figures it is also obvious how communication might be seen as the most important step in the analytics procedure to gain crucial insights regarding the inputs, outputs, profits, efficiency, and productivity. Unless and until the analysed data is not communicated properly to the concerned professionals or the common man in several cases, data seems to have no associated value. Figure 4 gives a visual overview of the steps performed in the data analytics procedure.

2.5 *E—Mobility*

Currently, the energy sector in transportation is majorly oil driven. Oil being a fossil fuel is a non-renewable source of energy hence limited in nature. In India, diesel and motor gasoline represented 90% of the final energy consumed in the transport sector, while jet kerosene represented 8% and electricity 2% in 2004. Diesel is the most used form of energy, with a share of 66%, and motor gasoline representing 24% (De la Rue du Can et al. 2009).

Apart from limited availability concerns, a major concern is the air pollution caused by the emission of harmful pollutants such as GHGs, VOCs (Volatile Organic Compounds), NOx, CO, etc. by combustion of oil in vehicles, which is also harmful to the health of living beings. In the United States, cars and trucks collectively emit around 24 pounds of carbon dioxide and other global warming gases for every gallon of gas (Union of Concerned Scientists 2014). About five pounds comes from the extraction, production, and delivery of the fuel, while the great bulk of heat-trapping emissions—more than 19 pounds per gallon—comes right out of a car's tailpipe (Union of Concerned Scientists 2014). Hence, the transportation advancement sector is now driven by the need for cleaner, RES such as electricity. Subsequently, transportation electrification is generally viewed as an appealing solution for reducing the reliance on oil for transportation and environmental impact of road transportation (Fig. 5).

The first step towards transportation electrification was the invention of electric vehicles (EVs). The number of PHEVs sold as on December 2020 in major global markets were about 13 million with China having the maximum sales every coming year, followed by Europe, USA, and other major countries (Wikipedia Contributors 2019a). India experienced sales of about 1.56 lakh units which is comparatively lower than sales in other leading countries but still an increase of about 20% from last year's sales (BloombergQuint, n.d.).

The numbers are huge, however not even close to taking over the ICE (Internal Combustion Engine) vehicle sales in the world. It is estimated that the world has surpassed the 1.5 billion mark in the number of vehicles on road (Chesterton 2018), which is quite a lot more than the total number of electric vehicles in the world.

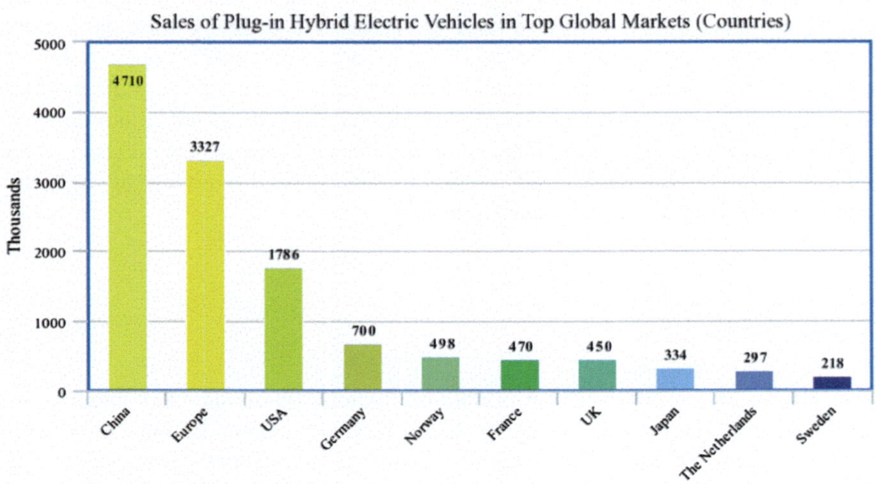

Fig. 5 Number of PHEVs sold till the year 2020 in major countries of the world

Thus, suggesting to us—'*the increasing but not yet enough*' EV sales in the world in order to achieve environmental sustainability. This largely directs us to the EV ecosystem which consists of several factors that determine the extent of adoption of EVs by users. Building a profitable EV ecosystem in a city implies building a network of service providers and businesses (Guardknox, n.d.) in the EV domain. An EV ecosystem is classified into four major service provider and business owner components based on their area of application (Fig. 6). Further, each component can be classified based on attributes of concern. The EV ecosystem is still on growth in the world and several current technologies and practices need to be on the radar of the government so as to popularize EVs among the users and make them think twice before they consider purchasing an ICE vehicle.

When an EV is manufactured, the technical component of the EV ecosystem plays an important role in determining the efficiency and benefits gained through an EV as a mode of transport as compared to oil-based modes of transport. Currently, lithium-ion batteries are the most popular type of EV batteries due to their higher specific energy (López-Ibarra et al. 2020) relative to the other battery types. However, it is found that batteries have much shorter lifetimes than electronic power systems with about 5–8 years of warranty (López-Ibarra et al. 2020) and (EDF Energy 2015). Shorter lifetimes imply investments in new batteries as much as after every 8–10 years of use. New battery investment costs along with the original total costs of ownership of the EVs (where batteries account up to around a quarter of the total EV cost López-Ibarra et al. 2020) are serious concerns for EV users worldwide. EV batteries also play a major role behind 'range anxieties' and 'charging time anxieties' of users. Such user anxieties can be solved by improving several battery features such as capacity, State of Charge (SOC), the integrated battery management system, and lifetime. However, manufacturers need to keep in mind another set of battery features while production. The batteries shall maintain an affordable cost range, should not be bulky or heavy as compared to the EV and finally have lesser harmful environmental impacts while disposal. The technical component also includes manufacturing of batteries with safer operations, in order to avoid risks.

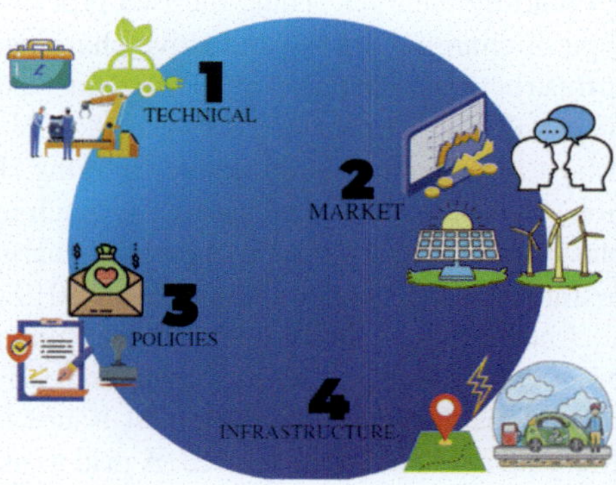

Fig. 6 Four major components of an EV ecosystem

Once, the EV is manufactured, it is time for its widespread acceptance by users through market trends and policies (Vikaspedia.in 2020). Market values along with policies play an economical role in popularizing EVs. The EV market is expected to assist EV buyers by answering queries related to consumer perceptions and emphasizing on long-term benefits through testing and certification, servicing and affordable overall costs. The Market is not just limited to factors directly related to an EV, but also include factors like quality of electricity in the region and market for storage of electricity. Several policies by the government regarding taxes, subsidies on sources of electricity, or electricity prices on a whole are expected to help lift the economic anxieties of citizens into purchasing EVs.

Finally, the last component of an EV ecosystem playing a crucial role in widespread EV acceptance is the EV charging infrastructure (Vikaspedia.in 2020). The charging infrastructure deals with appropriate locations of charging stations, allotment of charging points to charging stations and an adequate number of charging points. EV charging infrastructures are much more sensitive to the placement of chargers and charging stations as compared to fuel refilling stations for ICE vehicles due to one major issue—higher charging times. On average, even the fastest chargers require about at least 30 min to charge an EV and the charging times vary with the type of EV. Subsequently, higher charging times also require proper placement of charging stations for an EV user to be able to utilize his time while the EV is charging. Further, to assist issues related to high charging times, charging station service providers need to successfully implement the new-age technologies such as battery swapping and battery replacement. However, such practices require the batteries and EVs to undergo another set of standardization procedures.

Electrical Vehicle Supply Equipment (EVSE) commonly known as an EV charger, is not just a charger. An EVSE control aims to solve compatibility issues between EVs and different types of chargers. EVSE basically enables any EV user to charge from any outlet. It is equivalent to a settlement between the EV and the charger to keep the EV safe while charging using two-way communication between charger and EV (Green Car Reports, n.d.). At the same time, EV charging can be classified by the 'types of charging' based on the nature of contact/communication between EV and the charger. Thus, the two types of charging infrastructures based on the type of charging are—Conductive charging infrastructures and Inductive Charging infrastructures (Khalid et al. 2019). Since conductive charging involves direct metal to metal physical contact between the EV and the charger, it is more efficient & robust as compared to inductive charging (based on the principle of mutual induction). Hence, conductive charging is much more widely accepted and can be broadly classified based on the type of current, charging times, and voltage and current levels in Table 2 derived from (Khalid et al. 2019; Sharma et al. 2020).

Another major limitation in the charging infrastructure is the quality of electricity provided by the grid to the vehicles keep fluctuating as discussed earlier. These limitations thus tend to spark trust issues in the user about the time taken for EV charging and the cost generated. A grid management system might help in reviewing over-power flows in the grid so as to even out the fluctuations and avoid surges.

Table 2 Types of conductive EV chargers with specifications

Charger name	Type of current	Time taken to charge	Voltage and current ratings
Level—1 Charger	AC Slow	8–16 h for full charge	120 V, 15–20 Amp
Level—2 Charger	AC Slow	4–8 h for full charge	240–400 V, 40–80 Amp
Level—3 Charger	DC Fast with CHAdeMo technology	10–15 min for 80% charge, remaining 20% need constant voltage which consumes time	208–600 V, 200 Amp

Likewise, EV management can be given a more intelligent approach by several practices such as carpooling, ridesharing, self-driven cars, etc. A few studies insist on the large role of the influence of other consumer's perception in local EV adoption, i.e. 'word of mouth' effects (Cai et al. 2013) which simply encourages a non EV user to think about personal EV adoption. EV adoption in different regions or countries is also dependent on the average annual income of citizens of that country. For instance, most Americans and Chinese citizens can afford EVs and hence EV culture is soon to be predominant in those countries. However, EVs are still a foreign concept in developing countries like India. Having said that, income alone cannot be held responsible for EV adoption among citizens. For instance, the average annual income of citizens of the UK is among the highest in the world, but EV adoption is not exactly as predominant in the UK as in China. Hence, a customer service driven approach also plays an important role in transforming the EV market. It is basic to comprehend that customers resist change whenever introduced with a new technology or innovation. Thus, it is important that transportation electrification programmes offer genuine and tangible advantages to the EV drivers and afterwards plainly convey those advantages. Utilities can furnish both private and business clients with projects and administrations that energize and reward the adoption of EVs in their administration regions by leveraging best activities of the past and keeping in mind the client's point of view all through the advancement procedure (www.tdworld.com, n.d.).

3 Literature Review

It is crucial to study literature samples regarding recent advancements utilizing big data in several components of a smart city to not only analyse the extent up to which the big data is useful to us but to also raise relevant questions for future research directions.

3.1 Smart Cities

We begin our study by a general understanding of what urban areas under the umbrella idea of smart cities are expected to offer to its citizens.

Urban areas have held on to the Information and Communication Technologies (ICTs) as an advancement technique, hence bringing about the concept of a sustainable smart city. Through specific components of a smart city (Fig. 7), a big-data-driven smart city is expected to not only bring us closer to achieving the goals of sustainability and easing tasks for residents of the city, but also aims to address concerns related to socio-economic sustainability such as disaster detection and management, event detection, resources management, health monitoring, safety and security, profit maximization, etc.

A smart city can take several different definitions depending on the area of focus as classified in Fig. 7. For instance, the term smart city has been differently categorized in several pieces of literature such as 'intelligent cities' (Kominos 2002), 'wired cities' (Dutton et al. 1987), 'sentient cities' (Shepard 2011), or 'digital cities' (Gerhard Goos et al. 2000). (Kitchin 2013) studied the relationships of these terms with urbanization and concluded that these terms are utilized with a specific goal in mind to conceptualize the connection among ICT and contemporary urbanism, yet they mainly focus on the impacts of ICT on urban structure, procedures, and methods of living. (Gerhard Goos et al. 2000) further extensively classifies smart cities into two distinct yet related understandings with regards to what makes a city 'smart'. From one perspective, the idea of a smart city presented by (Khan et al. 2017) is—A city that monitors and integrates conditions of all of its critical infrastructures, including roads, bridges, tunnels, rails, subways, airports, seaports, communications, water,

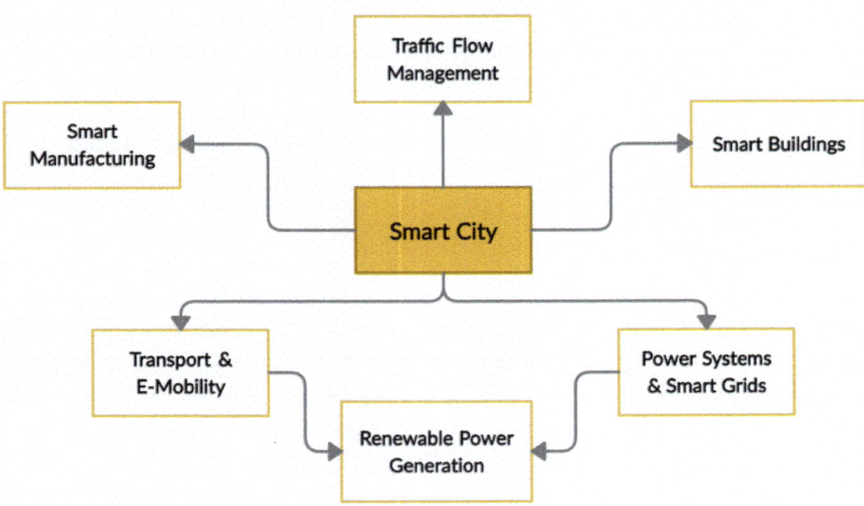

Fig. 7 Classification diagram of a smart city representing components of a smart city

power, even major buildings, can better optimize its resources, plan its preventive maintenance activities, and monitor security aspects while maximizing services to its citizens through data collected through smartphones, computers, GPS (Global Positioning System), sensors, cameras, and even people. But from the other perspective (Khan et al. 2017) has well defined the components of a smart city, such as mobility, governance, environment, and people as well as its applications and services such as healthcare, transportation, smart education, and energy systems.

The concept of IoT introduces the integration, monitoring and data collection processes (through sensors, actuators, smartphones, etc.) of several related objects working together for a specific task. While the IoT environment is vital for the development of a smart city, it is also responsible for the generation of huge amounts of real-time data in raw form which may be structured or unstructured, i.e. big data. (Rathore et al. 2018) in their study aim to collect real-time data concerning several domains of a smart city such as smart home, smart parking, smart vehicular traffic, weather and water system, surveillance and safety, and environment to harness and link the big data sets. Analysing and Visualization of these data sets is further expected to influence the decision-making processes in urban planning and hence in the development of a smart digital city.

Energy Management plays a very important role in the context of smart cities to substantially optimize the processes related to energy—generation, transmission, distribution, and utilization. It is of crucial importance to manage activities related to energy production and consumption not only to efficiently generate energy but also to leverage the generated energy. In the coming sections, we will emphasize on each of these aspects of energy management with the introduction of data sources and utilization of data-driven technologies.

3.2 *Power Systems and Smart Grids*

In this section, we will be emphasizing on 'Power Systems' that deal with the generation, distribution, and transmission of electricity through several different components. However, current power systems face a few distinct challenges associated with economic viewpoints (efficiency, productivity, affordability, system stability, and reliability) as well as in the successful implementation of big data analytics procedures (such as effective data acquisition and storage, data curation methodologies, how to exactly use the stored data to extract business value and how to reduce privacy issues). Eventually, once these challenges are dealt with big data analytics can again be utilized for the improvement in several operational sectors of power systems. Real-time optimization of power system network, precise forecast of load demand, consumption patterns analysis making way for new services, and pricing strategies are some of the areas that need assistance and big data seems to make its way (Zhou et al. 2016). Thus, the smart grid was conceptualized considering it to be a promising solution both from the consumer point of view and the industry point of view (Amaro and Pina 2017).

The introduction of smart grids employed advanced digital information and communication technologies which enabled the usage of information technologies like data analytics, forecasting, classification and clustering algorithms (Zhou et al. 2016) on the large amounts of smart grid big data. Apart from benefiting the industry owners, a smart grid also helps in rapid detection & restoration of faults and reducing the cost for consumers (Moradi et al. 2019). Smart grids are thus considered a promising solution both from the consumer point of view and the industry point of view.

A few intelligent devices currently being used in smart grids for efficient data collection and processing throughout the whole process of power flow include

- Phasor Measurement Units (PMUs) which have now replaced SCADA Systems due to higher sampling rates of PMUs (Junaidi and Shaaban 2018).
- BAS (Building Automation System)
- Sensors and Thermostats.

Similarly, major sources of data described in (Amaro and Pina 2017) are

- AMI data (smart metres)
- Distribution automation data (grid equipment)
- Third-party data (off-grid data sets)
- Asset management data (firmware for all smart devices and associated operating systems).

  Other sources also consist of
- GIS big data and weather big data (Voivontas et al. 1998; Jakubiec and Reinhart 2012).
- IED (Intelligent Electronic Devices): It integrates at least one microprocessor that can receive or send data/control from or to an outer source (Junaidi and Shaaban 2018).
- Electricity market data (Liang et al. 2018).

Further, the energy management system for smart grids is found to be more complex as compared to power systems. Ancillotti et al. (2013) in their study, described a smart energy management system as the new technologies and models that are fundamental for the well-off deployment of a smart grid and can be classified into three components

- Real-time wide-area situational awareness (WASA: a network of PMUs) of grid status through advanced metering and monitoring systems.
- Consumer's participation through home EMSs, demand response (DR) algorithms, and vehicle-to-grid (V2G) technology.
- Supervisory control through computer-based systems.

3.3 Smart Buildings

Out of the many energy-consuming aspects of smart cities, smart buildings turn out to be one of the biggest ones. Not only energy consumption rates but emission rates of GHG gases are also considerably high from smart buildings. Statistical studies have been performed and observed regarding energy consumed in the building sector. The building sector is one of the largest energy consumers, accounting for approximately 20–40% of the global energy usage (Chou et al. 2016; Linder et al. 2017) while generating 30% of all CO_2 emissions (Ancillotti et al. 2013); energy consumption by this sector is continuously increasing because of urbanization, rapid economic growth, rising income, and growing population. EIA's *International Energy Outlook 2017* (IEO 2017) projects that by 2040, the fastest growth in buildings energy consumption will occur in India as compared to all other countries. An average increase of about 2.7% per year delivered energy consumption for residential and commercial buildings in India is expected from 2015 to 2040 which is more than twice the global average increase. Figure 8 represents the country-wise annual percentage growth in the building energy sector (www.eia.gov, n.d.) (Fig. 9).

Major studies till date have concentrated on setting up a framework for anticipating the pattern of future power utilization through stochastic strategies and regression analysis to encourage effective usage of electricity by end-users (Lee et al. 2017). In this section, we will review some of the frameworks based on big data-driven technologies that have been described in the existing literature utilizing the context of an intelligent building. An intelligent system in the context of an intelligent building is composed of three levels: The infrastructure level of the input data, system infrastructure level and the level of services (Daissaoui et al. 2020). These three levels are going to form the basis of the majority of the big data models discussed.

A Smart Decision Support System (SDSS) proposed by (Chou et al. 2016) integrates smart grid big data analytics and cloud computing for building energy efficiency. Based on layered architecture the SDSS contains a data access layer followed

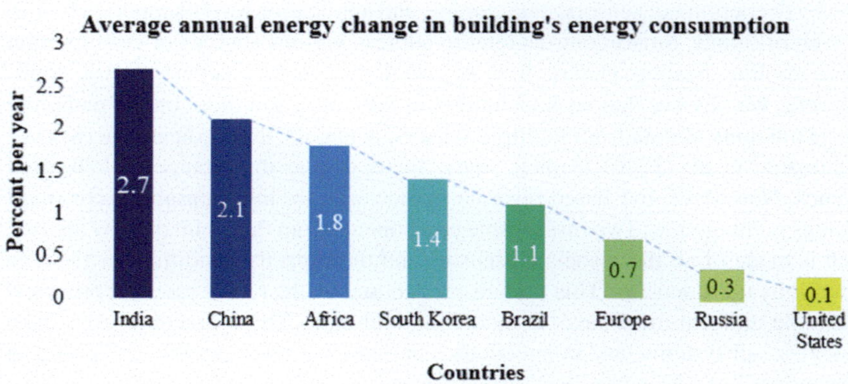

Fig. 8 Annual increase in building energy consumption for a few major countries/continents

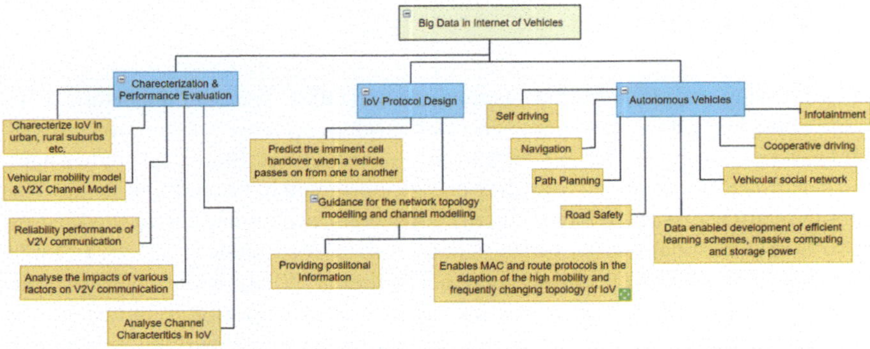

Fig. 9 Several domains of Big data analytics aided IoV

by an integrated analytics bench and a web-based portal. Advanced AI techniques and dynamic optimization algorithm are possibly coordinated behind the analytics bench to empower precise prediction and optimization of energy consumption. End users can minimize their costs by the best-case automatic operation of appliances by SDSS. The general flow of this framework includes four major steps

i. Collection of real-time data from AMI (smart metres).
ii. Wi-Fi and Bluetooth act as data communication platforms.
iii. AI & multi-objective algorithm for data analysing and
iv. Web-based tech visualizes the optimal operating schedule.

Additionally, (Linder et al. 2017) introduces Big Building Data (BBD) as a platform important for the paradigm shift from BMS to the web of buildings. BBD is ingestion, handling, and sharing framework ready to scale up to the Big Data desires for smart building situations. It provides data monitoring, anomaly detection and prediction, and renovation tools for users (building user, building owner, architects and Engineering consultancies). Another opportunity of this approach is BMS revolution by decoupling of sensors, data storage, and application levels supported by IoT and Web of things. Smart controls tend to address various sources of consumption such as cooling, heating, electric lighting, ventilation, solar shading, electric appliances, etc. We observe that such technologies rely more and more on the gathering of large amounts of data from multiple sensors, actuators, and dedicated networks.

Daissaoui et al. (2020) in their study, focus on the importance of IoT-based analytics. Nowadays, IoT-based big data frameworks are indispensable from smart buildings point of view. Two inseparable components from this point of view are IoT, which is made of all the associated sensors and the capacity condition for the data produced by these sensors. This framework is based on the recent research proposed to describe uniform meta-data for modelling buildings. These practices use sensor ontologies, subsystems, and connections, guaranteeing interoperable and compact applications.

3.4 Transport Systems

As of 2017, 1.43 million vehicles are registered in the world (Wikipedia Contributors 2019b) consuming a huge amount of crude oil derivatives (fossil fuels) and producing a considerable amount of air pollution which is also risky to human and animal health. While the solution relies on switching to alternative fuels such as biofuels, electrolytic hydrogen, and non-fossil fuel-based electricity (Wu and Chen 2017), there still exists a need to critically identify the current situation, evaluate the problems and to design appropriate strategies to achieve sustainable development goals. This is where big data analytics steps in.

The job is not just limited to reduction of emission but also to limit the usage of natural resources and to make the most out of the utilized resource. Wang et al. (2018) in his study elucidates into why and how big data and emerging technologies can help minimize both the harmful carbon emissions and the assets utilized in the transport sector. The report gives a general idea that the current gains in energy efficiency are much lesser than the efficiency gains observed in the Information Technology (IT) sector (as expressed in 'Moore's Law'). Thus, proposing that more utilization of IT in the transport sector will deliver enormous profits in sustainability.

Transport big data—One way to look at smart urban transportation management is to consider it as a multifaceted big data challenge which may incorporate tracking and management of every aspect of the vehicle body and engine/battery maintenance, warehousing, routing and mapping, communications, etc. by big data collection from all relevant sources such as road traffic index, metro operation data, road accident data, passenger card data, elevated off-ramp data, bus real-time data, microblogging traffic data, air quality status data, weather data, GPS data, FCD (Floating Car Data: to detect traffic flow speed) and LBS (location-based service) data (Miller and Harvey 2011; Wang et al. 2018).

Personal Travel Assistants (PTAs)—Currently, the most common platforms utilizing transport big data are PTAs. PTAs are expected to make use of transport big data to assist the user for the most optimal modes of transport and fuel-efficient routes based on basic user inputs. The PTA architecture generally consists of data ingestion, storage, analytics, and optimization modules to put forward a holistic as well as a sustainable approach for each trip (Miller and Harvey 2011; Wang et al. 2018).

Automated Vehicles (AVs)—The potential use of big data in AVs is particularly in object detection and deep learning approaches to make wiser and more accurate decisions.

Transport Technology and Mobility Assessment (TEMA)—A data processing platform expected to govern several road transport policies including the deployment of EVs, shift from oil to electrical energy, evaluating emissions, outflows, and so forth. To leverage the big data generated, a sample of vehicles was taken for analysis and it is found that half of the vehicles are parked for more than 90% of the time and 78% travel up to 50 km/day, 9% exceed 100 km/day and 3% exceed 150 km/day.

The figures also suggested a large potential for deploying battery electric vehicles (Gennaro et al. 2018).

Internet of Vehicles (IoV)—It is highly expected that very soon modern vehicles will have the ability to connect using radio technologies and exchange data with the surrounding environment under the dome of Internet of Vehicles (IoV) (Xu et al. 2018). Thus, this huge network of connected vehicles is not only consuming big data but also generating big data, hence explaining the reciprocal relationship between IoV and big data and how both assist each other in their respective domains. While IoV is solely assumed to be assisting in the generation of big data, it also supports in transmitting, storing, and computing big data. Similarly, (Xu et al. 2018) explored the role of big data in assisting IoV in several domains such as performance evaluation, characterization and communication protocol design utilizing measurement data, trace data, trajectory data, field data, vehicle movement data, senor data, traffic data, map data, etc.

A few models have also been developed in the recent past based on big data technologies to achieve sustainable environment goals in transportation. (Wu and Chen 2017) presents a model on hybrid data analytics to tackle critical issues of green transport systems as well as to generate valuable operational strategies for transport firms to achieve environmental sustainability. The model thus presented highlights three main stages

- Business analytics: Data collection and exploration, Data selection, Data pre-processing, Data mining, and Data transformation
- Hybrid data mining: Topic mining and association rules.
- Applies text mining of big data analytics.

3.5 *Electric Vehicles*

Electric Vehicles come with several significant benefits majorly—zero emissions and the capability to depend fully on RE sources. Despite such favourable outcomes of an EV, EVs are a rare sight in most countries of the world, especially India due to three major limitations of EVs. EVs are known to be much more expensive with shorter driving ranges and much greater charging times as compared to fuel vehicles. Thus, from an EV deployment point of view, it is necessary to review on the existing articles related to both—the technical aspect as well as the infrastructural aspect to tackle challenges associated with the mass adoption of EVs among users.

Technical Aspects—An EV battery plays the most important role in the technical aspect of an EV ecosystem. A Battery Management System (BMS) is an electronic system that controls a rechargeable EV battery (Yevgen Barsukov and Jinrong Qian 2013) by providing safety assurance to the battery, calculating and reporting data, monitoring the battery's state and balancing the environment (through several parameters such as State of Charge (SOC), State of Health (SOH), State of Power, State of Safety, temperature, voltage, current and coolant flow). The BMS not only manages the functions of a battery actively but also estimates the status of the battery and predicts several crucial variables such as SOH, Remaining useful time (RUL), SOC, etc.

First-hand, we will be reviewing and discussing battery ageing mechanisms and data-driven technologies proposed behind the health estimation and lifetime predictions for Li-ion batteries (Li et al. 2019). Li describes the health management unit of a battery consisting of two major steps—the SOH estimation (estimating the extent of failure) which is then followed by the health prediction (predicting the battery performance variation and identifying the time of failure). The main degradation of a Li-ion can be categorized into 3 main modes—the loss of lithium inventory (LLI), the loss of active material (LAM) in the electrodes, and the increase of cell internal resistance. These 3 modes of battery degradation are based on 9 ageing mechanisms which are then influenced by 7 factors based on temperature variations, pressure variations, SOC/Voltage, current rates and time, etc.

SOH estimation: Some of the most faithful data-driven SOH estimation approaches are based on Differential analysis (DA) and Machine learning (ML). While DA involves feature identification from differentiated curves of mechanical, thermal, or electrical parameters during battery cycling, ML methods train a model in accordance with the extracted input features to describe the cell ageing behaviour to estimate the SOH.

Battery health prognostics: Health prediction includes Remaining Useful time (RUL) prediction and Capacity forecasting. RUL prediction is done by pre-defining a failure threshold signal and then the modelled degradation signal is sent to compare it with the threshold value. RUL is obtained by

$$\textbf{RUL} = \textbf{estimated life of the training units} - \textbf{current life position of the test units}$$

Similarly, capacity forecasting is done by predicting future changes in SOH as a function of battery usage history. Both the above-mentioned frameworks for battery health prognostics are based on two methods namely—Analytical models (requires the development of an ageing model) and ML methods (model-free and learn from the ageing data to predict battery health).

The accuracy of RUL prediction can further be improved by utilizing a big data analytics framework that combines any one of the EV battery modelling methods as discussed previously with the driving pattern analysis (Karmavijaya et al. 2019). After collecting internal resistance and voltage data, the driving data of each trip is extracted from EV cloud platform and transformed into a consistent vector for pattern recognition. For analysis, an unsupervised clustering approach Growing Hierarchical Self-Organizing Maps (GHSOM) is utilized to cluster driving patterns and thus the analysis of energy consumption of each driving pattern is done for automakers to regulate the power consumption estimations according to different driving patterns. Finally, on the basis of route plans uploaded by EV users and SOH estimates, the system calculates RUL and advises the users through the user interface.

Table 3 Data collection nodes in the proposed methodology (Rahimi-Eichi and Chow 2014)

Data nodes		
Weather	Temperature—to adjust battery parameters	Wind speed and direction—to estimate aerodynamic power consumption due to opposite direction flowing winds
Route and Terrain	Points of decision (turns) for the driver	Latitude and longitude for each point available in polyline data format
Battery Manufacturer	Resistances, Capacitances, currents, SOCs and several more parameters for battery modelling	
EV manufacturer	Velocity of EV, mass, slope, acceleration, friction, and several other parameters to calculate EV power consumption—for EV modelling	
Driver History	Speed and location of EV from the GPS history of driver	

To solve the driving range anxieties in users, (Rahimi-Eichi and Chow 2014) proposes a range estimation framework to pre-inform users the maximum possible driving range, collecting data from five collecting nodes described in Table 3.

Cell to cell variations are known to accelerate the decay rate of battery life and hence reducing the performance and safety of an EV battery. Based on previous research, (Lu et al. 2020) found the five indicators determining cell to cell variations—time variation, voltage variation, current variation, resistance variation, and SOC variation. The first step involves collecting voltage, time, current, and temperature from charging cloud data. The five indicators are then estimated using obtained original signals. This paper suggests a weighted scoring mechanism to score the cell-to-cell variations of the battery pack. While the scoring mechanism for all five indicators are established based on thresholds and weight factors, the thresholds are determined on real applications and the weights are determined on an analytics hierarchy process. Thresholds contribute to individual unweighted scores whereas weight factors contribute to weighted scores, and together they both contribute to the total final score. Finally, 3 battery packs are evaluated using the proposed method, and it is observed that the pack with the highest score has less ageing whereas the pack with the least score has deep ageing.

Infrastructural Aspect—Using trajectory data analysis, (Li et al. 2015) proposes an Optimal Charging Station Deployment (OCSD) framework which takes a historical EV taxi trajectory data, road map data, and existing charging station information as input, and performs Optimal Charging Station Placement (OCSP) and Optimal Charging Point Assignment (OCPA). Both the mentioned frameworks are designed to optimize the Charging Pile Network (CPN) of the city by setting up 'K' new charging stations in the city, with 'M' new charging points distributed over the set of K new charging stations. In simple terms, these two optimization components are designed to limit the average time taken by an e-taxi to travel to charging station, and the average waiting time for an available charging point in the city.

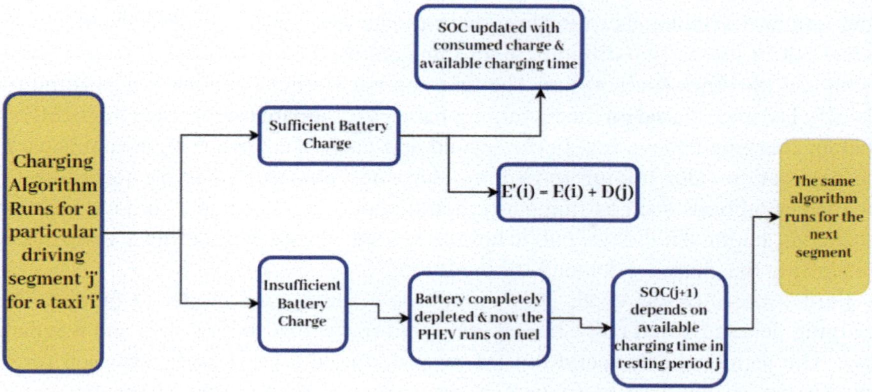

Fig. 10 Segment-wise charging algorithm proposed. Where, E(k) = electrified mileage of taxi 'k'. D(r) = total distance driven in segment 'r'. E'(k) = updated electrified mileage of taxi 'k'. SOC(r) = SOC at the beginning of segment 'r'

Trajectory data analysis is also utilized by (Cai et al. 2013) to characterize individual travel patterns by examining trajectory data of 10,375 taxis of Beijing by big data mining and further evaluates the impact of adopting plug-in hybrid EVs in the taxi fleet on GHG emissions, based on characterization of individual travel patterns. A Plug-in Hybrid EV charging algorithm for each segment is proposed in this study to analyse whether the total charge present in the battery is enough for the whole driving segment based on several inputs such as driving segments, resting periods, SOCs, and battery size. The charging algorithm is described in Fig. 10. Once all segments of a trip are analysed, the portion of the trip that can be electrified using a PHEV (with a given battery and charging conditions) is calculated. Next, several other factors are considered for simulations and modelling such as fuel cost saving and electrification rate, adoption, government subsidy, and GHG emissions to evaluate the extent to which each factor affects the PHEV adoption and lifecycle GHG emissions since lifecycle GHG emissions are dependent on the power grid, vehicle cycle, and the fuel cycle.

So far, it feels safe to assume that government subsidies could decrease lifecycle GHG emissions, however, that is not the case. With increased government subsidies, citizens are encouraged to adopt PHEVs and hence the electrification rate also increases. The emissions reduced by PHEVs are not enough to compensate or reduce the lifecycle GHG emissions. Hence, it can be concluded that lifecycle GHG emissions can only be reduced by switching to less carbon-intensive power grids (Cai et al. 2013).

An important step in aiming to popularize EVs is to analyse the current trends related to popularity. The most preferred charging times, the most preferred days, the most preferred locations, and how the charging behaviour is dependent on several other factors such as weather, holidays, etc. Wang et al. (2019) attempts to analyse the charging behaviour of users of a city utilizing the big data by making use of data mining technologies to explore the big data and further analyse the charging behaviour of users. Two different algorithms are used namely—the K-means algorithm and the Apriori algorithm. The first approach utilized K-means algorithm to classify EV users based on their charging habits to compare users with a huge difference in charging behaviours. In the second approach, the Apriori algorithm is used for association rules mining to explore charging behaviour of users according to weather conditions such as rain, snow, temperature, etc., and special days such as weekends, national holidays, public holidays. Analysis of EV charging also involves forecasting the charging demand for the future.

Furthermore, (Arias et al. 2016) proposes a forecasting model to predict EV charging demand based on historical real-word traffic distribution data and weather data. The technical architecture of the proposed model has four layers each for a specific target. The first two layers of data sources and storage extract data from local disks (data sources) on the computer and store them using the MATLAB store function. The stored data is provided in chunks to the third layer of data management which uses the MapReduce function of MATLAB for data handling. The fourth layer of data processing includes a cluster analysis as well as relational analysis for the identification of factors influencing traffic flow. Finally, a decision tree that classifies cars and buses was established using MATLAB. For charging demand forecasting, the user is expected to provide number of EVs, month and day as inputs to the model. The second step is to determine the weather and day type of the forecasting day from historical data sets. With these two forms of inputs, clusters and charging times (based on corresponding clusters) can be easily determined. A random sampling of data helps in determining the initial SOC. Finally, the EV charging demand can be calculated for the residential and commercial sectors.

Charging demand and Occupancy rates also play an important role in influencing the decision-making process of firms and agencies while determining potential locations of setting up new EV charging stations. Further examination of charging demand and occupancy rates give insights on electricity demand in particular regions. Lee et al. (2017) examines the charging demand in Jeju city utilizing open data sets of chargers across the city and open software components by performing a stream analysis to examine the number of occupied chargers over the city for as much as 500-h duration and increasing. Lee et al. (2017) also plots charging behaviour of the most frequent chargers and occupancy rates in different regions across the city.

3.6 Traffic Flow Predictions

Road transportation is the backbone of smart cities; it costs 1.35 million deaths (World Health Organisation 2020) and trillions of dollars to the global economy annually, damaging public health and the environment. The objective of traffic flow prediction is to provide traffic flow information that improves traffic efficiency, helps users to make better travel decisions, alleviate traffic congestion, reduce carbon emissions, etc. (Lv et al. 2014). Although traffic constitutes to be an integral part of a city, not much literature based on applications of big data-driven technologies in traffic exists. We begin our study by reviewing the sources of traffic data and then discussing the existing traffic flow prediction models.

Lv [31] lists the data sources related to traffic flow prediction as various sensor sources that include inductive loops, radars, cameras, mobile GPS (Geographical Positioning System), crowdsourcing, etc. Inductive loops are electromagnetic vehicle detection systems (Wikipedia 2021a), highly helpful in estimating the vehicle density on roads and the number of vehicles passing through a point location. The authors have thus proposed a deep-learning-based traffic flow prediction method, where a Stacked Auto-Encoder (SAE) model is used to learn generic traffic flow features, and it is trained in a layer-wise greedy fashion. Unlike the previously proposed ARIMA models and its variants that consider only the shallow structure of traffic data, the proposed model by (Lv et al. 2014) can successfully discover the latent traffic flow feature representation such as the non-linear spatial and temporal correlations.

Similar to traffic flow predictions, traffic congestion predictions also play an important role in providing better alternative routes to the drivers, thus reducing the time per trip and emissions as well as preventing future congestions. Two types of Long Short-Term Memory models (LSTMs) for congestion prediction based on data collected in a 5-min window are developed by (Majumdar et al. 2020). Model 1 is a univariate model based on the observed speed of the previous 5 min while Model 2 is a multivariate model based not only on previously observed speeds but also on traffic flow rates and vehicle headways. Since LSTM models are a subclass of Recurrent Neural Networks (RNNs) their training performances decrease with increasing data size and hence this study is primarily focussed on data analytics. In future, attempts can be made to utilize the big data generated using similar Machine Learning models for better congestion forecasting.

3.7 Smart Manufacturing

The manufacturing and production industry play a huge role in a company's economy, energy usage, and waste production. Almost all goods consume energy even before their work-life starts, as every manufacturing/production process needs energy. Similarly, the end of a product's economic life doesn't mean the end of its environmental impact. Hence, manufacturing methods need to be re-designed to use minimal energy and less environmental impacts. As laid down by (O'Donovan et al. 2015) smart manufacturing intends to convert data obtained over the product lifecycle into manufacturing intelligence so as to achieve positive impacts in all phases of manufacturing. Manufacturing frameworks are encountering an unstable growth of more than 1000 EB annually (Yin and Kaynak 2015) Hence, big data can prove to be an upgrade in the adequacy of smart manufacturing and there are several ways to go about it. Although not much literature exists to quantify the effects of utilizing manufacturing big data, we will be reviewing two pieces of literature that aim to utilize the manufacturing big data (1) to develop intelligent frameworks and streamline the manufacturing process (2) explore the relationship between big data and sustainable manufacturing.

Tao et al. (2018) proposes a very basic data-driven smart manufacturing framework where data from the manufacturing process is combined with data from orders and production plans. Intelligent algorithms and predictive models examine this data with the assistance of big data analytics so as to streamline the manufacturing process. The framework consists of four modules each for four such functions—input manufacturing data, data-driven analysis for information processing and generating recommendations, real-time monitoring to help encourage the efficient running of various types of equipments and problem processing by human operators or AI applications for a holistic approach.

Dubey et al. (2015) in their study aim to explore the relationship between big data and world-class sustainable manufacturing (WCSM). The methodology adopted consisted of identifying the constructs which impact upon sustainable manufacturing followed by principal component analysis (PCA) on the data collected. Constructs are formative and reflective and are studied by big data analytics perspective (measures, sampling design, data collection). Hence, the theoretical framework was generated, and it was tested using the gathered data.

3.8 Renewable Power Generation and Forecasting

An electric power generation system requires the adjustment of the generating and the loading components to be genuine and dependable and to operate safely at quality standards. Unfortunately, during the generation of power through RES, many issues are faced due to the fluctuations. Generation from RES results in uncertainty and variability to a high extent. The reason being changes in weather conditions due to which the intensity of ocean waves, winds, and sun rays might vary with time. Hence, it is important to develop frameworks supporting renewable power generation through a holistic approach. In this capacity, big data aids renewable energy power generation in mainly three capacities

- *Leveraging Renewable Energy (RE) Sources*: Newer power generating plants are now considering offshore renewables as the next potential source of clean energy after solar, and hence (Amaro and Pina 2017) specifically studied wave power as the target source to develop the largest maritime database through a multi-stage platform which is expected to integrate data of different velocity, variety, and volume under a trusted engine and finally produce a big data informative archive of value and veracity back to the participants and local communities.
- *Managing and analysing the performance of RE Systems*: Escobedo et al. (2017) designed a six-layer big data infrastructure for the management of PV Systems. The architecture of this management system is expected to perform general tasks of data ingestion, processing, storage, analysis, and monitoring of information in real time keeping in mind specific requirements of parallel processing, scalable

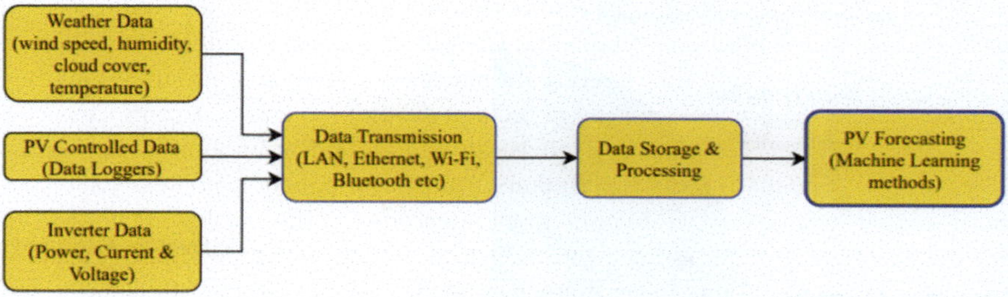

Fig. 11 Flowchart of PV forecasting (Preda et al. 2018)

infrastructure (horizontal, vertical, and fault-tolerant), and distributed storage. Additionally, the authors also perform a case study of power solar generation where a photovoltaic system is connected to a grid to estimate the working performances based on 22 essential factors (directly acquired by the collected data).

- *Forecasting the output renewable power*: Preda et al. (2018) proposed a methodology based on sensor applications where big data is captured from sensors and loggers, broken down and finally PV forecasting is done with the help of Support Vector Machine and Linear Regression algorithms (Fig. 11). By utilizing more parameters in the machine learning process, the root mean square error was observed to improve by considerable amounts.

4 Summary

Recently, a vast majority of the literature has contributed to research addressing challenges of sustainable development through big data-driven technologies and hence, the role of big data in several components of a smart city is pronounced, especially in the EV sector. One end that can be drawn from the examination of such big data aided methodologies is that embracing ICTs dependent on big data analytics and distributed computing storage facilities may support cooperation and correspondence between the various components of smart cities. With the growth and popularity of data-driven technologies, a significant amount of attention also goes to the advancements in the sector of data collection techniques, thus resulting in the growth of data collection devices such as AMIs, GIS, IEDs, sensors, actuators, etc. and finally all these combined result in the increased volume and variety of data. Integrating such huge amounts of heterogeneous data likewise assumes a significant job and a few arrangements have come up for the same, for example, virtual item and metadata. However, in any case, the source of data collected varies in each aspect. Similarly, big data tools (e.g. Hadoop, Cassandra, Hive) and techniques (Machine learning, Deep Learning, etc.) of big data have also evolved with time to address all kinds of data, matching data sets with the most suitable type of analysis models and visualization methods. Processing of data additionally varies in the few models delivered. While

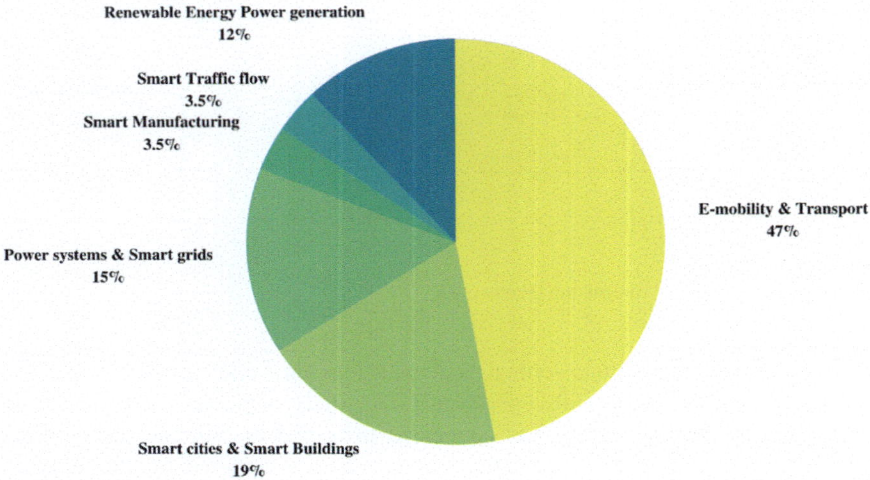

Fig. 12 Approximate quantitative analysis of the existing literature

all the above-mentioned frameworks use different data-driven technologies, they all aim to achieve the same goals in terms of sustainability with similar flow processes.

A common observation from the existing literature is that a significant amount of literature has been published for applications of big data in the transport sector, specifically towards E-mobility. Similarly, a considerable share of literature samples also focuses on the smart building and smart grid components of smart cities. A comparatively smaller amount of literature samples also discusses the forecasting and prediction of RES power generation as well as their monitoring and management systems. Smart manufacturing, traffic management, etc. have very limited research work at present (Fig. 12).

Concluding, Big data analytics sure has come up far in the digitalized era to ease sustainability-related challenges. However, big data techniques and methodologies need to be checked for suitability for the type of data, the application, and the desired knowledge to be obtained for their proper implementation. Some exploration holes despite everything exist in social and economic issues as well as in the proper implementation of technologies and there is an extension for research in those ways.

5 Limitations of Existing Research

The existing research surely does contribute a developing knowledge into existing theory, combines technologies and presents several new ideas in which the world can make use of the huge amounts of data generated to achieve goals of sustainability. However, some potential limitations shall be noted to overcome them in future research.

- The effect estimates in most of the models and frameworks presented above are just prospective observational studies and the results might differ in practical application, majorly due to inconsistency between the actual data and the procured data.
- IoV has a huge potential in creating more sustainable research opportunities due to the huge amounts of data generated. However, currently, very few IoV implementations are researched & several aspects such as safety, navigation, a social network on wheels, customer interests, etc. can be implemented. Safety issues also need to be addressed in network protocols (Xu et al. 2018).
- Few challenges associated with data-driven health estimation and prediction methods are
- ML methods are known to give less in-depth information regarding battery ageing & hence shall be combined with physical methods of battery decay such as DA to help in the identification of most sensitive indicators.
- Li-ion cells are also known to be environment sensitive hence compromising on the accuracy of models.
- Most of the existing research is done on the cellular level and the results might differ at a battery level.
- The sample size of the data collected needs to be increased and we need a more reliable method of analysing unstructured data (if collected by different methods) while enabling WCSM through big data techniques. Additionally, research needs to be done on data reduction techniques which would have helped to identify more enablers of WCSM (Dubey et al. 2015).
- Considering the evolution of sensors and loggers, a large number of PV systems exist without sensors. Hence, we are missing out on monitoring information while forecasting PV output. Even if the PV arrays are provided with data collection units, the accuracy of the collected data isn't guaranteed. For e.g. in Preda et al. (2018) the weather parameters are collected from weather stations for PV forecast which do not usually coincide with the location of PV array. Thus, the parameters are not accurate.
- While solar energy and wind energy are popular renewable sources of energy, ocean waves are still unpopular due to the several technological challenges associated with waves and hence extensive research is required in these fields. Some of the potential drawbacks of wave energy listed by Amaro and Pina (2017) are (1) Wave Energy Converters need to be designed to withstand extreme weather conditions (where the power may exceed over 100 times its mean value. (2) Energy storage means are required for the conversion of energy from slow random oscillations into useful motions to drive electric generators.
- Along with advantages in the application of big data, several issues need to be considered, such as the development of models that will give better accurate estimations for parameters in almost all climatic conditions and the development of reliable models for better prediction of weather data.
- Several confidential data sets and user data sets flow from one organization to the other in the process of procuring and analysing data. Several data protection steps need to be taken for the flow of such confidential data. In case of leaking of data,

it would be of crucial importance to identify from where the data got leaked & who was responsible for data leakage.

- The current literature doesn't discuss other monetary and social issues prompting atmosphere changes and contamination. Apart from smart cities and technical aspects of climate degradation, tourism is one such social issue. Tourism is heavily responsible for air, water, and land pollution due to stress on a particular land or water resource. Other factors may be due to more than usual traffic congestions and local shops becoming more expensive, stocks running out of shops and more generation of waste. These problems require predictive analysis and detailed forecasts utilizing big data techniques.
- Waste generation minimization and management are also important practices to achieve goals of sustainability. Big data analytics can not only help in the production of cleaner wastes (having less impact on the environment) but also help in the reduction of the amount of waste generated and manage them. Although no such study exists yet, there is scope for big data analytics and machine learning methods in the fields of waste segregation and management from homes to dumping grounds.
- Overall, most research articles aimed at achieving environmental sustainability through big data analytics indirectly aim to gain improvements in the business sectors as well. However, there is hardly any research addressing the social and economic aspects of sustainability directly. Although environmental sustainability sounds like a bigger challenge and rounds up the umbrella of sustainable development, the other two aspects can't be ignored since all three come hand in hand.

6 Scope of Future Work

Big data in itself is not a technology or a science field. Big data just enables us to get an actual idea of the real scenario taking place in the system to understand it better i.e. forming a basis for making technological and economic advancements. A better understanding of patterns, correlations, trends, etc. and deeper insights into the existing issues can then help in putting forward the technologies to overcome the issues. With the fast improvement of sensor technology, wireless transmission technology, network communication technology, cloud computing, distributed computing, smartphones (Zhou et al. 2016), etc. large amounts of data is generated every second. Previously, it was estimated that by 2020 1.7 MB data will be created every second for every person on earth by 2020 (Ahmad 2018). This data needs to be put to proper use for several issues of the modern world such as crisis response, disaster management and developing of smart management frameworks for transport systems, building systems, power systems, traffic systems, etc.

- Big data analytics aims to achieve the goals of clean power generation, efficient power transmission, dynamic power distribution, rational electricity consumption and innovative storage solutions by using one or many of the various data analysis techniques and algorithms, including optimization, forecasting, classification, and clustering, can be applied on the large amounts of smart grid big data (Sagiroglu et al. 2016)

- The efficient usage of building energy plays a vital role in the reduction of prices, minimizing harmful environmental effects, and brings ease for the users in terms of automation.
- The most predominant feature of this era is high-frequency generation and exchange of a large volume of data in real time that must be analysed to make proper decisions and to take appropriate actions as quickly as possible.
- Advances of big data in transport will help in large scale deployment of Battery Electric vehicles (BEVs) and Hybrid Electric Vehicles (HEVs) with de-carbonization as the main objective.
- Apart from the impact of GHG emissions on climate change, transport is also a significant source of noxious air pollutants. The data hence collected regarding these can help in improving the air quality for public health.
- Users are helped make better transport decisions by predicting traffic flow and monitoring traffic crowding by the huge data collected from various sensor sources, including inductive loops, radars, cameras, mobile Global Positioning, systems, crowdsourcing, social media (Lv et al. 2014).
- Heterogeneous network environment such as an industry or a service provider where delay in delivery of any action or goods will result in loss of the whole project it is necessary to have a system which can help to keep the record of item's location, current condition, and combining these details with inventory for faster delivery. Industrial IoT based on Big data analytics (IIoT) can be considered here (Lee et al. 2017). The same system can be used in a smart building where a user can wash clothes in a washing machine from anywhere just by connecting the washing machine to the internet. Of course, this technology is not just limited to washing machines but can help in all kinds of household chores. Thus, revolutionizing the concepts of household chores for several women across the world especially in developing Asian countries as well as for easing the tasks for students and bachelors across the world. However, the odds of achieving such a spread of machine intelligence in the near future is still debatable due to several risk factors and unreliability.
- As reviewed in several literature pieces, big data analytics plays an important role in the growth and development of an EV ecosystem for large scale deployment of EVs. Several technical data related to battery ageing and battery health play an important role in the development of efficient and cost-effective batteries, which solves major issues related to affordability, range anxieties, and charging time anxieties among users. Similarly, charging big data can be utilized for development in the market and infrastructural aspects of an EV ecosystem. Figure 13 represents a flow chart highlighting the potential usage of charging big data in the EV market and infrastructure.

- Big data can leverage the development of offshore renewables, particularly wave energy conversion systems. Potential locations for wave power development

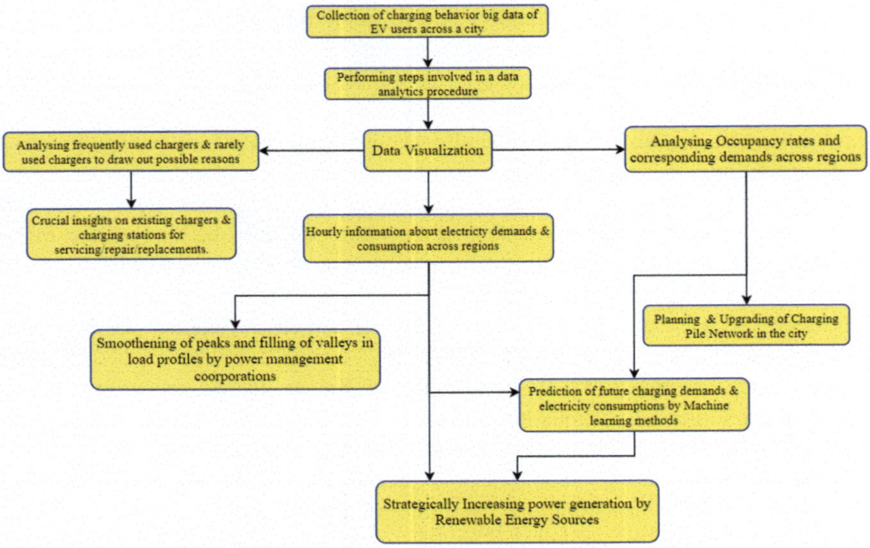

Fig. 13 Flowchart representing the usage of an EV user charging behaviour big data in future planning of several aspects of an EV ecosystem

in India are coasts of Maharashtra, Goa, Karnataka, and Kerala (Kanyakumari has the highest power owing to strong winds) (Wikipedia 2021b). On a whole, renewable energy generation systems are known to have several drawbacks limiting their popularity. Uncertainty of the source and less efficiency of the power generating system top the list. However, solar energy, wind energy, wave energy, and other forms of RES are infinitely available. Machine learning models based on big data proves to be the biggest achiever in this case for forecasting and prediction.

- Government records and surveys are an incredible source of basic information about each and every citizen and hence, the government agencies can utilize the citizen data for the identification of the needy to provide them with the necessary assistance. From the business point of view, it is highly expected that an adjustment in the viewpoint at business issues through big data analytics will provide a wide scope of modifications for a wide range of associations and supply chains, in turn helping their business grow. Business owners can leverage the big data generated to understand their customers better and improving business operations answering both environmental and economic concerns. Since organizations need to hire a special team of data engineers, scientists, and analysts to assess the real case scenario of a business, the introduction of big data analytics also generates new income opportunities. Hence, through the business sector, big data can be expected to have a hand in dissolving issues related to the economic sustainability of a country on a larger scale.

7 Case Study

7.1 Test System

From the EV charging station placement point of view, it is crucial to monitor and hence realize the power supply demands pressure in a region and the subsequent heavy loading on distribution transformer due to both—residential charging stations and commercial charging stations. However, in this study, we focus mainly on the analysis of charging demand and occupancy of commercial charging stations because residential chargers are very unlikely to happen for any densely populated Indian city in near future. The proposed strategy not only helps in relieving the burden on distribution transformer but also in decreasing the cost of EV charging by making it more economical.

The purpose of a case study is to provide new information, trends, and results to EV charging station management firms by the analysis and examination of traffic and charging big data sets (Fig. 3). Thus, we perform a case study by assuming charging demands in the city of Mumbai, India based on route traffic congestion analysis with reference to the studies performed in Li et al. (2015) and Wang et al. (2019). The methodology used in this case study is described in Fig. 14.

The development of an EV ecosystem and deployment of EVs are both cause and effect of each other. Due to limited on road EVs in Mumbai and due to less popularity among Indian users as well as charging station management firms there exists a lack of sensors and other data collection devices at EV charging stations.

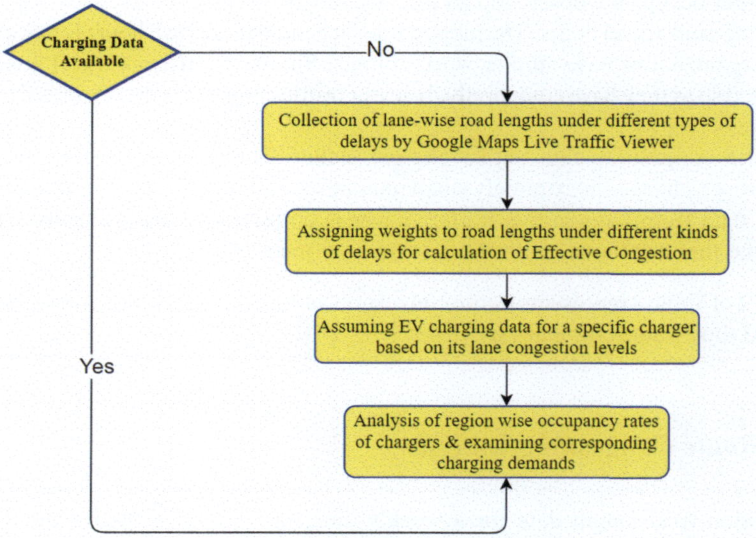

Fig. 14 Route Congestion analysis methodology utilized in the case study

Hence, the historical time stamp data of chargers is unavailable. With the same background reasoning, we attempted to collect the remaining data related to the EV infrastructure for Mumbai.

- Mumbai suburban and Mumbai city combined have around only 27 eV charging stations till date.
- Most of the charging stations are equipped with one or two chargers at max while data regarding number of chargers is not available for the rest.
- Types of chargers available are mainly AC slow chargers and DC fast chargers.

7.2 Route Selection

Most of the EV charging stations can be categorized under two major expressways in Mumbai—Eastern Expressway and the Western Expressway, hence Set of n routes $R = \{R_1, R_2\}$ where $n = 2$. Part 1 of Table 4 gives additional first-hand information about features of both routes considered in our study. Out of 27 charging stations, 3 of them either do not come under any route or are a part of Navi-Mumbai and hence, are excluded from our analysis. Figure 15 shows the Map of Mumbai with locations of charging stations situated along its major express highways.

7.3 Selection of Time Frames

- Charger occupancy observational time frames are divided mainly into rush hours and off-peak travel hours. The biggest reason behind these rush hours is that almost all organizations have similar work timings. Similarly, off-peak travel hours are named so as they have comparatively lesser traffic congestion due to no reasonable cause.
- Traffic congestion data unlike charger status data changes at every minute. However, changes at extremely small intervals are negligible. Thus, traffic congestion data from Google live traffic viewer is collected for the previously stated 7-time frames of a day with shorter one-hour intervals.

Part 2 of Table 4 represents traffic congestion and charger occupancy observational selected time frames for analysis.

7.4 Route Congestion Analysis

Congestion road length data for a week's time, i.e. 3 August 2020 (Monday)–9 August 2020 (Sunday) was collected from Google Maps Live traffic viewer. The

Table 4 Represents important set of information considered in the performed case study

Part 1—Information of Selected Routes

Features	Eastern express highway (R_1)	Western express highway (R_2)
Number of charging stations	8 + 2 (Shared with R_2)	14 + 2 (Shared with R_1)
Route length under observation	17.4 kms	30.5 kms
Start—Destination	Parel to Vikhroli	Borivali to Bandra + Worli Sea link
Part of Highway	National Highway—3	National Highway—8

Part 2—Time Frames

Charger occupancy observational time frames		Traffic congestion observational time frames
Rush Hours	Off-Peak Travel Hours	
7–9 AM	1 AM – 4 PM	7–8 AM
		8–9 AM
		1–2 PM
5–7 PM	8–9PM	3–4 PM
		5–6 PM
		6–7 PM
		8–9 PM

Part 3—Route Lengths highlighted under different kinds of delay in Google maps

Colour of Highlighted part	Congestion caused delay
Green	No delay
Orange	Slight delay (S)
Red	Medium delay (M)
Maroon	Heavy delay (H)

road lengths are highlighted under different kinds of delay on the Google Maps live Traffic viewer as represented in Part 3 of Table 4.

- *Lane Considerations*—The total distance highlighted under each colour is measured and numerically added for each time frame. Lane-wise congestion analysis plays an important role in determining charging behaviour as an EV user is likely to in a particular direction via lane 1 is much more likely to search and spot for an EV charging station lying adjacent to Lane 1 than the ones lying adjacent to Lane 2A route generally comprises of two lanes owing to vehicles flowing in the opposite directions.
- *Assigning weights*—Proper weights are assigned to each road length depending on the type of delay it has. Assuming the speed of the vehicle to be ‘x’ for heavy delay in one of the many road types, the vehicle speeds for medium and slight

Fig. 15 Charging stations represented in Blue lie under Route 1 whereas stations in Red lie under Route 2. Orange highlighted stations lie equidistant between both routes whereas the charging stations in black are situated in Navi Mumbai or do not lie in the area surrounding any route

delays can be assumed to be '2x' and '3x', respectively for the same road type. Through lane-wise congestions of both routes we derive the graph in Fig. 16 which represents the effective congestion of both routes after assimilating all kinds of delays by assignment of weights (Figs. 15 and 16).

Therefore, effective congestion in metres $= \frac{S}{3} + \frac{M}{2} + \frac{H}{1}$

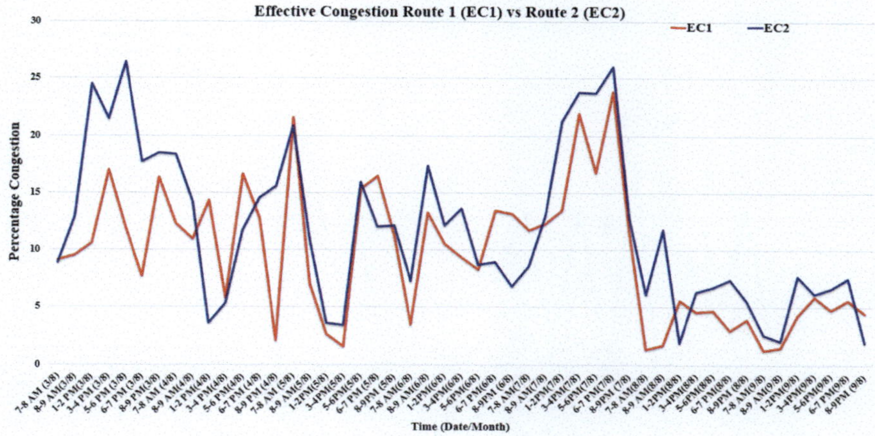

Fig. 16 Comparison of route congestions of route 1 and route 2

7.5 Charging Data Assumption

Occupancy of chargers across the city at a particular time frame was assumed on the basis of several constructive reasons adding up, lane congestion levels being the strongest of them all since the number of vehicles on road is directly proportional to the charging demand especially in case of EVs majorly due to the need of frequent charging requirements. EVs require frequent charging due to many reasons

- EVs need to be charged before the battery reaches below 20% to avoid cycles of deep discharge which adds up to the limited storage capacities of batteries.
- Charging less than its capacity—To avoid damage to the battery due to generated heat. Further, research suggests that charging an EV battery to its full capacity can decrease the range of the EV (InsideEVs n.d.).
- Lesser driving range—As compared to an ICE car, most EV batteries can store charge capable of running much lesser than ICE cars due to the battery capacities.

Also, different EV users can be expected to charge their EVs differently, some might prefer charging for shorter intervals regularly while others might charge for full capacity at once less frequently.

7.6 Route -Wise Charging Data Analysis

With the previously explained basis of charging data assumption, charging data is assumed. Further, plots are generated so as to identify the peaks, valleys, and points of comparison (Fig. 17).

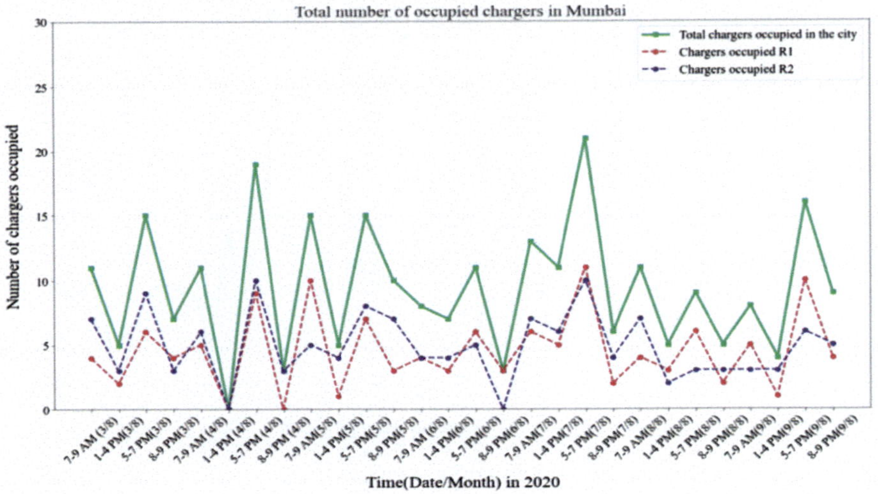

Fig. 17 Represents the number of chargers occupied on a whole for both routes and total chargers occupied in the city

7.7 Region Wise Charging Data Analysis

Mumbai mainly comprises of two districts of Maharashtra—Mumbai Suburban and Mumbai City. Mumbai Suburban is further divided into three sub districts/regions—Andheri, Borivali, and Kurla as in Fig. 19a. Thus, we have four major regions of importance—Borivali, Andheri, Kurla, and City. Table 5 describes the spread of EV charging stations and chargers across the city.

'Charging Demand' and 'Occupancy Rates' are two terms which are often used interchangeably. However, there is a slight difference in between both the terms which also accounts for the logical relation between them. The relation is crucial from business point of view as it helps firms with deeper insights while aiming to optimize a charging pile network in a city. The relation can be explained by Fig. 18.

Bar graphs are plotted with the assumed charging data to examine occupancy rates in each of the four districts/regions of Mumbai. Figure 19b is a bar graph representation of percentage occupancies at each time frame for each region. However, it doesn't deliver a clear picture of overall occupancies of regions in order to make out which region has the most occupancy or charging demand. Figure 19c represents effective region wise occupancy of a week for all regions.

Table 5 Region wise allocation of charging stations and chargers

	Borivali	Andheri	Kurla	City
Charging stations	3	9	5	7
Chargers	4	13	10	9

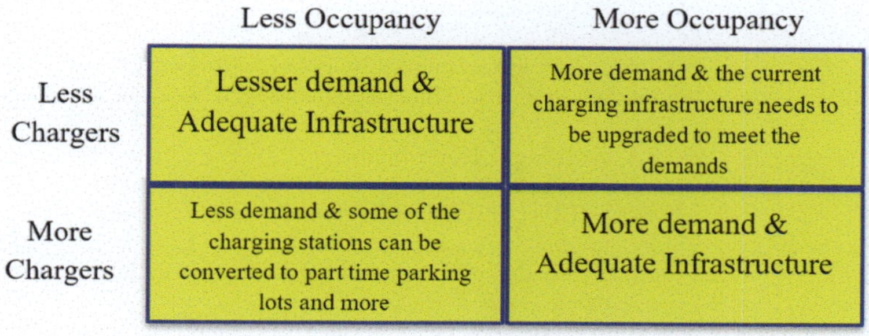

Fig. 18 Relationship between demand and occupancy

We intend to draw out conclusions utilizing the above proposed relations from the generated graphs. It would be safe to assume that Borivali region draws out most of the EV users, while it still has very few chargers. Hence, it is safe to conclude that the charging infrastructure in Borivali needs to be upgraded to meet the future charging demands. Similarly, Andheri represents a situation close to Case 2. Even after having the maximum number of chargers in Andheri, occupancy rates remain the lowest. And hence, it is important for firms to analyse the least occupied charging stations to convert them into part time parking lots or for other purposes.

7.8 Predication of Congestion Data—Without Pandemic Conditions

Real time traffic congestion levels for Mumbai is collected from the live traffic viewer website of *TomTom* (www.tomtom.com n.d.) for the same time frame during the year 2019 to assess the conditional changes without the effect of pandemic. Further, this traffic data is not comparable to the traffic congestion levels previously calculated for Pandemic conditions and is made comparable by converting time-based congestion into distance-based congestion.

7.9 Route-Wise Congestion Analysis for Without Pandemic Conditions

Similar graphs as in the previous case of Pandemic conditions are plotted for normalized conditions of the city, to compare the distance-based congestions of pandemic and without pandemic conditions and thus make predictions about the charging demand. Figure 19 is an overall representation of congestion data comparison of 2019 versus 2020 which is approximated to suit our congestion observation time frames. The peaks and valleys of congestion line plots are incident at the same time

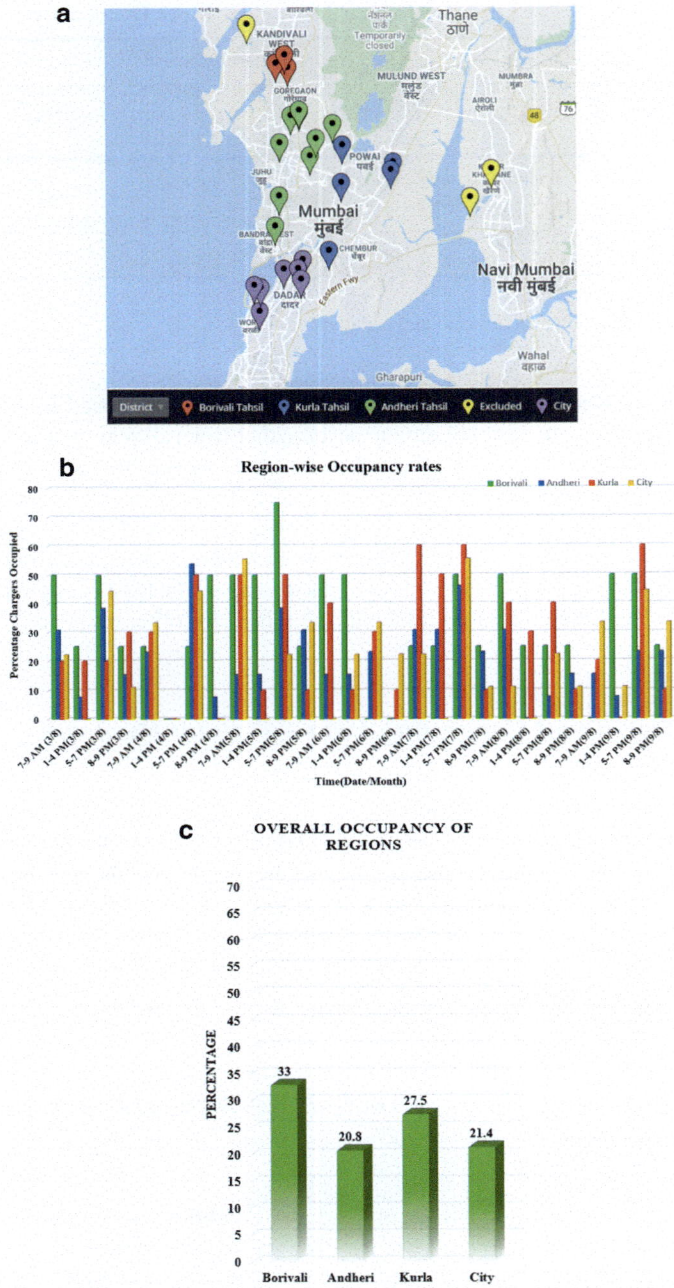

Fig. 19 **a** District-wise division of charging stations in Mumbai. **b** Region wise Occupancy rates of chargers in Mumbai. **c** Overall percentage occupancy of regions over a week

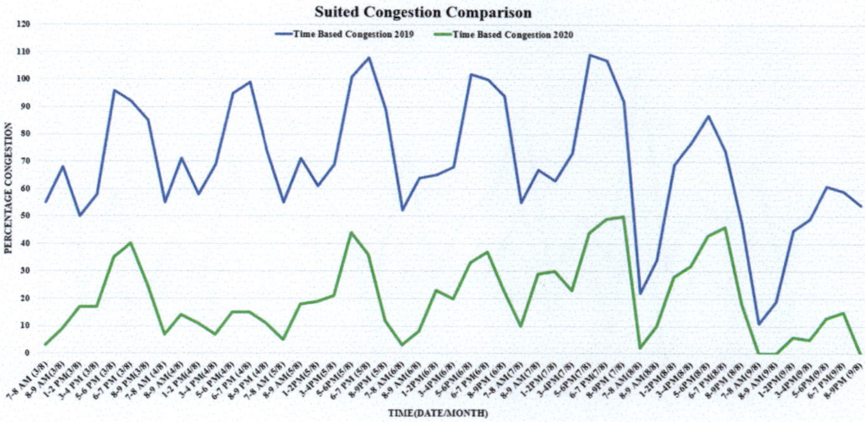

Fig. 20 Overall congestion comparison for the city of Mumbai by the obtained congestion data from TomTom N.V

frames in Fig. 20, thus, suggesting to us the travel patterns of users (rush hours and off-peak hours) remain the same, but differ in magnitude for both cases.

7.10 Charging Data Assumption—Without Pandemic Conditions

With the predicted congestion rates of 2020 without pandemic conditions, charging behaviour was assumed and plotted for both lanes of both routes as well as for all four regions. Further, Fig. 21a, b compare the occupancy rates of the without pandemic conditions with the pandemic conditions.

Figure 21b suggests to us similar charging occupancy ranks of regions in 'without pandemic' conditions as in the 'pandemic conditions' since the factors determining charging behaviour in a particular region remain the same, and only the amount of EV users seeking charging increases. However, in 'without pandemic' conditions Kurla region seems to have highest occupancy rates followed by Borivali. Hence, it is safe to conclude that the region of Kurla might come under Case 2, i.e. the current infrastructure is suitable for demands. Also, a considerable increase in the number of occupied chargers in the city resulted in slight region wise occupancies, which suggests to us the denounced effects of overall city occupancy rate in region wise occupancies. Further, more visual graphs need to be examined in a similar way and cases need to be identified for different regions in a city for the successful deployment of Charging Pile Network in the city.

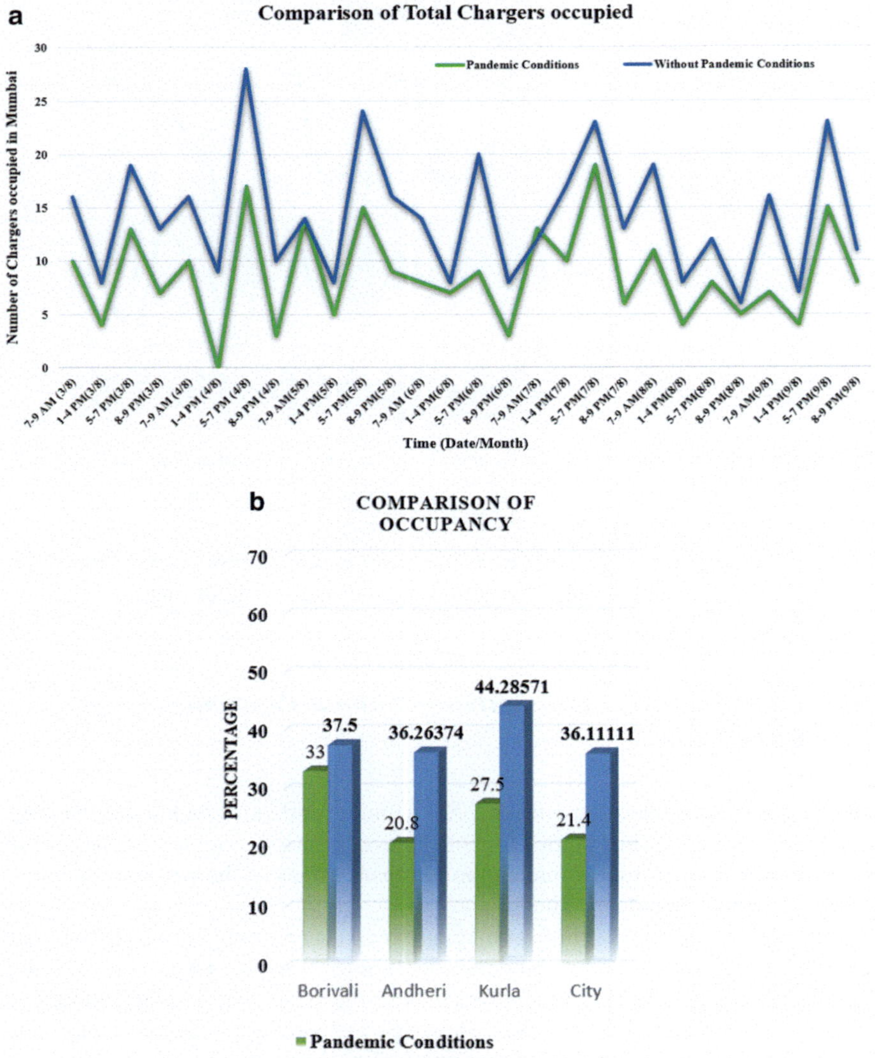

Fig. 21 **a** Comparison of total number of chargers occupied during pandemic conditions and without pandemic conditions. **b** Comparison of Region wise overall percentage occupancies in Pandemic conditions versus Without Pandemic Conditions

8 Discussion and Conclusion

Big data is produced at a fast rate every second through a huge number of data generating and recording sources. With the increasing population, one can only imagine the amount of data generated in the coming years. This chapter aims at providing a state-of the-art literature review of the recent research advances in achieving goals of sustainable development through big data-driven technologies majorly through components of a smart city. Although the goal is to cover all three aspects of sustainable development, this chapter largely focuses on achieving goals of environmental sustainability mainly because of two reasons—(1) Most existing research articles focus on environmental sustainability-driven business and economic viewpoints (maximizing profit through the utilization of renewable sources and implementation of environmentally sustainable practices (2) We believe environmental sustainability is a prerequisite; opening doors for citizens, organizations, and governments to achieve socio-economic aspects of sustainable development. Having said that, the need for research in economic and business viewpoints need to be examined in depth for a holistic picture of the data being generated.

The case study performed revolves around the 'most-raw utilization of data'—analytics. Examining and exploring the data to influence the decision-making tasks of organizations is one of the most beneficial and wise uses of data sets generated. With the same mindset, we attempt to analyse the current Charging Pile Network in the city of Mumbai. The E-Mobility sector is expected to grow exponentially in India once the EV technologies address the cost and time anxieties for better user preferences. Thus, it is important to understand the current charging infrastructure, study the electricity demands, and monitor the EV sales in order to prepare the city for mass adoption of EVs.

With the necessary background, readers are encouraged to research focussing on overcoming one of the many limitations of the existing literature articles thus help in increasing the domain of reach of data-driven sustainable development. We also hope our research opens new research directions for researchers to explore the usage and potential of data-driven technologies for a wider range of tasks including analytics, optimization, forecasting, management, risk analysis, location of faults, and many more in the future.

References

Ahmad I (2018) How much data is generated every minute? [Infographic]. Social Media Today. https://www.socialmediatoday.com/news/how-much-data-is-generated-every-minute-infographic-1/525692/

Amaro N, Pina JM (2017) Big data in power systems leveraging grid optimization and wave energy integration. In: 2017 international conference on engineering, technology and innovation (ICE/ITMC). Funchal, pp 1046–1054. https://doi.org/10.1109/ICE.2017.8279997

Ancillotti E, Bruno R, Conti M (2013) The role of communication systems in smart grids: architectures, technical solutions and research challenges. Comput Commun 36:1665–1697. https://doi.org/10.1016/j.comcom.2013.09.004
Apache.org (2019) Apache Hadoop. http://hadoop.apache.org/
Arias M, Bae S (2016) Electric vehicle charging demand forecasting model based on big data technologies. Appl Energy 183.https://doi.org/10.1016/j.apenergy.2016.08.080
BloombergQuint (n.d.) Electric vehicle sales in India up 20% in 2019–20, industry body says. https://www.bloombergquint.com/business/electric-vehicle-sales-in-india-up-20-in-2019-20-industry-body-says. Accessed 31 May 2021
Cai H, Xu M (2013) Greenhouse gas implications of fleet electrification based on big data-informed individual travel patterns. Environ Sci Technol 47.https://doi.org/10.1021/es401008f
Chesterton A (2018) How many cars are there in the world? CarsGuide. https://www.carsguide.com.au/car-advice/how-many-cars-are-there-in-the-world-70629
Chou J-S, Ngo N-T, Chong WK, Gibson GE (2016) Big data analytics and cloud computing for sustainable building energy efficiency. In: Start-up creation, pp 397–412
Daissaoui A, Boulmakoul A, Karim L, Lbath A (2020) IoT and big data analytics for smart buildings: a survey. Procedia Comput Sci 170:161–168
Datadog (2016) Hadoop architectural overview. https://www.datadoghq.com/blog/hadoop-architecture-overview/. Accessed 31 May 2021
De la Rue du Can S, Letschert V, Mcneil M, Zhou N, Sathaye J, Orlando E, Berkeley L (2009) Residential and transport energy use in India: past trend and future outlook. https://doi.org/10.2172/951788
Dittrich J, Quiané-Ruiz J-A (2012) Efficient big data processing in Hadoop MapReduce. In: Proceedings of the VLDB endowment, vol 5, no 12, pp 2014–2015
DoSomething.org (2021) 11 Facts about global poverty. https://www.dosomething.org/us/facts/11-facts-about-global-poverty#fnref7. Accessed 31 May 2021
Dubey R, Gunasekaran A, Childe SJ, Wamba SF, Papadopoulos T (2015) The impact of big data on world-class sustainable manufacturing. The Int J Adv Manuf Technol. 84(1–4):631–645. https://link.springer.com/article/https://doi.org/10.1007/s00170-015-7674-1
Dutton WH, Blumler JG, Kraemer KL (1987) Wired cities: shaping future communication. Macmillan, New York
EDF Energy (2015) EV: Batteries. EDF Energy. https://www.edfenergy.com/electric-cars/batteries
Escobedo G, Jacome N, Arroyo-Figueroa G (2017) Big data & analytics to support the renewable energy integration of smart grids—case study: power solar generation. In: Proceedings of the 2nd international conference on internet of things, big data and security
Gennaro M, Paffumi E, Martini G (2018) Big data for supporting low-carbon road transport policies in Europe: applications, challenges and opportunities", case studies on transport policy 6(4)
Goos G, Hartmanis J, Van Leeuwen J, Ishida T, Isbister K (2000) Digital cities. Heidelberg Springer, Berlin Heidelberg, Berlin
Green Car Reports (n.d.) What Is EVSE and why does your electric car charger need it? https://www.greencarreports.com/news/1050948_what-is-evse-and-why-does-your-electric-car-charger-need-it. Accessed 31 May 2021
Growth of electricity sector in India from 1947–2020 government of India ministry of power central electricity authority New Delhi (2020). https://cea.nic.in/wp-content/uploads/pdm/2020/12/growth_2020.pdf
Guardknox (n.d.) The EV charging ecosystem. https://www.guardknox.com/electric-vehicle-charging-ecosystem/. Accessed 31 May 2021
Indian consumers: second place. (n.d.) https://www.nationalgeographic.com/greendex/assets/dl/pdf/Indians.pdf. Accessed 31 May 2021
InsideEVs (n.d.) You should set EV Charge limit to just 60 percent: here's why. https://insideevs.com/news/368097/video-60-percent-ev-charge-limit-benefits/. Accessed 31 May 2021

Jakubiec JA, Reinhart CF (2012) A method for predicting city-wide electric production from photovoltaic panels based on LiDAR and GIS data combined with hourly DAYSIM simulations. Solar Energy 93(C):127–43

Johnson J (2021) Global digital population 2019|Statistic. Statista. https://www.statista.com/statistics/617136/digital-population-worldwide/

Junaidi N, Shaaban M (2018) Big data applications in electric energy systems. In: 2018 international conference on computational approach in smart systems design and applications (ICASSDA)

Karmawijaya MI, Nashirul Haq I, Leksono E, Widyotriatmo A (2019) Development of big data analytics platform for electric vehicle battery management system. In: 2019 6th international conference on electric vehicular technology (ICEVT)

Khalid MR, Alam MS, Sarwar A, Jamil Asghar MS, A comprehensive review on electric vehicles charging infrastructures and their impacts on power-quality of the utility grid. eTransportation. https://doi.org/10.1016/j.etran.2019.100006

Khan M, Babar M, Ahmed SH, Shah SC, Han K (2017) Smart city designing and planning based on big data analytics. Sustain Cities Soc 35:271–279

Kitchin R (2013) The real-time city? Big Data and Smart Urbanism. GeoJournal 79(1):1–14

Komninos N (2002) Intelligent cities: innovation. Routledge, Knowledge systems and digital spaces

Kuhlman T, Farrington J (2010) What is sustainability? Sustainability 2(11):3436–3448

Lee J, Park G-L, Han Y, Yoo S (2017) Big data analysis for an electric vehicle charging infrastructure using open data and software. In: Proceedings of the eighth international conference on future energy systems

Li Y, Luo J, Chow C-Y, Chan K-L, Ding Y, Zhang F (2015) Growing the charging station network for electric vehicles with trajectory data analytics. In: 2015 IEEE 31st international conference on data engineering

Li Y, Liu K, Foley AM, Zülke A, Berecibar M, Nanini-Maury E, Van Mierlo J, Hoster HE (2019) "Data-driven health estimation and lifetime prediction of lithium-ion batteries: a review. Renew Sustain Energy Rev, Elsevier, 113(C):1–1

Liang F, Yu W, An D, Yang Q, Fu X, Zhao W (2018) A survey on big data market: pricing, trading and protection. IEEE Access 6:15132–15154

Linder L, Vionnet D, Bacher J-P, Hennebert J (2017) Big building data—a big data platform for smart buildings. Energy Procedia. 122:589–594. https://doi.org/10.1016/j.egypro.2017.07.354

López-Ibarra JA, Goitia-Zabaleta N, Herrera VI, Gazta ñaga H, Camblong H (2020) Battery aging conscious intelligent energy management strategy and sensitivity analysis of the critical factors for plug-in hybrid electric buses. eTransportation 5:100061

Lu Y, Li K, Han X, Feng X, Chu Z, Lu L, Huang P, Zhang Z, Zhang Y, Yin F, Wang X, Dai F, Ouyang M, Zheng Y (2020) A method of cell-to-cell variation evaluation for battery packs in electric vehicles with charging cloud data. eTransportation 6:100077

Lv Y, Duan Y, Kang W, Li Z, Wang F-Y (2014) Traffic flow prediction with big data: a deep learning approach. IEEE Trans Intell Transp Syst 1–9

Majumdar S, Subhani MM, Roullier B, Anjum A, Zhu R (2020) Congestion prediction for smart sustainable cities using IoT and machine learning approaches. Sustain Cities Soc 102500

Miller H (2011) Collaborative mobility: using geographic information science to cultivate cooperative transportation systems. Procedia—Soc Behavioral Sci 21:24–28. https://doi.org/10.1016/j.sbspro.2011.07.005

Months TG 9, Weeks 4 (2020) Terabytes of data were buried in the Arctic. We ask the person in charge, why? Mashable SEA. https://sea.mashable.com/tech/11810/terabytes-of-data-were-buried-in-the-artic-we-ask-the-person-in. Accessed 31 May 2021

Moradi J, Shahinzadeh H, Nafisi H, Marzband M, Gharehpetian GB (2019) Attributes of big data analytics for data-driven decision making in cyber-physical power systems. In: 2020 14th international conference on protection and automation of power systems (IPAPS)

Ndukwu M, Onwude D, Bennamoun L, Abam F, Simo-Tagne M, Horsfall I, Briggs T (2020) Nigeria's energy deficit: the challenges and eco- friendly approach in reducing the energy gap. Int J Sustain Eng 1–18

Nitesh Jain (2015) Big data and hadoop essentials. Udemy. https://www.udemy.com/course/big-data-and-hadoop-essentials-free-tutorial/. Accessed 6 Nov 2019
O'Donovan P, Leahy K, Bruton K, O'Sullivan DTJ (2015) An industrial big data pipeline for data-driven analytics maintenance applications in large-scale smart manufacturing facilities. J Big Data 2(25):1–26
Onya (2021) Greendex ranks 18 countries based on sustainability. https://www.onyalife.com/greendex-ranks-18-countries-based-on-sustainability/. Accessed 31 May 2021
Preda S, Oprea S-V, Bâra A, Belciu (Velicanu) A (2018) PV forecasting using support vector machine learning in a big data analytics context. Symmetry 10(12):748
Rahimi-Eichi H, Chow M-Y (2014) Big-data framework for electric vehicle range estimation. In: Presented at the 40th annual conference of the IEEE industrial electronics society (iecon2014). IEEE, Dallas, TX
Rathore MM, Paul A, Hong W-H, Seo H, Awan I, Saeed S (2018) Exploiting IoT and big data analytics: defining smart digital city using real-time urban data. Sustain Cities Soc 40:600–610. https://www.sciencedirect.com/science/article/abs/pii/S2210670717309782
Sagiroglu S, Terzi, R, Canbay Y, Colak I (2016) Big data issues in smart grid systems. In: 2016 IEEE international conference on renewable energy research and applications (ICRERA)
Sanila S, Subramanian D, Sathyalakshmi S (2017) Real- time mining techniques: a big data perspective for a smart future. Indian J Sci Technol 10(42):1–7
Science (2018) Cities emit 60% more carbon than thought in embodied consumption, C40 report says. https://www.nationalgeographic.com/science/article/city-consumption-greenhouse-gases-carbon-c40-spd. Accessed 31 May 2021
Selecthub.com (2021) Big data versus business intelligence versus data mining|the differences. https://www.selecthub.com/business-intelligence/bi-vs-big-data-vs-data-mining/. Accessed 31 May 2021
Sharma G, Sood VK, Alam MS, Shariff SM (2020) Comparison of common DC and AC bus architectures for EV fast charging stations and impact on power quality. eTransportation 5:100066
Shepard M (2011) Sentient city: ubiquitous computing, architecture, and the future of urban space. MIT Press, Cambridge, Mass
Snowden S (2021) Melting of greenland's ice sheets has passed the point of no return, scientists say. Forbes. https://www.forbes.com/sites/scottsnowden/2020/08/14/greenlands-melting-ice-sheets/?sh=74244bba2449. Accessed 31 May 2021
Softwaretestinghelp.com (2019) Top 15 big data tools (big data analytics tools) in 2019. https://www.softwaretestinghelp.com/big-data-tools/
Solability (2021) The global sustainable competitiveness index. https://solability.com/the-global-sustainable-competitiveness-index/the-index. Accessed 31 May 2021
Tao F, Qi Q, Liu A, Kusiak A (2018) Data-driven smart manufacturing. J Manuf Syst 48:157–169
Team B (2021) First economic contraction in 4 decades: India's GDP shrinks 23.9% in Q1, FY21. Business-standard.com. https://www.business-standard.com/article/economy-policy/first-economic-contraction-in-4-decades-india-s-gdp-shrinks-x-in-q1-fy21-120083101022_1.html. Accessed 31 May 2021
The Conversation (2021) Siberia heatwave: why the Arctic is warming so much faster than the rest of the world. https://theconversation.com/siberia-heatwave-why-the-arctic-is-warming-so-much-faster-than-the-rest-of-the-world-141455. Accessed 31 May 2021
The Guardian (2021) Earth has lost 28 trillion tonnes of ice in less than 30 years. https://www.theguardian.com/environment/2020/aug/23/earth-lost-28-trillion-tonnes-ice-30-years-global-warming. Accessed 31 May 2021
The Hindu (2021) Data|An estimated 12.2 crore Indians lost their jobs during the coronavirus lockdown in April: CMIE. https://www.thehindu.com/data/data-over-12-crore-indians-lost-their-jobs-during-the-coronavirus-lockdown-in-april/article31520715.ece. Accessed 31 May 2021
The Indian Express (2020) Explained: How does the Rewa Solar Power Plant match up to similar plants in India and abroad? https://indianexpress.com/article/explained/rewa-solar-power-plant-madhya-pradesh. Accessed 31 May 2021

Tien J (2013) Big data: unleashing information. J Syst Sci Syst Eng 22(2):127–151
Udemy (n.d.) Free Hadoop tutorial—Hadoop starter kit. https://www.udemy.com/course/hadoop starterkit/. Accessed 31 May 2021
Union of Concerned Scientists (2014) Car emissions and global warming. Union of concerned scientists. https://www.ucsusa.org/resources/car-emissions-global-warming
Vanani IR, Majidian S (2019) Literature review on big data analytics methods. Social media and machine learning. https://www.intechopen.com/books/social-media-and-machine-learning/literature-review-on-big-data-analytics-methods
Vikaspedia.in (2020) vikaspedia domains. https://vikaspedia.in/energy/energy-efficiency/electric-vehicles-in-india
Voivontas D, Assimacopoulos D, Mourelatos A, Corominas J (1998) Evaluation of renewable energy potential using a GIS decision support system. Renew Energy 13:333–344. https://doi.org/10.1016/S0960-1481(98)00006-8
Wang S, Moriarty P (2018) The potential for big data for urban sustainability. https://doi.org/10.1007/978-3-319-73610-5_3
Wang Y, Cai S, Shen Y, Shao D, Gong X, Zhou C, Chu D (2019) The analysis of electrical vehicles charging behavior based on charging big data 63–67. https://doi.org/10.1109/ICCCBDA.2019.8725776
web.archive.org (2018) Physical progress (achievements)|ministry of new and renewable energy|government of India. https://web.archive.org/web/20180503151432, http://mnre.gov.in/physical-progress-achievements. Accessed 31 May 2021
Wikipedia Contributors (2019a) Electric car use by country. Wikipedia https://en.wikipedia.org/wiki/Electric_car_use_by_country
Wikipedia Contributors (2019b) Motor vehicle. Wikipedia. https://en.wikipedia.org/wiki/Motor_vehicle
Wikipedia (2021a) Induction loop. https://en.wikipedia.org/wiki/Induction_loop
Wikipedia (2021b) Wave power in India. https://en.wikipedia.org/wiki/Wave_power_in_India. Accessed 31 May 2021
World Economic Forum (2021) The plastic pandemic is only getting worse during COVID-19. https://www.weforum.org/agenda/2020/07/plastic-waste-management-covid19-ppe/. Accessed 31 May 2021
World Health Organization (2020) Road traffic injuries. who.int. https://www.who.int/news-room/fact-sheets/detail/road-traffic-injuries
Worldometers.info (2021) COVID live update: 171,029,617 cases and 3,556,677 deaths from the coronaviru—worldometer. https://www.worldometers.info/coronavirus/?utm_campaign=homeAdvegas1?. Accessed 31 May 2021
Wu P-J, Chen Y-C (2017) Big data analytics for transport systems to achieve environmental sustainability. In: 2017 international conference on applied system innovation (ICASI)
www.power-technology.com (n.d.) The 100% club: the cities going all-out on renewables. https://www.power-technology.com/features/100-club-cities-going-renewables/. Accessed 31 May 2021
www.tdworld.com. (n.d.) StackPath. https://www.tdworld.com/electrification/article/20972652/transportation-electrification-how-to-do-it-right. Accessed 31 May 2021
www.tomtom.com (n.d.) Mumbai traffic report|TomTom Traffic Index. https://www.tomtom.com/en_gb/traffic-index/mumbai-traffic/
Xu W, Zhou H, Cheng N, Lyu F, Shi W, Chen J, Shen X (2018) Internet of vehicles in big data era. IEEE/CAA J Automatica Sinica 5(1):19–35
Yevgen B, Jinrong Q (2013) Battery power management for portable devices. Artech House, Cop, Boston

Yin S, Kaynak O (2015) Big data for modern industry: challenges and trends [Point of View]. In: Proceedings of the IEEE, vol 103, no 2, pp 143–146. https://ieeexplore.ieee.org/abstract/document/7067026/
Zhou K, Fu C, Yang S (2016) Big data driven smart energy management: From big data to big insights. Renew Sustain Energy Rev 56(January):215–225

Chapter 10
Inventory Models with Lot-Size Dependent Discount for Deteriorating Items: Pricing and Inventory Policies

Mohammad Abdul Halim

Abstract Two inventory models are studied for disintegrating products whose market demand be contingent upon the selling price. The retailer enjoys an opportunity on the per-unit purchasing charge of the products based on the order amount so that a lower purchase cost per unit is possible only for a sufficiently large enough order size. Under this lot-size related discount, the inventory models for without ending situation and a fully backlogged shortages situation are formulated mathematically. Deterioration starts instantly when the products are kept in the warehouse at a known certain rate. The ultimate intention is to find the best inventory strategy and price of the product to maximize both situations' profit. The entire problem is formulated geometrically and solved by proposing an efficient algorithm. The optimum of the profit function is examined theoretically and also graphically by using MATLAB software. Finally, the proposed models are validated by using two numerical examples for solving each case. Sensitivity investigation is done by transforming each parameter in turn while keeping fixed the others.

Keywords EOQ model · Price discount based on the order amount · Non-zero terminal inventory · Fully backlogged shortage

1 Introduction

During the most recent couple of decades, numerous analysts/researchers who contemplated inventory models extensively with various inventory features demonstrated by accepting interest rate as steady. In any case, in authenticity, the demand of a thing has been thought about forever in a unique state. This grabs the eye of analysts to feel concern regarding the fluctuation of the interest rate. Many researchers have been carrying their research considering price-dependent demand with deterioration in present days. The primary target of inventory analysis is to satisfy the clients'

M. A. Halim (✉)
Department of General Educational Development, Daffodil International University, Dhaka, Bangladesh
e-mail: halim.ged@diu.edu.bd

I. Ali et al. (eds.), *Computational Modelling in Industry 4.0*,
https://doi.org/10.1007/978-981-16-7723-6_10

interest and optimize the benefit. The clients' interest relies upon different boundaries, for example, stock, time, cost, and so on. In the current combative market circumstance, the impact of marketing strategies such as the price varieties of an item boosts its market interest among the potential customers. The market cost of an item is one of the unequivocal factors in picking a thing to burn through. A portion of the explored outcomes shows that a smaller market price source increment sought after a larger market price has the opposite impact. Thus, it tends to be inferred that a thing's interest is a component of showed stock in a showroom and market price of a product. A substantial number of researchers explored the effects of varieties of the price. Kotler (1972) forced showcasing procedures into stock choices to examine the connection between optimum order amount and pricing policy. Harris (1913) has first introduced an inventory model taking constant demand while Ladany and Sternlieb (1974) described the consequence deviation of the charge on EOQ besides on marketing.

In the inventory investigation, deterioration is one of the essential aspects of a real-life situation. In any case, it has an outcome to the stock framework because breaking down item are not useable; for example, it is ultimately harmful. So, we cannot disregard this outcome to the stock investigation. In the present combative market circumstance, it is really challenging to convince any potential customer and the entrepreneurs, therefore, allowing different genres of offers to boost the market demands (for instance, price concession, quantity concession, delay payment scheme, and others).

1.1 Literature Review

Schrader and Ghare (1963) familiarized an EOQ model accepting a consistent weakening and steady rate of demand over a limited arranging skyline for the first time. Covert and Philip (1973) changed the model of Ghare and Schrader's unchanging disintegration frequency to a two-boundary Weibull distribution. Later, Pal et al. (1993) delineated a dependent stock demand model with constant decline and holding cost. A model consuming a stock-sensitive consumption rate, shortage, and deterioration was delineated by Padmanabhan and Vrat (1995). Later, Panda et al. (2012), Khan et al. (2019b), Shaikh et al. (2019b), Panda et al. (2019), and Khan et al. (2020c), the decaying items assuming the fixed capacity of the retailer's warehouse were further examined. Incorporating the advertisement's consequence on the customers' demand, Khan et al. (2020d) found the optimal advertising policy assumed as a discrete decision variable.

Not a wide range of stock products disintegrated in the real world when they got by the retailer. During a particular time, interval, deterioration does not appear, and all products hold their initial quality. Later, Shaikh et al. (2017), Mashud et al. (2018) and Khan et al. (2020c) described the effect of non-instantaneous decay on the retailer's optimal inventory scheme. An EPQ (Economic production quantity) model about decaying products was studied by Mandal and Phaujdar (1989), taking linear stock-sensitive demand and steady production rate into account. Then the model of Baker and Urbans (1988) for deteriorating items was extended by Pal et al. (1993). Some of the examples related to this characteristic comprise Sarker et al. (1997), Ray and Chaudhuri (1997), Ray et al. (1998), Dye and Ouyang (2005), Bhunia and Shaikh (2011), Lee and Dye (2012), Meher et al. (2012), Panda et al. (2012), Avinadav et al. (2013), Taleizadeh et al. (2013), Shaikh et al. (2017), Sahu et al. (2017), and Shaikh et al. (2019c). On the other hand, Panda et al. (2013) expressed the models noticing characteristics of ameliorating items.

Often to increase the order, they need discounts on all of the units or incremental. One scaled-down value applies to all the units in the provided request in the entirety of the unit's limits through a progressive decrease of value applies to subsets of units in the provided order amount inside specific cutoff points. Weng (1995) established a model making an allowance for optimal quantity discounts depending on market price. Li et al. (1996) established a lot-for-lot optimum pricing policy. Burwell et al. (1997) introduced another EOQ model along with reliant price demand and constant carrying cost without shortage. Considering the deterioration of items with constant demand, Wee (1998) established a discount scheme on price. Goh and Sharafali (2002) improved the model of Wee (1998) incorporating the price discount opportunity in their model. Later, Alferes (2016) incorporated linear time-dependent holding cost and market price-dependent demand under all units' discount opportunities to achieve the optimal inventory decision. Khan et al. (2019a) further improved Alferes (2016) model for decaying items with a certain lifetime. Afterwards, Shaikh et al. (2019a) and Khan et al. (2020a) studied the discount prospect on the purchase price for marketing price-dependent demand and the current stock level dependent. Recently, Khan et al. (2020b) imposed the discount facilities on the total purchase price from the supplier to the retailer according to the prepayment amount.

Hu and Munson (2002) introduced another heuristic model for investigating the effect of the incremental quantity discounts while Hu et al. (2004) proposed an improvement of the classical heuristic Silver Meal model to boost the Hu and Munson (2002) 's outcomes. Integrating both all-units and gradual amount rebate structures, Mendoza and Ventura (2008) described another model with two distinct transportation methods: full-truckload and not exactly truckload transporters. San-José and García-Laguna (2009) studied another model with ultimately accumulated deficiencies and all-units amount limits. In contrast, Valliathal and Uthayakumar (2011) presented an EPQ model with both current amount and time-subordinate interest and non-linear stock ward holding cost with shortage somewhat accumulated.

Alfares (2015) developed a model considering the most outstanding benefit stock with stock amount subordinate market interest, time-subordinate holding cost, and discount based on the order amount. Again, Alfares and Ghaithan (2016), adjusted Alfares (2015)'s model with market price value subordinate interest, time-differing holding cost, and discount based on the order amount. Our principal commitment has been presented in Table 1, comparing published studies.

One of the essential assumptions used in the above model is that the replenishment period must stop at zero stock amounts. Urban (1992) first proposed that it is more productive to use higher stock levels bringing about more prominent interest by

Table 1 Principal attainment of the proposed model

Literature	Model		Demand		Deterioration	Non-zero ending inventory	Shortage	Discount
	EPQ	EOQ	Price-dependent	Stock-dependent				
Panda and Sukla (2013)	√				√		√	
Alfares (2015)	√			√				√
Singh and Panda (2015)		√	√		√		√	
Alfares (2016)		√	√					√
Khan et al. (2019a)		√	√		√		√	√
Shaikh et al. (2019a)		√		√	√		√	√
Shaikh et al. (2020)		√			√		√	
Khan et al. (2020a)		√	√		√			√
Das et al. (2021)	√		√		√			
Khan et al. (2021)		√					√	√
Rahman et al. (2021)		√	√	√	√		√	
This Work		√	√		√	√	√	√

utilizing a loosened up terminal state of zero-consummation stock. Giri et al. (1996) broadened the Urban's (1992) model for stock things having consistent crumbling. Chang et al. (2004) improved Giri and Chaudhuri's (1998) model by altering the target to the augmentation of benefit and loosening up a zero-closure stock limitation for falling apart things. Teng et al. (2005) expanded Urban's (1992) model and proposed a calculation to acquire the ideal recharging process duration and request non-zero closure stock. As of late, Chang et al. (2010) modified the Wu et al. (2006) model to ease the constriction of zero completion stock where shortcomings are not allowed. The zero-consummation stock stage's terminal condition is lost, and the deficiencies are allowed and fully multiplied. Table 1 exposes that very few studies incorporated the concession policy based on the purchase amount. However, any study for decay items with non-zero closure stock has not been described in an all-units concession environment. The current effort attempts to accomplish this gap in inventory management.

This study describes two inventory models for declining goods with price demand in a model (i) a model without an end to inventory (ii) an inventory model with a wholly backlogged shortage. We consider the price rate of the purchase price belongings, and it depends on order size in both models. Deterioration is taken into account, and it is started instantly when the goods are received, and this rate is constant. This work's focal theme is to explore the best response to the selling price of the goods and the order quantity and the profit in both cases. An algorithm is proposed to tackle this issue while the concavity of the target work 3D plot is deliberated by utilizing MATLAB software. At last, we validate our proposed inventory models by solving two numerical examples for both cases.

2 Assumptions and Notations

Notations

To flourish a model the following symbols are considered.

Notations	Illustration
$\boldsymbol{A}$	Expenditure to place an order ($)
$\boldsymbol{c_i}$	Per unit purchase price ($)
$\boldsymbol{c_h}$	EXPENSE to hold a unit product for a year ($)
$\boldsymbol{c_b}$	Charge for a unit backordered item per year ($)
θ	Deterioration proportion ($\theta \in (0, 1)$)
$\boldsymbol{T}$	The cycle length in year
t_1	The moment at which gets to zero (for case II)
t_2	Time duration in which shortages appear (for case II)
$\boldsymbol{TC_i}\ (p, \boldsymbol{Q}, \boldsymbol{B})$	The total cost per year ($)

(continued)

(continued)

Notations	Illustration
$\boldsymbol{TP_i}$ $(p, \boldsymbol{Q}, \boldsymbol{B})$	The total profit per year ($)
Decision variables	
p	Selling price per unit ($)
$\boldsymbol{Q}$	Order size per cycle
$\boldsymbol{B}$	Remaining stock level (for case I) and maximum shortage (for case II)

The models are delineated on the grounds of the pursuit suppositions.

1. The inventory coordination's replenishment scale and preparation horizon are both infinite; moreover, lead time is minimal to nil.
2. The market demand D is connected with the price linearly, i.e. $D(p) = m - pn$.
3. Deterioration rate θ is constant but independent of stock level.
4. Any substitution or restoration of deteriorated goods within the time under review is not allowed.
5. The stock reduces to zero after meeting up customers' demand exactly when t_1 and shortage appears in t_2 time length. (for case II)
6. Shortages are permitted (for case II), and during the shortage period, it will be partially backlogging with a constant rate δ.

3 Problem Definition

Consider the circumstance wherein a practitioner purchase a decaying thing from a manufacturer. Also, because of the amount markdown, it has concurred that there is a compromise connection between the unit purchase price and the requested amount, and the unit purchase price is a declining advance ability of the requested amount. The product's market demand is associated with the price, and the relationship is a linearly declining function of price. The stock amount diminishes because of satisfying the clients' needs and decay as well. As per the impacts of interest and decay on stock, the accompanying two cases are recognized and examined.

Case I: Inventory model for without ending inventory

Case II: Inventory model with fully backlogged shortage

4 Mathematical Modeling

We have developed two inventory models based on the above-stipulated assumptions.

4.1 Case I: Inventory Model for Without Ending Inventory

Initially, a retailer purchased of goods Q units. Due to the demand D and the deterioration θ as well, the stock decreases linearly period T. After satisfying customers' appeal, the retailer has a remaining stock level B at time T (Fig. 1).

The stock has decreased from Q to B through the customers' demand D and also for the constant deterioration rate θ. Hence the accumulated stock decreasing rate is $(D+\theta)$ in unit time. Now in the case of the retailer, the total cycle time is given by the following relationships:

$$T = \frac{Q - B}{D + \theta} \tag{1}$$

Here we have described inventory-related cost:

(a) *Ordering cost per year:*

$$\frac{A}{T} = \frac{A(D+\theta)}{Q - B} \tag{2}$$

(b) *Holding cost per year:*

$$c_h \frac{\frac{1}{2}(Q+B)\frac{Q-B}{D+\theta}}{T} = \frac{c_h(Q+B)}{2} \tag{3}$$

(c) *Purchase cost per year:*

$$\frac{c_i Q}{T} = \frac{c_i Q(D+\theta)}{Q - B} \tag{4}$$

Therefore, the total cost per year is

$$TC_i(p, Q) = \frac{A(D+\theta)}{Q - B} + \frac{c_h(Q+B)}{2} + \frac{c_i Q(D+\theta)}{Q - B}. \tag{5}$$

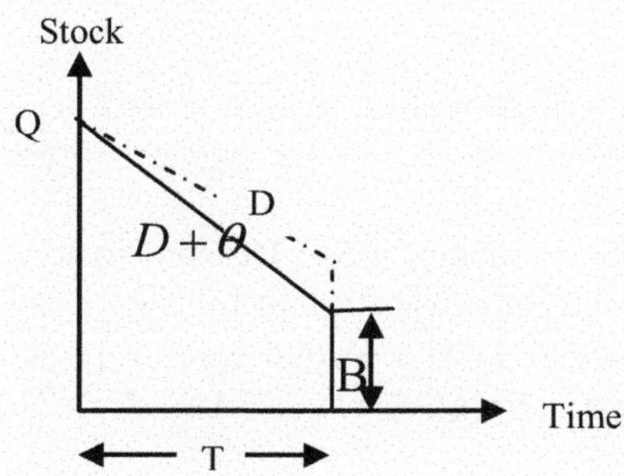

Fig. 1 Current stock amount versus time for non-zero ending situation

In the projected model, maximization of the total profit per year $TP_i(p, Q)$ is the objective function. The objective function, i.e. the benefit function, consists of two measurements: the overall income from sales and the total expense, while the total expense consists of three types of expenditures, namely the charge of ordering per year, the charge of keeping per year and the expenditure of purchasing per year. The sales revenue can be calculated by multiplying the unit selling price (p) with the number of units ended per year, which is precisely equal to the number of demand units per year, i.e. $m - pn$. The total profit function per year can be calculated by subtracting the total cost per year from the sales revenue per year. Therefore,

$$TP_i(p, Q) = p(m - pn) - \frac{A(D + \theta)}{Q - B} - \frac{c_h(Q + B)}{2} - \frac{c_i Q(D + \theta)}{Q - B} \tag{6}$$

That is,

$$TP_i(p, Q) = p(m - pn) - \frac{A(m - pn + \theta)}{Q - B} - \frac{c_h(Q + B)}{2} - \frac{c_i Q(m - pn + \theta)}{Q - B} \tag{7}$$

For the requisite conditions for optimizing total benefit, fix partial derivatives in the first order concerning p and Q of $TP_i(p, Q)$ as zero.

$$m - 2np + \frac{An}{Q - B} + \frac{c_i Qn}{Q - B} = 0 \tag{8}$$

$$\frac{A(m - pn + \theta)}{(Q - B)^2} - \frac{c_h}{2} + \frac{c_i B(m - pn + \theta)}{(Q - B)^2} = 0 \tag{9}$$

4.1.1 Concavity of the Profit Function for Case I

We calculate the Hessian matrix to verify the optimality of the overall profit (7). $H_{kk} = \begin{bmatrix} \frac{\partial^2 TP_i(p,Q)}{\partial p^2} & \frac{\partial^2 TP_i(p,Q)}{\partial Q \partial p} \\ \frac{\partial^2 TP_i(p,Q)}{\partial p \partial Q} & \frac{\partial^2 TP_i(p,Q)}{\partial Q^2} \end{bmatrix}$ and the expression of the Hessian matrix is:

$$H_{kk} = \begin{bmatrix} -2n & -\frac{An}{(Q-B)^2} - \frac{c_i nB}{(Q-B)^2} \\ -\frac{An}{(Q-B)^2} - \frac{c_i nB}{(Q-B)^2} & -\frac{2A(m-pn+\theta)}{(Q-B)^3} - \frac{2c_i QB(m-pn+\theta)}{(Q-B)^3} \end{bmatrix}. \tag{10}$$

Form the above Hessian matrix H_{kk}, without any calculation; it can be seen straightforwardly that for any p and Q, the first leading principal minor is strictly negative. The second leading principal minor $H(p, Q)$ will be positive if $4(m - pn + \theta)(Q - B) > n(A + Bc_i)$. This can be examined as follows:

$$H_{22} = \begin{vmatrix} -2n \quad amp; -\frac{An}{(Q-B)^2} - \frac{c_i nB}{(Q-B)^2} \\ -\frac{An}{(Q-B)^2} - \frac{c_i nB}{(Q-B)^2} \; amp; -\frac{2A(m-pn+\theta)}{(Q-B)^3} - \frac{2c_i QB(m-pn+\theta)}{(Q-B)^3} \end{vmatrix}$$
$$= \frac{4nA(m-pn+\theta)}{(Q-B)^3} + \frac{4nc_i QB(m-pn+\theta)}{(Q-B)^3} - \left\{-\frac{An}{(Q-B)^2} - \frac{c_i nB}{(Q-B)^2}\right\}^2$$
$$= \frac{4nA(m-pn+\theta)}{(Q-B)^3} + \frac{4nc_i QB(m-pn+\theta)}{(Q-B)^3} - \frac{n^2(A+c_i B)^2}{(Q-B)^4}$$
$$= \frac{1}{(Q-B)^4}[\{4nA(m-pn+\theta)(Q-B) + 4nc_i QB(m-pn+\theta)(Q-B)\}$$
$$-\{n^2(A+c_i B)^2\}]. \tag{11}$$

Since $(Q-B) > 0$, to be positive H_{22}, the numerator part in the last expression should be positive. It is possible only for.

$$4nA(m-pn+\theta)(Q-B) + 4nc_i QB(m-pn+\theta)(Q-B) > n^2(A+c_i B)^2. \tag{12}$$

This can be rewritten as $4n(m-pn+\theta)(Q-B)(A+c_i B) > n^2(A+c_i B)^2$. Since $n(A+c_i B)$ is always positive, so finally, we can write,

$$4(m-pn+\theta)(Q-B) > n(A+c_i B). \tag{13}$$

4.2 Case II: Inventory Model with Full Backorder

In this scenario, the practitioner often buys the product at the commencement of each Q units cycle. The entire cycle length T consists of two parts t_1 and t_2 where t_1 is the moment at which the stock amount grasps to zero and t_2 stands for the time duration of shortages appearance. The stock decreases linearly in the account of the resultant effect of demand and deterioration till during $[0, t_1]$ and reduces to zero after meeting up customers demand exactly at $t = t_1$. Shortly after the shortage occurs, the rate D is completely backlogged with shortages. The total shortage amounts, at the ending moment of the cycle, is R (Fig. 2).

Within $t = t_1$ the stock decreases from $Q - R$ for the account of D and θ, i.e. the accumulated decreasing rate $(D + \theta)$ per unit time. So,

$$t_1 = \frac{Q-R}{D+\theta}. \tag{14}$$

Since the shortage length is t_2 and the maximum shortage level is R,

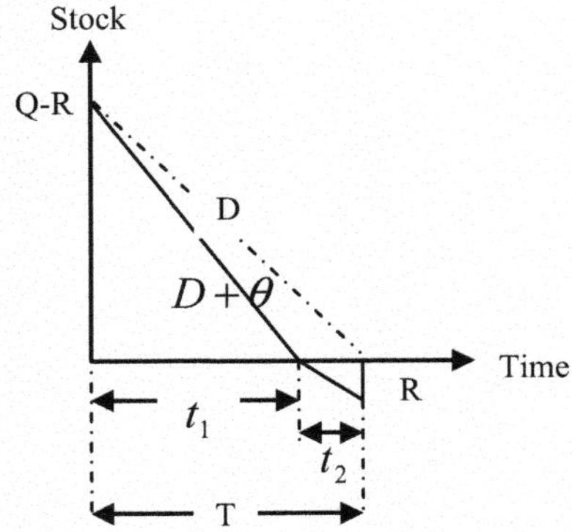

Fig. 2 Current stock amount versus time for full backorder situation

$$t_2 = \frac{R}{D}. \tag{15}$$

Therefore, the total cycle time of the retailers can be exposed by the following relations:

$$T = t_1 + t_2 = \frac{Q-R}{D+\theta} + \frac{R}{D} = \frac{DQ + R\theta}{D(D+\theta)}. \tag{16}$$

Here we have described inventory-related cost:

(a) *Ordering cost per year:*

$$\frac{A}{T} = \frac{AD(D+\theta)}{DQ + R\theta} \tag{17}$$

(b) *Holding cost per year:*

$$c_h \frac{\frac{1}{2}(Q-R)\frac{Q-R}{D+\theta}}{T} = c_h \frac{\frac{1}{2}(Q-R)\frac{Q-R}{D+\theta}}{\frac{DQ+R\theta}{D(D+\theta)}} = \frac{c_h D(Q-R)^2}{2(DQ + R\theta)} \tag{18}$$

(c) *Purchase cost per year:*

$$\frac{c_i Q}{T} = \frac{c_i QD(D+\theta)}{DQ + R\theta} \tag{19}$$

(d) *Shortage cost per year:*

$$c_b \frac{\frac{1}{2}.Rt_2}{T} = \frac{\frac{1}{2}.R\frac{R}{D}}{\frac{DQ+R\theta}{D(D+\theta)}} = \frac{c_b R^2(D+\theta)}{2(DQ + R\theta)} \tag{20}$$

Consequently, the total cost per year is

$$TC_i = \frac{AD(D+\theta)}{DQ+R\theta} + \frac{c_h D(Q-R)^2}{2(DQ+R\theta)} + \frac{c_i QD(D+\theta)}{DQ+R\theta} + \frac{c_b R^2(D+\theta)}{2(DQ+R\theta)}. \quad (21)$$

As far as, it may be written in the accompanying form in terms of p and Q:

$$\begin{aligned} TC_i(p,Q) = {} & \frac{A(m-pn)\{(m-pn)+\theta\}}{Q(m-pn)+R\theta} + \frac{c_h(m-pn)(Q-R)^2}{2\{Q(m-pn)+R\theta\}} \\ & + \frac{c_i Q(m-pn)\{(m-pn)+\theta\}}{\{Q(m-pn)+R\theta\}} + \frac{c_b R^2\{(m-pn)+\theta\}}{2\{Q(m-pn)+R\theta\}}. \end{aligned} \quad (22)$$

We have to deduct the net cost per year $TC_i(p,Q)$ from yearly income $p(m-pn)$ for attaining the profit in a year $TP_i(p,Q)$ (as the amount of goods sold per year is exactly equal to the demand per year $m-pn$ and the selling price per unit is p).

$$\begin{aligned} TP_i(p,Q) = p(m-pn) & - \frac{A(m-pn)\{(m-pn)+\theta\}}{Q(m-pn)+R\theta} - \frac{c_h(m-pn)(Q-R)^2}{2\{Q(m-pn)+R\theta\}} \\ & - \frac{c_i Q(m-pn)\{(m-pn)+\theta\}}{\{Q(m-pn)+R\theta\}} - \frac{c_b R^2\{(m-pn)+\theta\}}{2\{Q(m-pn)+R\theta\}}. \end{aligned} \quad (23)$$

For the important states of augmentation of the all-out benefit just as acquiring the ideal estimation of choice variable, set the main request fractional subordinates for the choice factors p and Q of $TP_i(p,Q)$ are equivalent to zero.

$$\text{i.e., } \frac{\partial TP_i(p,Q)}{\partial p} = 0$$

or

$$\left\{ \begin{aligned} & m - 2np - \frac{c_h n Q(m-pn)(Q-R)^2}{2\{Q(m-pn)+R\theta\}^2} - \frac{(c_i Q + A)nQ(m-pn)(m-pn+\theta)}{\{Q(m-pn)+R\theta\}^2} \\ & - \frac{c_b n Q R^2(m-pn+\theta)}{2\{Q(m-pn)+R\theta\}^2} + \frac{(c_i Q + A)n(m-pn)}{Q(m-pn)+R\theta} \\ & + \frac{c_h n(Q-R)^2 + c_b n R^2}{2\{Q(m-pn)+R\theta\}} + \frac{(c_i Q + A)n(m-pn+\theta)}{Q(m-pn)+R\theta} = 0 \end{aligned} \right. \quad (24)$$

and

$$\text{i.e., } \frac{\partial TP_i(p,Q)}{\partial Q} = 0$$

or

$$\begin{cases} \dfrac{c_h(m-pn)^2(Q-R)^2}{2\{Q(m-pn)+R\theta\}^2}+\dfrac{(c_iQ+A)(m-pn)^2(m-pn+\theta)}{\{Q(m-pn)+R\theta\}^2} \\ +\dfrac{c_b(m-pn)R^2(m-pn+\theta)}{2\{Q(m-pn)+R\theta\}^2}-\dfrac{c_h(m-pn)(Q-R)}{Q(m-pn)+R\theta} \\ -\dfrac{c_i(m-pn)(m-pn+\theta)}{Q(m-pn)+R\theta}=0 \end{cases} \tag{25}$$

4.2.1 Concavity of the Profit for Case II

To check the optimality of the profit (23), we calculate its Hessian matrix

$$H(p,Q)=\begin{bmatrix} \frac{\partial^2 TP_i}{\partial p^2} & \frac{\partial^2 TP_i}{\partial Q\partial p} \\ \frac{\partial^2 TP_i}{\partial p\partial Q} & \frac{\partial^2 TP_i}{\partial Q^2} \end{bmatrix}$$

.

Please see the detail calculations in Appendix 1.

Lemma 1 $TP_i(p,Q)$, has the maximum value for those values of p and T which satisfied the following equations

(i)

$$m-2np-\frac{c_hnQ(m-pn)(Q-R)^2}{2\{Q(m-pn)+R\theta\}^2}-\frac{(c_iQ+A)nQ(m-pn)(m-pn+\theta)}{\{Q(m-pn)+R\theta\}^2}$$
$$-\frac{c_bnQR^2(m-pn+\theta)}{2\{Q(m-pn)+R\theta\}^2}+\frac{(c_iQ+A)n(m-pn)}{Q(m-pn)+R\theta}$$
$$+\frac{c_hn(Q-R)^2+c_bbR^2}{2\{Q(m-pn)+R\theta\}}+\frac{(c_iQ+A)n(m-pn+\theta)}{Q(m-pn)+R\theta}=0$$

(ii)

$$\frac{c_h(m-pn)^2(Q-R)^2}{2\{Q(m-pn)+R\theta\}^2}+\frac{(c_iQ+A)(m-pn)^2(m-pn+\theta)}{\{Q(m-pn)+R\theta\}^2}$$
$$+\frac{c_b(m-pn)R^2(m-pn+\theta)}{2\{Q(m-pn)+R\theta\}^2}-\frac{c_h(m-pn)(Q-R)}{Q(m-pn)+R\theta}$$
$$-\frac{c_i(m-pn)(m-pn+\theta)}{Q(m-pn)+R\theta}=0$$

provided the following conditions are satisfied

(iii)

$$\frac{\partial^2 TP_i}{\partial p^2} = \left\{ \begin{array}{l} -2n - \frac{c_h n^2 Q^2 (m-pn)(Q-R)^2}{\{Q(m-pn)+R\theta\}^3} - \frac{2(c_i Q+A)n^2 Q^2 (m-pn)(m-pn+\theta)}{\{Q(m-pn)+R\theta\}^3} \\ -\frac{c_b n^2 Q^2 R^2 (m-pn+\theta)}{\{Q(m-pn)+R\theta\}^3} - \frac{2(c_i Q+A)n^2}{Q(m-pn)+R\theta} \\ +\frac{2(c_i Q+A)n^2 Q(m-pn)}{\{Q(m-pn)+R\theta\}^2} + \frac{n^2 Q\{c_h (Q-R)^2 + c_b R^2\}}{\{Q(m-pn)+R\theta\}^2} \\ +\frac{2(c_i Q+A)n^2 Q(m-pn+\theta)}{\{Q(m-pn)+R\theta\}^2} \end{array} \right\} < 0$$

i.e.,
$$\left\{ \begin{array}{l} \frac{2(c_i Q + A)n^2 Q(m - pn)}{\{Q(m - pn) + R\theta\}^2} + \frac{n^2 Q\{c_h (Q - R)^2 + c_b R^2\}}{\{Q(m - pn) + R\theta\}^2} \\ + \frac{2(c_i Q + A)n^2 Q(m - pn + \theta)}{\{Q(m - pn) + R\theta\}^2} \end{array} \right\}$$
$$> \left\{ \begin{array}{l} 2n + \frac{c_h n^2 Q^2 (m-pn)(Q-R)^2}{\{Q(m-pn)+R\theta\}^3} \\ + \frac{2(c_i Q+A)n^2 Q^2 (m-pn)(m-pn+\theta)}{\{Q(m-pn)+R\theta\}^3} \\ + \frac{c_b n^2 Q^2 R^2 (m-pn+\theta)}{\{Q(m-pn)+R\theta\}^3} + \frac{2(c_i Q+A)n^2}{Q(m-pn)+R\theta} \end{array} \right\}$$

(iv)

$$\frac{\partial^2 TP_i}{\partial Q^2} = \left\{ \begin{array}{l} -\frac{c_h (m-pn)^3 (Q-R)^2}{\{Q(m-pn)+R\theta\}^3} - \frac{2(c_i Q+A)(m-pn)^3 (m-pn+\theta)}{\{Q(m-pn)+R\theta\}^3} \\ -\frac{c_b R^2 (m-pn)^2 (m-pn+\theta)}{\{Q(m-pn)+R\theta\}^3} + \frac{2c_h (m-pn)^2 (Q-R)}{\{Q(m-pn)+R\theta\}^2} \\ +\frac{2c_i (m-pn)^2 (m-pn+\theta)}{\{Q(m-pn)+R\theta\}^2} - \frac{c_h (m-pn)}{Q(m-pn)+R\theta} \end{array} \right\} < 0$$

i.e.,
$$\left\{ \frac{2c_h (m - pn)^2 (Q - R)}{\{Q(m - pn) + R\theta\}^2} + \frac{2c_i (m - pn)^2 (m - pn + \theta)}{\{Q(m - pn) + R\theta\}^2} \right\}$$
$$> \left\{ \begin{array}{l} \frac{c_h (m - pn)^3 (Q - R)^2}{\{Q(m - pn) + R\theta\}^3} + \frac{2(c_i Q + A)(m - pn)^3 (m - pn + \theta)}{\{Q(m - pn) + R\theta\}^3} \\ + \frac{c_b R^2 (m - pn)^2 (m - pn + \theta)}{\{Q(m - pn) + R\theta\}^3} + \frac{c_h (m - pn)}{Q(m - pn) + R\theta} \end{array} \right\}$$

Proof Please see Appendix 2.

Theorem 1 If the objective functions satisfy the conditions (iii) and (iv) in Lemma 1, the following condition $\left(\frac{\partial^2 TP_i}{\partial p^2}\right)\left(\frac{\partial^2 TP_i}{\partial Q^2}\right) - \left(\frac{\partial^2 TP_i}{\partial p \partial Q}\right)^2 > 0$, then TP_i has the maximum value for the respective variables.

Proof Please see Appendix 3.

5 Solution Algorithm

If the supplier defines $n + 1$ quantity breaks to define n price discounts on purchase price per unit, in the following steps (below), the algorithm to achieve the best inventory policy would be summarized, where q_i, $i = 1, 2, ..., n + 1$ $(q_1 < q_2 <$

.... $< q_n < q_{n+1} = \infty$) are the quantity breaks which define the n per unit purchase price c_i, $i = 1, 2, ..., n$ $(c_1 > c_2 > .... > c_n)$. Additionally, using Mathematica-5.0 software, the values of p and Q are obtained from the Eqs. (8) and (9) for Case I while that of from Eqs. (24) and (25) for Case II. It is noted that (i) if the value of any one of p and Q is either negative or complex number, then that (p, Q) will be ignored (ii) if both the values of p and Q are positive but the objective function's value is negative, then that (p, Q) will be ignored (iii) if we get more than one set of (p, Q) for which the objective function's value is positive, then we will choose that (p, Q) for which the objective function's value is maximum. Now the algorithms for computing optimal price and inventory policy for cases I and II.

5.1 Algorithm to Achieve Optimal Price and Inventory Policy for Case I

Step 1: Initialize $TP_{\max}(p, Q) = 0$ and $i = n$

Step 2: For p and Q, solve Eqs. (8) and (9) by placing all known values $(A, m, n, \theta, c_h, B)$ and c_i

  i) When Q maintains the restriction $q_n \leq Q < q_{n+1}$, the solution, then, is achievable. Calculate $TP_i(p, Q)$ from (7) by utilizing the inferred estimations of p and Q and if $TP_i(p, Q) > TP_{\max}(p, Q)$, set $TP_{\max}(p, Q) = TP_i(p, Q)$. GOTO, Step 5
  ii) When Q fails to maintain the restriction $q_n \leq Q < q_{n+1}$, the solution breaks the feasibility and GOTO Step 3

Step 3: Substitute c_i, $Q = q_n$ and all the other given values $(A, m, n, \theta, c_h, B)$ into the Eq. (8) and solve for p. Calculate $TP_i(p, Q)$ from (7) by utilizing the inferred estimations of p and Q and if $TP_i(p, Q) > TP_{\max}(p, Q)$, set $TP_{\max}(p, Q) = TP_i(p, Q)$. GOTO, Step 4

Step 4: When $i \geq 2$, GOTO Step 2 modify i by $i - 1$; else, GOTO Step 5

Step 5: With p and Q, Calculate $TP_{\max}(p, Q)$

5.2 Algorithm to Achieve Optimal Price and Inventory Policy for Case II

Step 1: Initialize $TP_{\max}(p, Q) = 0$ and $i = n$

Step 2: For p and Q, solve Eqs. (24) and (25) by placing all known values $(A, m, n, \theta, c_h, B)$ and c_i

  i) When Q maintains the restriction $q_n \leq Q < q_{n+1}$, the solution, then, is achievable. Calculate $TP_i(p, Q)$ from (23) by utilizing the inferred estimations of p and Q and if $TP_i(p, Q) > TP_{\max}(p, Q)$, set $TP_{\max}(p, Q) = TP_i(p, Q)$. GOTO, Step 5

ii) When Q fails to maintain the restriction $q_n \leq Q < q_{n+1}$, the solution breaks the feasibility and GOTO Step 3

Step 3: Substitute c_i, $Q = q_n$ and all the other given values $(A, m, n, \theta, c_h, B)$ into the Eq. (24) and solve for p. Calculate $TP_i(p, Q)$ from (23) by utilizing the inferred estimations of p and Q and if $TP_i(p, Q) > TP_{\max}(p, Q)$, set $TP_{\max}(p, Q) = TP_i(p, Q)$. GOTO, Step 4

Step 4: When $i \geq 2$, GOTO Step 2 modify i by $i - 1$; else, GOTO Step 5

Step 5: With p and Q, Calculate $TP_{\max}(p, Q)$

To investigate the applicability of the algorithms, two numerical examples have solved in the next section.

6 Numerical Illustration

Example 1. For Case I

Let study the example deliberately,

Ordered quantity	$0 = q_1 \leq Q < q_2 = 200$	$200 = q_2 \leq Q < q_3 = 400$	$400 = q_3 \leq Q < q_4 = \infty$
Purchase price (per unit)	$c_1 = 5.5$	$c_2 = 5$	$c_3 = 4.5$

Along with $A = 220$, $m = 100$, $n = 1.5$, $c_h = 0.2$, $B = 10$, $\theta = 0.02$.

With the dint of Algorithm 5.1, the maximum total profit for this problem can be defined through the following steps:

Step 1: Initialize $TP_{\max}(p, Q) = 0$ and $i = 3$.

Iteration 1: $i = 3$.

Step 2: $c_3 = 4.5(400 \leq Q < \infty)$.

Utilizing (8) and (9) compute p and Q with $c_3 = 4.5$ and all the known values, one can obtain

$p = 35.96$, $Q = 359$ and from (7) $TP_3(p, Q) = 1377.07$. This is not a feasible solution because Q

does not have in the following range $(400 \leq Q < \infty)$ GOTO, Step3.

Step 3: $c_3 = 4.5$, $Q = 400$.

Utilizing (8) and (8) compute p by using the lowest cost $c_3 = 4.5$ for $Q = 400$, one can find $p = 35.92$.

From (7), $TP_3(p, Q) = 1376.65$. Since $TP_3(p, Q) = 1376.65 > 0 = TP_{\max}(p, Q)$, set

$TP_{\max}(p, Q) = 1376.65$. GOTO, Step 4.

Step 4: Update i by 2 and GOTO Step 2.

Iteration 2: For i $= 2$.

Step 2: $c_2 = 5(200 \leq Q < 400)$.

Plugging $c_2 = 5$ and the remaining known values into the Eqs. (8) and (9) for p

and Q, one can obtain $p = 36.22$, $Q = 361$ and $TP_2(p, Q) = 1353.47$ use Eq. (7). As Q

maintains the range $(200 \leq Q < 400)$, Consequently, this solution p, Q is feasible and

$TP_2(p, Q) = 1353.47 < TP_{max}(p, Q) = 1376.65$. Therefore, $TP_{max}(p, Q) = 1376.65$ which has

found in Step 3 of iteration 1. GOTO, Step 5.

Step 5: The best price and inventory policy is $p = 35.92, Q = 400, TP_3(p, Q) = 1376.65$.

Therefore, the optimal price and inventory policy are $Q = 400, p = 35.92, T = 8.45$, $TC_3(p, Q) = 279.98, TP_3(p, Q) = 1376.65$. The total profit per year versus selling price per unit and order size is presented in Fig. 3. The total profit versus the selling price per unit is supplied in Fig. 4, and the total profit versus the order size is presented in Fig. 5.

Example 2. For Case II

Let us deliberate the example

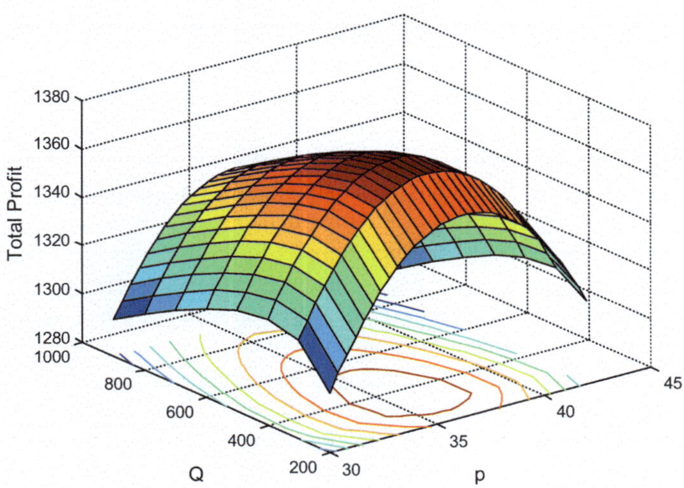

Fig. 3 Total profit versus selling price per unit and order size, for Example 1

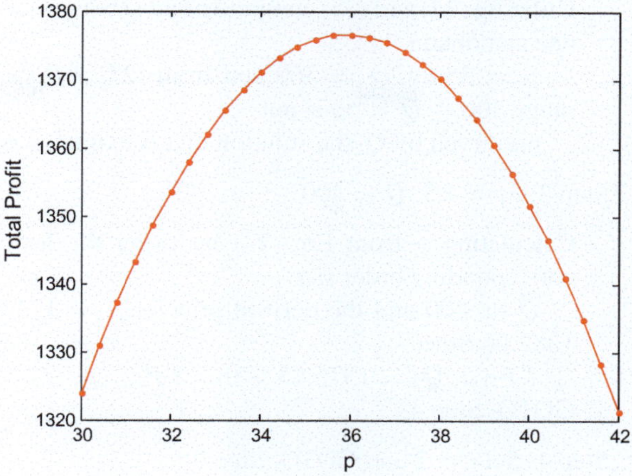

Fig. 4 Total profit versus selling price per unit, for Example 1

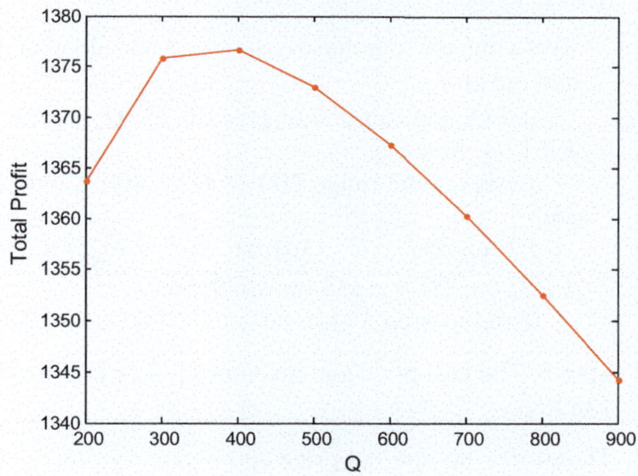

Fig. 5 Total profit versus order size, for Example 1

Ordered quantity	$0 = q_1 \leq Q < q_2 = 200$	$200 = q_2 \leq Q < q_3 = 400$	$400 = q_3 \leq Q < q_4 = \infty$
Purchase price (per unit)	$c_1 = 5.5$	$c_2 = 5$	$c_3 = 4.5$

Along with $A = 220$, $m = 100, n = 1.5$, $c_h = 0.2$, $c_b = 6$, $R = 30$, $\theta = 0.02$.

With the dint of Algorithm 5.2, the maximum total profit for this problem can be defined through the following steps:

Step 1: Initialize $TP_{\max}(p, Q) = 0$ and $i = 3$.

Iteration 1: $i = 3$.

Step 2: $c_3 = 4.5(400 \leq Q < \infty)$.

Utilizing (24) and (25) compute p and Q with $c_3 = 4.5$ and all the given values, one can obtain

$p = 35.89$, $Q = 360$ and from (23) $TP_3(p, Q) = 1383.06$. Since the range $400 \leq Q < \infty$ is not

preserved by Q, the solution that is extracted is not viable. GOTO, Step 3.

Step 3: $c_3 = 4.5$, $Q = 400$.

Calculating p from Eq. (24) by using the lowest cost $c_3 = 4.5$ and the corresponding order size

$Q = 400$ and the derived price is $p = 35.86$. From (23),$TP_3(p, Q) = 1382.66$.Since

$TP_3(p, Q) = 1376.65 > 0 = TP_{\max}(p, Q)$, set $TP_{\max}(p, Q) = 1382.66$. GOTO, Step 4.

Step 4: Set $i = 2$ and GOTO, Step 2.
Iteration 2: For $i = 2$.
Step 2: $c_2 = 5(200 \leq Q < 400)$.

Replacing unit purchase cost $c_2 = 5$ and all other given values into the Eqs. (24) and (25) for p

and Q, one can obtain $p = 36.14$, $Q = 359$ and from (23) $TP_2(p, Q) = 1361.94$. Since Q

preserves the range $200 \leq Q < 400$, Consequently, this solution p, Q is and

$TP_2(p, Q) = 1361.94 < TP_{\max}(p, Q) = 1382.66$. Therefore, $TP_{\max}(p, Q) = 1382.66$ which has

found in Step 3 of iteration 1. GOTO, Step 5.

Step 5: The best price and inventory policy is $p = 35.86$,$Q = 400$,$TP_3(p, Q) = 1382.66$.

Therefore, the optimal price and inventory policy are $Q = 400$,$p = 35.86$,$T = 8.65$, $TC_3(p, Q) = 274.43$,$TP_3(p, Q) = 1382.66$.

To cheek, the concavity of the total profit function (23), calculate its Hessian matrix $H(p, Q) = \begin{bmatrix} \frac{\partial^2 TP_i}{\partial p^2} & \frac{\partial^2 TP_i}{\partial Q \partial p} \\ \frac{\partial^2 TP_i}{\partial p \partial Q} & \frac{\partial^2 TP_i}{\partial Q^2} \end{bmatrix}$ at $(p, Q) = (35.86, 400)$. Hence,

$H(p, Q) = \begin{bmatrix} -3.0000033 & -0.00206228 \\ -0.0020623 & -0.0004049 \end{bmatrix}$. In the Hessian matrix, we observe that the first leading principal minor is strictly negative (–3.0000033) and the second leading principal minor is the determinant of $H(p, Q)$, i.e. $\begin{vmatrix} -3.0000033 & -0.00206228 \\ -0.0020623 & -0.0004049 \end{vmatrix} = 0.0012104 > 0$. Thus $TP_i(p, Q)$ is maximum at $(p, Q) = (35.86, 400)$. The total profit per year versus selling price per unit and order size is shown in Fig. 6. Furthermore, the total profit versus selling price per unit is shown in Fig. 7, and also the total profit versus the order size is shown in Fig. 8.

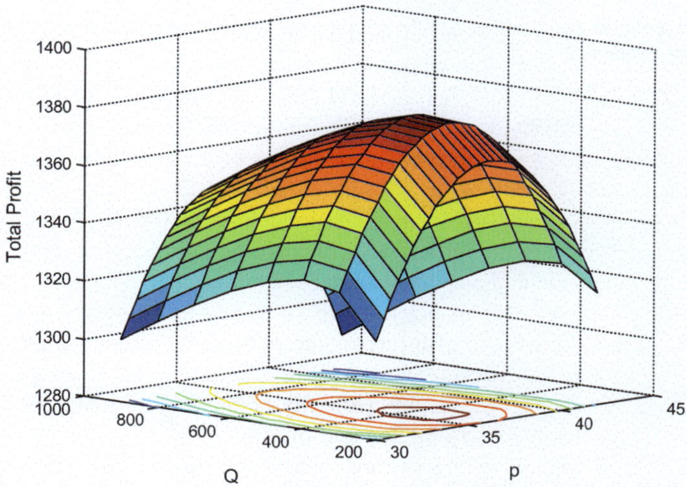

Fig. 6 Total profit versus selling price per unit and order size, for Example 2

Fig. 7 Total profit versus selling price per unit, for Example 2

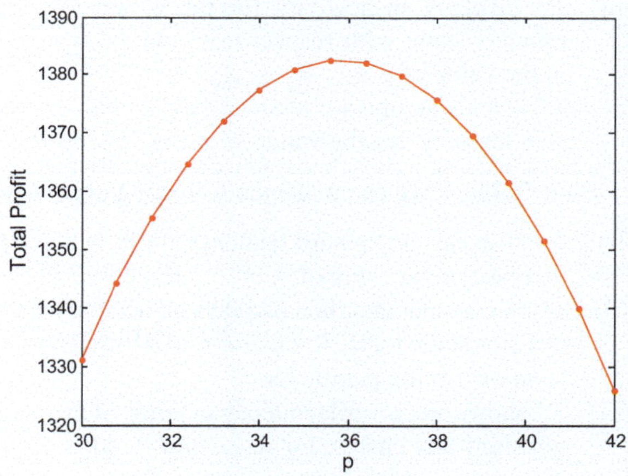

7 Sensitivity Inspection

The above examples are given in this section to explore the consequence of the known parameters on the solution quality sensitivity analysis.

The upcoming findings can be seen from Table 2 by the following:

(i) TP^* reveals an acclivity for the parameter m while TP^* reveals a declivity in regard to the values of A, n, c_h, B *and* θ.

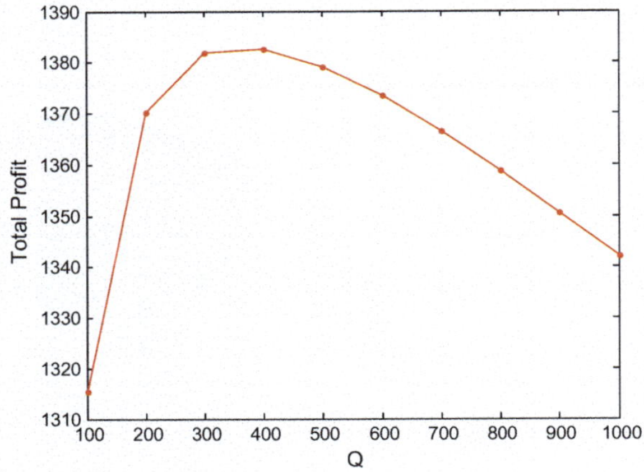

Fig. 8 Total profit versus order size, for Example 2

(ii) Q^* shows an upward tendency for the parameters A, m, however, Q^* remains static for n, B *and* θ while Q^* displays a downward tendency for the parameters c_h.

(iii) P^* displays an acclivity for the parameters A, m, c_h *and* B, however, P^* remains static with respect to θ and P^* reveals a declivity with an increase in the value of n.

(iv) T^* reveals an upward tendency for A *and* n, however, T^* shows a downward characteristic for the values of m, c_h, B *and* θ.

From Table 3, we can make the following observation which stated bellow:

(i) TP^* reveals an upward tendency for m while TP^* reveals a declivity with an increase in the value of A, n, c_h, R, c_b *and* θ.

(ii) Q^* is constant and then displays an acclivity for parameters A *and* m and Q^* remains static for n, R, c_b *and* θ. Additionally, Q^* decreases and then becomes level off for the parameter c_h.

(iii) P^* shows an upward tendency in terms of the parameters A, m, c_h and P^* does not show any change for R, c_b *and* θ. Also, P^* reveals a declivity for n.

(iv) T^* reveals an acclivity in terms of the parameters A *and* n, however, T^* upsurges and becomes constant for R. Again, T^* gives constant value for c_b while T^* displays a downward tendency for m, c_h *and* θ.

(v) t_1^* displays an acclivity for the parameters A *and* n while t_1^* remains static for the values of c_b. On the other hand, t_1^* exhibits a declivity for the parameters m, c_h, R *and* θ.

(vi) t_2^* reveals an upward inclination for A, n *and* R, however, t_2^* increases and thence becomes constant for c_h. Moreover, t_2^* reveals a downward inclination for m while t_2^* becomes constant and thence decreases for θ.

Table 2 Sensitivity Inspection for without ending a model (Case I)

Parameter	Original value	New values	TP_i^*	p^*	Q^*	T^*
A	220	132	1387.0780	35.8103	400	8.4225
		176	1381.8590	35.8667	400	8.4379
		264	1371.4490	35.9795	400	8.4689
		308	1366.3000	36.0214	413	8.7612
m	100	60	413.5720	22.5897	400	14.9223
		80	828.4437	29.2564	400	10.7928
		120	2058.1870	42.5897	400	6.9475
		140	2873.2700	49.2331	429	6.3284
n	1.5	0.9	2483.7360	58.1453	400	8.1779
		1.2	1791.3040	44.2564	400	8.3134
		1.8	1100.8830	30.3675	400	8.5982
		2.1	904.4825	26.3993	400	8.7480
c_h	0.2	0.12	1393.6320	35.8766	462	9.7769
		0.16	1384.8490	35.9223	401	8.4734
		0.24	1368.4490	35.9231	400	8.4534
		0.28	1360.2490	35.9231	400	8.4534
B	10	6	1379.4760	35.8968	400	8.5328
		8	1378.0680	35.9099	400	8.4931
		12	1375.2170	35.9364	400	8.4137
		14	1373.7740	35.9499	400	8.3740
θ	0.02	0.012	1376.6900	35.9231	400	8.4548
		0.016	1376.6700	35.9231	400	8.4541
		0.024	1376.6280	35.9231	400	8.4526
		0.03	1376.5970	35.9231	400	8.4515
c_1, c_2, c_3	5.5, 5, 4.5	3.3, 3, 2.7	1463.1000	35.0000	400	8.2071
		4.4, 4, 3.6	1419.5550	35.4615	400	8.3284
		6.6, 6, 5.4	1334.3820	36.3846	400	8.5822
		7.7, 7, 6.3	1292.7540	36.8462	400	8.7149

8 Conclusion

In this work, two inventory models for decaying items (i) non-zero ending inventory (ii) shortage situations, as well as price break situations of the purchase amount of the product, are studied. The problem is formulated for the retailer and solved by Mathematica software. Considering all the possibilities of the optimal purchase quantity under all-units concession scheme, two simple and efficacious algorithms are supplied to attain the best market price and stock amount for the retailer. The concavity of the objective functions is examined by dint of 3D plots. Also, two-line

Table 3 Sensitivity analysis for filly backlogged shortage (Case II)

Parameter	Original value	New values	TP_i^*	p^*	Q^*	t_1^*	t_2^*	T^*
A	220	132	1392.8490	35.7483	400	7.9746	0.6466	8.6214
		176	1387.7500	35.8033	400	7.9888	0.6477	8.6368
		264	1377.5790	35.9133	400	8.0173	0.6501	8.6677
		308	1372.5440	35.9570	412	8.2910	0.6510	8.9423
m	100	60	416.9922	22.5250	400	14.1046	1.1436	15.2491
		80	833.1597	29.1916	400	10.2118	0.8280	11.0403
		120	2065.49	42.5250	400	6.5798	0.5335	7.1135
		140	2881.7300	49.1807	417	5.8368	0.4528	6.2897
n	1.5	0.9	2489.9460	58.0805	400	7.7491	0.6283	8.3777
		1.2	1797.4140	44.1917	400	7.8740	0.6384	8.5127
		1.8	1106.7950	30.3028	400	8.1363	0.6597	8.7963
		2.1	910.2949	26.3345	400	8.2742	0.6709	8.9453
c_h	0.2	0.12	1396.9680	35.8202	464	9.3826	0.6481	10.0310
		0.16	1389.5060	35.8568	402	8.0503	0.6489	8.6994
		0.24	1375.8150	35.8583	400	8.0030	0.6489	8.6522
		0.28	1368.9710	35.8583	400	8.0030	0.6489	8.6522
R	30	18	1384.7220	35.8583	400	8.2626	0.3893	8.6521
		24	1383.9700	35.8583	400	8.1328	0.5191	8.6521
		36	1380.7920	35.8583	400	7.8732	0.7787	8.6523
		42	1378.3650	35.8583	400	7.7435	0.9085	8.6523
c_b	6	3.6	1385.3610	35.8583	400	8.0030	0.6489	8.6522
		4.8	1384.0110	35.8583	400	8.0030	0.6489	8.6522
		7.2	1381.3100	35.8583	400	8.0030	0.6489	8.6522
		8.4	1379.9590	35.8583	400	8.0030	0.6489	8.6522
θ	0.02	0.012	1382.6980	35.8583	400	8.0044	0.6490	8.6536
		0.016	1382.6790	35.8583	400	8.0037	0.6490	8.6529
		0.024	1382.6410	35.8583	400	8.0023	0.6488	8.6515
		0.03	1382.6130	35.8583	400	8.0013	0.6488	8.6505
c_1, c_2, c_3	5.5, 5, 4.5	3.3, 3, 2.7	1467.0910	34.9583	400	7.7760	0.6305	8.4067
		4.4, 4, 3.6	1424.5720	35.4083	400	7.8879	0.6396	8.5277
		6.6, 6, 5.4	1341.3560	36.3083	400	8.1216	0.6585	8.7804
		7.7, 7, 6.3	1300.6590	36.7583	400	8.2437	0.6684	8.9125

diagrams of price versus total profit and order size versus total profit are deliberated for each case. The executed sensitivity analysis reveals that the discount on the purchase price based on the ordered amount has an integral and indispensable role in enhancing not only the retailer's profit but also purchase quantity.

For future research, this model extends in several ways such as demand can be assumed as a powerful form of time and non-linear marketing price-dependent. One may also extend this model for the products with a maximum span of time with several credit starts.

Appendix 1

Differentiate (23) against p, one has

$$\frac{\partial TP_i(p,Q)}{\partial p}=\left\{\begin{array}{l} m-2np-\frac{c_h nQ(m-pn)(Q-R)^2}{2\{Q(m-pn)+R\theta\}^2}-\frac{(c_iQ+A)nQ(m-pn)(m-pn+\theta)}{\{Q(m-pn)+R\theta\}^2}\\ -\frac{c_b nQR^2(m-pn+\theta)}{2\{Q(m-pn)+R\theta\}^2}+\frac{(c_iQ+A)n(m-pn)}{Q(m-pn)+R\theta}\\ +\frac{c_h n(Q-R)^2+c_b nR^2}{2\{Q(m-pn)+R\theta\}}+\frac{(c_iQ+A)n(m-pn+\theta)}{Q(m-pn)+R\theta} \end{array}\right\} \tag{A1}$$

Differentiate of the Eq. (23) with respect to Q, one finds

$$\frac{\partial TP_i(p,Q)}{\partial Q}=\left\{\begin{array}{l} \frac{c_h(m-pn)^2(Q-R)^2}{2\{Q(m-pn)+R\theta\}^2}+\frac{(c_iQ+A)(m-pn)^2(m-pn+\theta)}{\{Q(m-pn)+R\theta\}^2}\\ +\frac{c_b(m-pn)R^2(m-pn+\theta)}{2\{Q(m-pn)+R\theta\}^2}-\frac{c_h(m-pn)(Q-R)}{Q(m-pn)+R\theta}\\ -\frac{c_i(m-pn)(m-pn+\theta)}{Q(m-pn)+R\theta} \end{array}\right\} \tag{A2}$$

Second-order partial derivatives against p and Q of $TP_i(p,Q)$ are

$$\frac{\partial^2 TP_i(p,Q)}{\partial p^2}=\left\{\begin{array}{l} -2n-\frac{c_h n^2Q^2(m-pn)(Q-R)^2}{\{Q(m-pn)+R\theta\}^3}\\ -\frac{2(c_iQ+A)n^2Q^2(m-pn)(m-pn+\theta)}{\{Q(m-pn)+R\theta\}^3}\\ -\frac{c_b n^2Q^2R^2(m-pn+\theta)}{\{Q(m-pn)+R\theta\}^3}+\frac{2(c_iQ+A)n^2Q(m-pn)}{\{Q(m-pn)+R\theta\}^2}\\ +\frac{n^2Q\{c_h(Q-R)^2+c_bR^2\}}{\{Q(m-pn)+R\theta\}^2}+\frac{2(c_iQ+A)n^2Q(m-pn+\theta)}{\{Q(m-pn)+R\theta\}^2}\\ -\frac{2(c_iQ+A)n^2}{Q(m-pn)+R\theta} \end{array}\right\} \tag{A3}$$

$$\frac{\partial^2 TP_i(p,Q)}{\partial Q^2}=\left\{\begin{array}{l} -\frac{c_h(m-pn)^3(Q-R)^2}{\{Q(m-pn)+R\theta\}^3}-\frac{2(c_iQ+A)(m-pn)^3(m-pn+\theta)}{\{Q(m-pn)+R\theta\}^3}\\ -\frac{c_bR^2(m-pn)^2(m-pn+\theta)}{\{Q(m-pn)+R\theta\}^3}+\frac{2c_h(m-pn)^2(Q-R)}{\{Q(m-pn)+R\theta\}^2}\\ +\frac{2c_i(m-pn)^2(m-pn+\theta)}{\{Q(m-pn)+R\theta\}^2}-\frac{c_h(m-pn)}{Q(m-pn)+R\theta} \end{array}\right\} \tag{A4}$$

$$\frac{\partial^2 TP_i(p, Q)}{\partial p \partial Q} = \left\{ \begin{array}{l} \frac{c_h n(m-pn)^2 Q(Q-R)^2}{\{Q(m-pn)+R\theta\}^3} + \frac{2n(c_i Q+A)Q(m-pn)^2(m-pn+\theta)}{\{Q(m-pn)+R\theta\}^3} \\ + \frac{c_b n Q R^2 (m-pn)(m-pn+\theta)}{\{Q(m-pn)+R\theta\}^3} - \frac{(c_i Q+A)n(m-pn)^2}{\{Q(m-pn)+R\theta\}^2} \\ - \frac{c_h n(m-pn)(Q-R)(2Q-R)}{\{Q(m-pn)+R\theta\}^2} - \frac{c_b n R^2(2m-2np+\theta)}{2\{Q(m-pn)+R\theta\}^2} \\ - \frac{n(3c_i Q+2A)Q(m-pn)(m-pn+\theta)}{\{Q(m-pn)+R\theta\}^2} + \frac{c_i n(2m-2np+\theta)}{Q(m-pn)+R\theta} \\ + \frac{c_h n(Q-R)}{Q(m-pn)+R\theta} \end{array} \right\} \tag{A5}$$

Appendix 2

For maximization of the objective function, one has

$$\frac{\partial^2 TP_i(p, Q)}{\partial p^2} = \left\{ \begin{array}{l} -2n - \frac{c_h n^2 Q^2 (m-pn)(Q-R)^2}{\{Q(m-pn)+R\theta\}^3} \\ - \frac{2(c_i Q+A)n^2 Q^2 (m-pn)(m-pn+\theta)}{\{Q(m-pn)+R\theta\}^3} \\ - \frac{c_b n^2 Q^2 R^2 (m-pn+\theta)}{\{Q(m-pn)+R\theta\}^3} + \frac{2(c_i Q+A)n^2 Q(m-pn)}{\{Q(m-pn)+R\theta\}^2} \\ + \frac{n^2 Q\{c_h(Q-R)^2 + c_b R^2\}}{\{Q(m-pn)+R\theta\}^2} \\ + \frac{2(c_i Q+A)n^2 Q(m-pn+\theta)}{\{Q(m-pn)+R\theta\}^2} - \frac{2(c_i Q+A)b^2}{Q(m-pn)+R\theta} \end{array} \right\} < 0$$

$$\text{i.e.,} \left\{ \begin{array}{l} \frac{2(c_i Q+A)n^2 Q(m-pn)}{\{Q(m-pn)+R\theta\}^2} + \frac{n^2 Q\{c_h(Q-R)^2 + c_b R^2\}}{\{Q(m-pn)+R\theta\}^2} \\ + \frac{2(c_i Q+A)n^2 Q(m-pn+\theta)}{\{Q(m-pn)+R\theta\}^2} \end{array} \right\}$$

$$> \left\{ \begin{array}{l} 2n + \frac{c_h n^2 Q^2 (m-pn)(Q-R)^2}{\{Q(m-pn)+R\theta\}^3} + \frac{2(c_i Q+A)n^2 Q^2 (m-pn)(m-pn+\theta)}{\{Q(m-pn)+R\theta\}^3} \\ + \frac{c_b n^2 Q^2 R^2 (m-pn+\theta)}{\{Q(m-pn)+R\theta\}^3} + \frac{2(c_i Q+A)n^2}{Q(m-pn)+R\theta} \end{array} \right\}$$

and

$$\frac{\partial^2 TP_i(p, Q)}{\partial Q^2} = \left\{ \begin{array}{l} -\frac{c_h(m-pn)^3(Q-R)^2}{\{Q(m-pn)+R\theta\}^3} - \frac{2(c_i Q+A)(m-pn)^3(m-pn+\theta)}{\{Q(m-pn)+R\theta\}^3} \\ - \frac{c_b R^2 (m-pn)^2 (m-pn+\theta)}{\{Q(m-pn)+R\theta\}^3} + \frac{2c_h(m-pn)^2(Q-R)}{\{Q(m-pn)+R\theta\}^2} \\ + \frac{2c_i(m-pn)^2(m-pn+\theta)}{\{Q(m-pn)+R\theta\}^2} - \frac{c_h(m-pn)}{Q(m-pn)+R\theta} \end{array} \right\} < 0$$

$$\text{i.e.,} \left\{ \frac{2c_h(m-pn)^2(Q-R)}{\{Q(m-pn)+R\theta\}^2} + \frac{2c_i(m-pn)^2(m-pn+\theta)}{\{Q(m-pn)+R\theta\}^2} \right\}$$

$$> \left\{ \begin{array}{l} \dfrac{c_h(m-pn)^3(Q-R)^2}{\{Q(m-pn)+R\theta\}^3} + \dfrac{2(c_iQ+A)(m-pn)^3(m-pn+\theta)}{\{Q(m-pn)+R\theta\}^3} \\ + \dfrac{c_bR^2(m-pn)^2(m-pn+\theta)}{\{Q(m-pn)+R\theta\}^3} + \dfrac{c_h(m-pn)}{Q(m-pn)+R\theta} \end{array} \right\}$$

Appendix 3

According to Lemma 1

$\frac{\partial^2 TP_i(p,Q)}{\partial p^2} < 0$ and $\frac{\partial^2 TP_i(p,Q)}{\partial Q^2} < 0$ then obviously the condition $\left(\frac{\partial^2 TP_i}{\partial p^2}\right)\left(\frac{\partial^2 TP_i}{\partial Q^2}\right) - \left(\frac{\partial^2 TP_i}{\partial p \partial Q}\right)^2 \geq 0$.

and the objective function gives maximum value due to the optimal value of p and Q.

References

Alfares HK (2015) Maximum-profit inventory model with stock-dependent demand, time-dependent holding cost, and all-units quantity discounts. Math Model Anal 20(6):715–736

Alfares HK, Ghaithan AM (2016) Inventory and pricing model with price-dependent demand, time-varying holding cost, and quantity discounts. Comput Ind Eng 94:170–177

Avinadav T, Herbon A, Spiegel U (2013) Optimal inventory policy for a perishable item with demand function sensitive to price and time. Int J Prod Econ 144(2):497–506

Baker RA, Urban TL (1988) A deterministic inventory system with an inventory-level-dependent demand rate. J Oper Res Soc 39(9):823–831

Bhunia A, Shaikh A (2011) A deterministic model for deteriorating items with displayed inventory level dependent demand rate incorporating marketing decisions with transportation cost. Int J Ind Eng Comput 2(3):547–562

Burwell TH, Dave DS, Fitzpatrick KE, Roy MR (1997) Economic lot size model for price-dependent demand under quantity and freight discounts. Int J Prod Econ 48(2):141–155

Chang CT, Teng JT, Goyal SK (2010) Optimal replenishment policies for non-instantaneous deteriorating items with stock-dependent demand. Int J Prod Econ 123(1):62–68

Covert RP, Philip GC (1973) An EOQ model for items with Weibull distribution deterioration. AIIE Trans 5(4):323–326

Chang WJ, Hung CH, Dye CY (2004) An inventory model for deteriorating items with stock-dependent demand and time-value of money when credit period is provided. J Inf Optim Sci 25(2):237–254

Dye CY, Ouyang LY (2005) An EOQ model for perishable items under stock-dependent selling rate and time-dependent partial backlogging. Eur J Oper Res 163(3):776–783

Das S, Khan MAA, Mahmoud EE, Abdel-Aty AH, Abualnaja KM, Shaikh AA (2021) A production inventory model with partial trade credit policy and reliability. Alex Eng J 60(1):1325–1338

Ghare PM (1963) A model for an exponentially decaying inventory. J Ind Engng 14:238–243

Giri BC, Goswami A, Chaudhuri KS (1996) An EOQ model for deteriorating items with time varying demand and costs. J Oper Res Soc 47(11):1398–1405

Giri BC, Chaudhuri KS (1998) Deterministic models of perishable inventory with stock-dependent demand rate and nonlinear holding cost. Eur J Oper Res 105(3):467–474
Goh M, Sharafali M (2002) Price-dependent inventory models with discount offers at random times. Prod Oper Manag 11(2):139–156
Harris FW (1913) How many parts to make at once factory. Mag Manag 10(2):135–136
Hu J, Munson CL (2002) Dynamic demand lot-sizing rules for incremental quantity discounts. J Oper Res Soc 53(8):855–863
Hu J, Munson CL, Silver EA (2004) A modified Silver-Meal heuristic for dynamic lot sizing under incremental quantity discounts. J Oper Res Soc 55(6):671–673
Khan MAA, Shaikh AA, Panda GC, Konstantaras I, Taleizadeh AA (2019a) Inventory system with expiration date: pricing and replenishment decisions. Comput Ind Eng 132:232–247
Khan MAA, Shaikh AA, Panda GC, Konstantaras I (2019b) Two-warehouse inventory model for deteriorating items with partial backlogging and advance payment scheme. RAIRO-Oper Res 53(5):1691–1708
Khan MAA, Ahmed S, Babu MS, Sultana N (2020) Optimal lot-size decision for deteriorating items with price-sensitive demand, linearly time-dependent holding cost under all-units discount environment. Int J Syst Sci Oper Logist:1–14
Khan MAA, Shaikh AA, Panda GC, Konstantaras I, Cárdenas-Barrón LE (2020b) The effect of advance payment with discount facility on supply decisions of deteriorating products whose demand is both price and stock dependent. Int Trans Oper Res 27(3):1343–1367
Khan MAA, Shaikh AA, Panda GC, Bhunia AK, Konstantaras I (2020) Non-instantaneous deterioration effect in ordering decisions for a two-warehouse inventory system under advance payment and backlogging. Ann Oper Res: 1–33
Khan MAA, Shaikh AA, Konstantaras I, Bhunia AK, Cárdenas-Barrón LE (2020) Inventory models for perishable items with advanced payment, linearly time-dependent holding cost and demand dependent on advertisement and selling price. Int J Prod Econ 230:107804
Khan MAA, Shaikh AA, Cárdenas-Barrón LE (2021) An inventory model under linked-to-order hybrid partial advance payment, partial credit policy, all-units discount and partial backlogging with capacity constraint. Omega: 102418
Kotler P (1972) Marketing decision making: a model building approach
Ladany S, Sternlieb A (1974) The interaction of economic ordering quantities and marketing policies. AIIE Trans 6(1):35–40
Lee YP, Dye CY (2012) An inventory model for deteriorating items under stock-dependent demand and controllable deterioration rate. Comput Ind Eng 63(2):474–482
Li SX, Huang Z, Ashley A (1996) Inventory, channel coordination and bargaining in a manufacturer-retailer system. Ann Oper Res 68(1):47–60
Mandal BA, Phaujdar S (1989) An inventory model for deteriorating items and stock-dependent consumption rate. J Oper Res Soc 40(5):483–488
Mashud A, Khan M, Uddin M, Islam M (2018) A non-instantaneous inventory model having different deterioration rates with stock and price dependent demand under partially backlogged shortages. Uncertain Supply Chain Manag 6(1):49–64
Meher MK, Panda GC, Sahu SK (2012) An inventory model with Weibull deterioration rate under the delay in payment in demand decling market. Appl Math Sci 6(23):1121–1133
Mendoza A, Ventura JA (2008) Incorporating quantity discounts to the EOQ model with transportation costs. Int J Prod Econ 113(2):754–765
Padmanabhan G, Vrat P (1995) EOQ models for perishable items under stock dependent selling rate. Eur J Oper Res 86(2):281–292
Pal S, Goswami A, Chaudhuri KS (1993) A deterministic inventory model for deteriorating items with stock-dependent demand rate. Int J Prod Econ 32(3):291–299
Sheikh SR, Patel R (2017) Two-warehouse inventory model with different deterioration rates under time dependent demand and shortages. Glob J Pure Appl Math 13(8):3951–3960
Panda GC, Sahoo S, Sukla PK (2012) Analysis of constant deteriorating inventory management with quadratic demand rate. Am J Oper Res 2(6):98–103

Panda GC, Sukla PK (2013) Production-inventory management model for a Weibull deteriorating item with linear demand and shortages. LogForum 9(2)
Panda GC, Sahoo S, Sukla PK (2013) A note on inventory model for ameliorating items with time dependent second order demand rate. LogForum 9(1)
Panda GC, Khan MAA, Shaikh AA (2019) A credit policy approach in a two-warehouse inventory model for deteriorating items with price-and stock-dependent demand under partial backlogging. J Indus Eng Int 15(1):147–170
Rahman MS, Khan MAA, Halim MA, Nofal TA, Shaikh AA, Mahmoud EE (2021) Hybrid price and stock dependent inventory model for perishable goods with advance payment related discount facilities under preservation technology. Alex Eng J 60(3):3455–3465
Ray J, Chaudhuri KS (1997) An EOQ model with stock-dependent demand, shortage, inflation and time discounting. Int J Prod Econ 53(2):171–180
Ray J, Goswami A, Chaudhuri KS (1998) On an inventory model with two levels of storage and stock-dependent demand rate. Int J Syst Sci 29(3):249–254
Sahu S, Panda G, Das AK (2017) A fully backlogged deteriorating inventory model with price dependent demand using preservation technology investment and trade credit policy. Int J Eng Tech Res 6(6):851–858
San-José LA, García-Laguna J (2009) Optimal policy for an inventory system with backlogging and all-units discounts: application to the composite lot size model. Eur J Oper Res 192(3):808–823
Sarker BR, Mukherjee S, Balan CV (1997) An order-level lot size inventory model with inventory-level dependent demand and deterioration. Int J Prod Econ 48(3):227–236
Shaikh AA, Mashud AHM, Uddin MS, Khan MAA (2017) Non-instantaneous deterioration inventory model with price and stock dependent demand for fully backlogged shortages under inflation. Int J Bus Forecast Mark Intell 3(2):152–164
Shaikh AA, Khan MAA, Panda GC, Konstantaras I (2019a) Price discount facility in an EOQ model for deteriorating items with stock-dependent demand and partial backlogging. Int Trans Oper Res 26(4):1365–1395
Shaikh AA, Das SC, Bhunia AK, Panda GC, Khan MAA (2019b) A two-warehouse EOQ model with interval-valued inventory cost and advance payment for deteriorating item under particle swarm optimization. Soft Comput 23(24):13531–13546
Shaikh AA, Panda GC, Sahu S, Das AK (2019c) Economic order quantity model for deteriorating item with preservation technology in time dependent demand with partial backlogging and trade credit. Int J Logist Syst Manag 32(1):1–24
Shaikh AA, Panda GC, Khan MAA, Mashud AHM, Biswas A (2020) An inventory model for deteriorating items with preservation facility of ramp type demand and trade credit. Int J Math Oper Res 17(4):514–551
Singh SP, Panda GC (2015) An inventory model for generalized weibull deteriorating items with price dependent demand and permissible delay in payments under inflation. LogForum 11(3)
Taleizadeh AA, Pentico DW, Jabalameli MS, Aryanezhad M (2013) An EOQ model with partial delayed payment and partial backordering. Omega 41(2):354–368
Teng JT, Chang CT, Goyal SK (2005) Optimal pricing and ordering policy under permissible delay in payments. Int J Prod Econ 97(2):121–129
Urban TL (1992) Deterministic inventory models incorporating marketing decisions. Comput Ind Eng 22(1):85–93
Valliathal M, Uthayakumar R (2011) Simple approach of obtaining the optimal pricing and lot-sizing policies for an EPQ model on deteriorating items with shortages under inflation and time-discounting. İstanbul Üniversitesi İşletme Fakültesi Dergisi 40(2):304–320
Wee HM (1998) Optimal buyer-seller discount pricing and ordering policy for deteriorating items. Eng Econ 43(2):151–168
Weng ZK (1995) Modeling quantity discounts under general price-sensitive demand functions: optimal policies and relationships. Eur J Oper Res 86(2):300–314

Wu KS, Ouyang LY, Yang CT (2006) An optimal replenishment policy for non-instantaneous deteriorating items with stock-dependent demand and partial backlogging. Int J Prod Econ 101(2):369–384

Chapter 11
Intuitionistic Fuzzy Inventory Model with Pre-payment Scheme for Damageable Item

Puja Supakar, Sanat Kumar Mahato, and Pintu Pal

Abstract The purpose of this chapter is to present a model for maximizing retail profit through an economic order quantity inventory model. First, we derive the deterministic model for the damageable item under the assumption that advance payment is allowed by the distributor to increase the sales of the items in lots. Since every parameter involves some sort of uncertainty, we developed imprecise model and crispified model in commensurate with the deterministic one and represent this impreciseness in terms of linear triangular intuitionistic fuzzy numbers and crispified these with a well-known signed distance method. We solved both the deterministic and crispified models and results are compared for the numerical experiments. To interpret managerial insights, we carried out the sensitivity studies of the controlling parameters for the crispified model.

Keywords Pre-payment · Damageable item · Intuitionistic fuzzy number · EOQ model · Crispification · Signed distance method

1 Introduction

In the present market scenario, it is commonly believed that large stockpiles of the products that are kept for display in the mall, shopping centre or street market or the amassed stress of the stocked products will attract the customers to buy more. Due to this storage the products break or damage or this can reduce the actual utility after some time period.

Majumder et al. (2000) have established an imperfect production model with time-varying demand for multi-items. Saha et al. (2010) have proposed a stock dependent EOQ models for breakable products. Rahman et al. (2021) have also evolved a stock and price dependent demand for deteriorating EOQ model. Maragatham and

P. Supakar · S. K. Mahato (✉)
Department of Mathematics, Sidho-Kanho-Birsha University, Purulia, W.B 723104, India

P. Pal
Department of Computer Application, Asansol Engineering College, Paschim Burdwan, W.B 713305, India

I. Ali et al. (eds.), *Computational Modelling in Industry 4.0*,
https://doi.org/10.1007/978-981-16-7723-6_11

Gnanvel (2017) have presented a model for purchasing breakable inventory items with allowable delay in payments. Kundu et al. (2018) have discussed a production model for breakable and imperfect products.

The effect of pre-payment financing scheme is observed as an everyday life. Also, it is one of the effective business tools to encourage the direct customers or to confirm the order or to start new business for the small vendors. It is not possible to arrange full amount at the time of order or supply for the beginners or small vendors. The vendor or customer receives fixed percentage of the total amount as pre-payment during the order and the remainder is adjusted after the delivery. It is further noted that the seller offers a certain percentage discount on the total price depending on the amount purchased. The discount rate may be fixed or it is influenced by the amount of pre-payment, the time period when pre-payment is given and many other factors. However, the customers or small vendors who are not able to pay the pre-payment part of the amount immediately, have to take a loan from a bank or some agency with some rate of interest. In this case, if they are benefited after paying the loan interest, then only they choose the pre-payment financing scheme. Gupta et al. (2009) have suggested a pre-payment EOQ model with uniform demand. Priyan et al. (2014) have expanded a pre-payment EOQ model including fuzzy uncertainties. Zhang et al. (2014) have also presented EOQ model incorporating pre-payment financing scheme. For perishable goods, Tsao et al. (2019) have proposed an optimal pricing and ordering policy using the advance-cash-credit financing scheme. Teng et al. (2016) have devised a deteriorating EOQ model under advance payments scheme. Supakar and Mahato (2018) have discussed a fuzzy-stochastic advance payment EOQ model with linear time varying demand. Khan et al. (2020) have evolved a two-warehouse inventory system with pre-payment financing and demand component based on selling price. Khan et al. (2021) have extended the previous research into an EOQ model considering advance and delay in payments.

Many researchers have considered the parameters to be precise, i.e. every parameter is perfectly determinable. However, in real-life situations, due to insufficient information, lack of evidence, fluctuating financial market, weather change, etc. parameters may not be a fixed value always which can be expressed by several aspects such as interval-valued, fuzzy, intuitionistic fuzzy, stochastic, or combination of these.

Saha et al. (2012) have established a breakable EOQ models with imprecise environment. Supakar and Mahato (2020) have established a pentagonal fuzzy pre-payment EPQ model with time varying deterioration and demand. Chakrabortty et al. (2011) have established intuitionistic fuzzy EOQ model and IF programming technique for solution.

1.1 Identification of Research Gaps

After careful literature review, we have summarized the research findings of some renowned researchers related to this work and presented in a table given below. From this table, the research gaps can easily be viewed. Also, the attempts made in this work are shown in this table from which it can be easily distinguished.

References	Model	Item type	Payment scheme	Demand rate	Uncertainty
Majumder et al. (2000)	EPQ	Imperfect	–	Time varying	–
Chiang et al. (2005)	–	–	–	–	Fuzzy
Gupta et al. (2009)	EOQ	–	Advance	Uniform	Interval valued
Saha et al. (2010)	EOQ	Breakable	–	Stock dependent	Trapezoidal Fuzzy
Chakrabortty et al. (2011)	EOQ	–	–	Uniform	Intuitionistic fuzzy
Saha et al. (2012)	EOQ	Breakable	–	Stock and selling price-dependent	Fuzzy cost and resources
Priyan et al. (2014)	EOQ	–	Advance	Constant	Trapezoidal and Triangular fuzzy
Zhang et al. (2014)	EOQ	–	Advance		–
Teng et al. (2016)	EOQ	Deteriorating	Advance	Constant	–
Maragatham and Gnanvel (2017)	Purchasing Inventory Model	Breakable	Permissible delay	Linear time-dependent	–
Supakar and Mahato (2018)	EOQ	–	Advance	Linear time-dependent	Fuzzy-Stochastic
Kundu et al. (2018)	EPQ	Breakable Item, Imperfect Production	–	Time sensitive	Triangular fuzzy
Tsao et al. (2019)	EPQ	Deterioration	Advance	Price-dependent	-
Supakar and Mahato (2020)	EPQ	Deteriorating	Advance	Ramp type	Pentagonal fuzzy Parameters
Khan et al. (2020)	two-warehouse inventory system	Deterioration	Advance	Selling price-dependent	–

(continued)

(continued)

References	Model	Item type	Payment scheme	Demand rate	Uncertainty
Khan et al. (2021)	EOQ	–	Advance and delay payment	–	–
Rahman et al. (2021)	EOQ	Deterioration	Advance	Price and stock-dependent	–
This chapter	EOQ	Damageable	Advance	Selling price-dependent	Intuitionistic fuzzy

1.2 Objectives of This Chapter

The objectives of this work are:

i. to develop realistic models and then find the maximum profit from this model.
ii. to handle the uncertainty from the real data with suitable intuitionistic fuzzy number.
iii. to make the computation easy, use crispified method to convert the intuitionistic fuzzy numbers to a fixed real number.
iv. to present actual market demand, that is not always constant.
v. to consider damageable units in the model, which is a common fact because of the amassed stress of the stocked products kept in heaped or displayed in the mall, supermarket.

1.3 Outcomes of the Chapter

The outcomes of the present work are given below

i. Designing of an EOQ model in crisp and intuitionistic fuzzy environments involving pre-payment scheme, selling price-dependent demand for damageable item.
ii. The output value of the profit function for crispified input data is better than the crisp input data.
iii. To compute intuitionistic fuzzy model is very hard. So we convert the intuitionistic fuzzy model to crispified model using signed distance method.
iv. The profit function is increasing when the parameters $\tilde{t}_c$, $\tilde{o}_c$ and n_0 are increasing and $\tilde{P}_c$ and α are decreasing.
v. The profit function does not behave uniformly with respect to the parameters $\tilde{h}_c$ and t_1 for whole cycle length.

Schematic diagram of proposed study

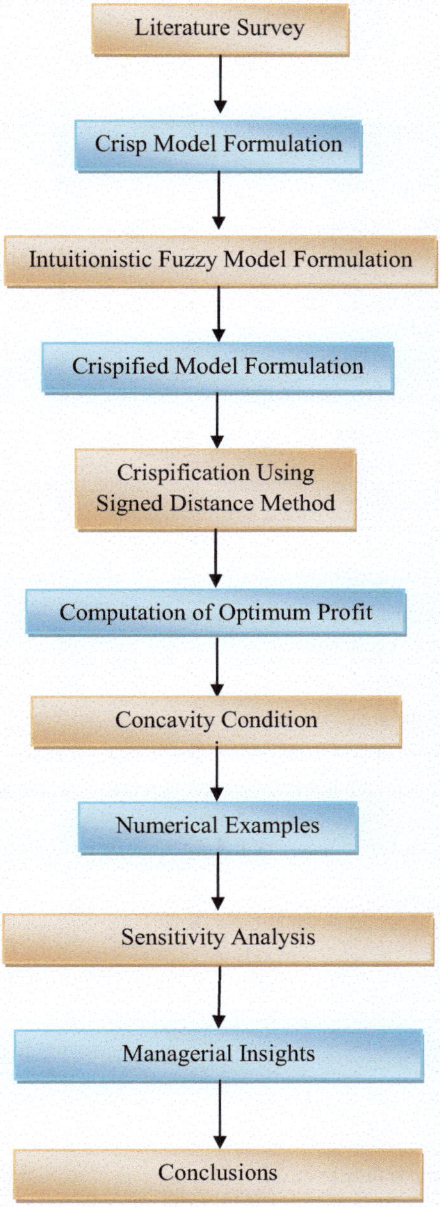

2 Nomenclatures and Assumptions

2.1 Nomenclatures

Symbol	Meaning	Unit
O_c	Ordering cost per order	\$
$\tilde{O}_c$	Intuitionistic fuzzy ordering cost per order	\$
$d_{avg}(\tilde{O}_c, 0)$	Crispified ordering cost per order	\$
P_c	Purchasing cost per unit item	\$
$\tilde{P}_c$	Intuitionistic fuzzy purchasing cost per unit item	\$
$d_{avg}(\tilde{P}_c, 0)$	Crispified purchasing cost per unit item	\$
t_c	Transportation cost per unit item	\$
$\tilde{t}_c$	Intuitionistic fuzzy transportation cost per unit item	\$
$d_{avg}(\tilde{t}_c, 0)$	Crispified transportation cost per unit item	\$
h_c	Holding cost per order	\$
$\tilde{h}_c$	Intuitionistic fuzzy holding cost per order	\$
$d_{avg}(\tilde{h}_c, 0)$	Crispified holding cost per order	\$
n_0	The mark up of selling price for damaged item	year
I_e	Interest charged per annum	\$
$\tilde{I}_e$	Intuitionistic fuzzy interest charged per annum	\$
$d_{avg}(\tilde{I}_e, 0)$	Crispified interest charged per annum	\$
t_1	The length of time during which the pre-payments are paid,$0 \leq t_1 \leq 1$	Year
t_2	The length of time after that system face to damageable item, $t_2 > t_1$	Year
α	Prior to the delivery date, the amount of purchase costs to be prepaid, $0 \leq \alpha \leq 1$	Unitless
N	The number of equal pre-payments before receiving the order quantity	Unitless

Decision variables

S_p	Selling price per unit item	*\$*
T	Length of cycle time	*year*

Functions

$D(S_p)$	Annual demand rate depends on unit selling price S_p
$Q(t)$	Inventory level in units at time t
$TP(T, S_p)$	Average of total profit

(continued)

(continued)

$B(q)$	The number of damaged units per unit of time and is based on initial inventory level q

2.2 Assumptions

To formulate the crisp and the intuitionistic fuzzy EOQ model, the following assumptions are under consideration:

i. There is a single item under consideration.
ii. Pre-payment is allowed with n equal instalments.
iii. The products break or damage after some time period. The number of damaged units per unit of time is a function of initial inventory level q, i.e. $B(q) = a_1 q^r T$, a_1 is a constant.
iv. The demand rate is selling price-dependent, i.e. $D(S_p) = a - bS_p$, $a > bS_p$, a and b are constants.
v. Intuitionistic fuzzy numbers are used to express the uncertain parameters for the imprecise model.
vi. Transportation cost is applicable.
vii. Time horizon is finite.
viii. Finite numbers of replenishments are allowed.

3 Some Important Definitions

3.1 Intuitionistic Fuzzy Number (IFN)

An intuitionistic fuzzy number $\tilde{A}$ with the membership function is $\sigma_{\tilde{A}}(x)$ and non-membership function $\delta_{\tilde{A}}(x)$ is:

(i). An intuitionistic fuzzy subset of the real line.
(ii). Normal, i.e., there is a $x_0 \in X$ such that $\sigma_{\tilde{A}}(x_0) = 1$ and $\delta_{\tilde{A}}(x_0) = 0$.
(iii). Convex for the membership function, i.e., for all $x_1, x_2 \in X$, $\sigma_{\tilde{A}}(\lambda x_1 + (1 - \lambda)x_2) \geq \min\{\sigma_{\tilde{A}}(x_1), \sigma_{\tilde{A}}(x_2)\}$, where $\lambda \in [0, 1]$.
(iv). Concave for non-membership function $x_1, x_2 \in X$, $\delta_{\tilde{A}}(\lambda x_1 + (1 - \lambda)x_2) \leq \max\{\delta_{\tilde{A}}(x_1), \delta_{\tilde{A}}(x_2)\}$, where $\lambda \in [0, 1]$ (Fig. 1).

3.2 Triangular Intuitionistic Fuzzy Number (TIFN)

A triangular intuitionistic fuzzy number (TIFN) $\tilde{A}$ is an intuitionistic fuzzy set in R with following membership function ($\sigma_{\tilde{A}}(x)$) and non-membership function ($\delta_{\tilde{A}}(x)$)

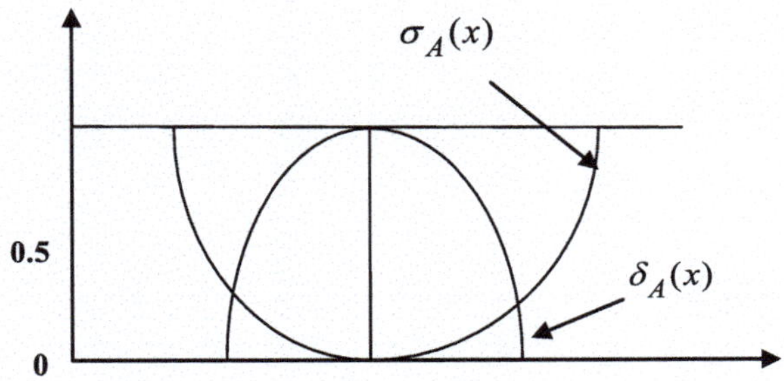

Fig. 1 Membership and Non-membership functions of IFN

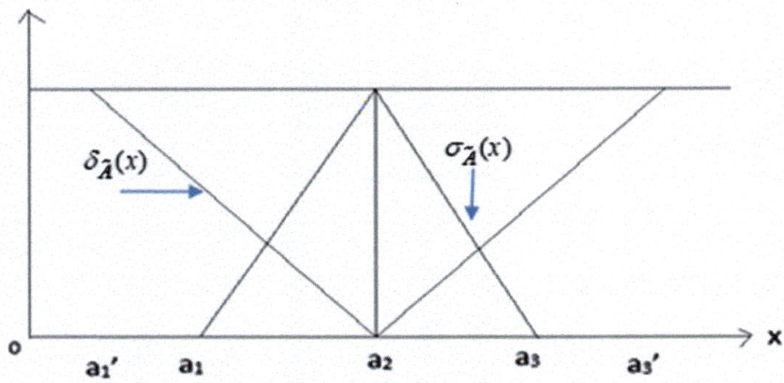

Fig. 2 Graphical representation of TIFN

$$\sigma_{\tilde{A}}(x) = \begin{cases} \dfrac{x - a_1}{a_2 - a_1}, \text{ for } a_1 \le x \le a_2 \\ \dfrac{a_3 - x}{a_3 - a_2}, \text{ for } a_2 \le x \le a_3 \\ 0, \text{ otherwise} \end{cases} \text{ and } \delta_{\tilde{A}}(x) = \begin{cases} \dfrac{a_2 - x}{a_2 - a_1'}, \text{ for } a_1' \le x \le a_2 \\ \dfrac{x - a_2}{a_3' - a_2}, \text{ for } a_2 \le x \le a_3' \\ 1, \text{ otherwise} \end{cases}$$

where, $a_1' \le a_1 \le a_2 \le a_3 \le a_3'$ and $0 \le \sigma_{\tilde{A}}(x), \delta_{\tilde{A}}(x) \le 1$, for $\sigma_{\tilde{A}}(x) = 1 - \delta_{\tilde{A}}(x), \forall x \in R$ (Fig. 2).

This TIFN is denoted by $\tilde{A} = (a_1, a_2, a_3; a_1', a_2, a_3')$.

4 Signed Distance Method of Crispification (Chiang et al. 2005)

Suppose $x, 0 \in R$, then $d_0(x, 0) = x$ is stated as the signed distance of x determined from the origin 0. If $x > 0$, the distance between x and 0 is $d_0(x, 0) = x$. Accordingly, if $x < 0$, the distance between x and 0 is $-d_0(x, 0) = -x$. Due to this, $d_0(x, 0) = x$ is defined as the signed distance of x to 0.

Therefore, the signed distance of $A_1(\alpha)$ and $A_2(\alpha)$ determined from 0 are $d_0(A_1(\alpha), 0) = A_1(\alpha)$ and $d_0(A_2(\alpha), 0) = A_2(\alpha)$, respectively.

Similarly, signed distance of $A'_1(\beta)$ and $A'_2(\beta)$ determined from 0 are $d_0(A'_1(\beta), 0) = A'_1(\beta)$ and $d_0(A'_2(\beta), 0) = A'_2(\beta)$.

Thus, the signed distance of the interval $[A_1(\alpha), A_2(\alpha)]$ determined from origin 0 is given by

$$d_0([A_1(\alpha), A_2(\alpha)], 0) = \frac{1}{2}[d_0(A_1(\alpha), 0) + d_0(A_2(\alpha), 0)] = \frac{1}{2}[A_1(\alpha) + A_2(\alpha)].$$

Similarly, the signed distance of the interval $[A'_1(\beta), A'_2(\beta)]$ determined from origin 0 is given by

$$d_0([A'_1(\beta), A'_2(\beta)], 0) = \frac{1}{2}[d_0(A'_1(\beta), 0) + d_0(A'_2(\beta), 0)] = \frac{1}{2}[A'_1(\beta) + A'_2(\beta)].$$

Signed distance of the TIFN $\tilde{A} = (a_1, a_2, a_3; a'_1, a_2, a'_3)$ determined from origin 0 with respect to the membership function is

$$d_\sigma(A, 0) = \frac{1}{2}\int_0^1 [A_1(\alpha) + A_2(\alpha)]d\alpha$$
$$= \frac{1}{2}\int_0^1 a_1 + \alpha(a_2 - a_1) + a_3 - \alpha(a_3 - a_2)d\alpha$$
$$= \frac{1}{4}(a_1 + 2a_2 + a_3)$$

Again, the signed distance of the TIFN $\tilde{\mathrm{A}}$ determined from origin 0 with respect to the non-membership function is

$$d_\delta(A, 0) = \frac{1}{2}\int_0^1 [A'_1(\beta) + A'_2(\beta)]d\beta$$
$$= \frac{1}{2}\int_0^1 a_2 - \beta(a_2 - a'_1) + a_2 + \beta(a'_3 - a_2)d\beta$$
$$= \frac{1}{4}(a'_1 + 2a_2 + a'_3)$$

Therefore, the average signed distance of the TIFN $\tilde{A}$ measured from origin 0 is given by

$$d_{\text{avg}}(A, 0) = \frac{d_\sigma(A, 0) + d_\delta(A, 0)}{2} = \frac{1}{8}(a_1 + 4a_2 + a_3 + a'_1 + a'_3).$$

5 Model Formulation

A single item EOQ model has been developed for damageable item in crisp and intuitionistic fuzzy environments considering allowable pre-payment and selling price-dependent demand. To make the computation easier we have converted the intuitionistic fuzzy model by using very well-known signed distance method.

5.1 Crisp Model

Let us assume that there is no breaking/damage within the period $0 \leq t \leq t_2$. After, that damage is started and continued throughout the whole cycle. If $Q(t)$ indicates the inventory level at any instant t, thus the EOQ model can be represented by the differential equation stated hereunder:

$$\begin{aligned} \frac{dQ(t)}{dt} &= -D(S_p),\ 0 \leq t \leq t_2 \\ \frac{dQ(t)}{dt} &= -D(S_p) - B(q),\ t_2 \leq t \leq T \end{aligned} \tag{1}$$

with the boundary conditions $Q(T) = 0,\ Q(0) = q$.

By solving the differential Eq. (1), we can obtain,

$$Q(t) = \begin{Bmatrix} q - D(S_p)t, 0 \leq t \leq t_2 \\ (D(S_p) + B(q))(T - t),\ t_2 \leq t \leq T \end{Bmatrix} \tag{2}$$

Since $Q(t)$ is continuous at $t = t_2$, utilizing the result of Eq. (2) we get,

$$t_2 = \frac{(D(S_p) + B(q))T - q}{B(q)} \tag{3}$$

Each cycle of profit is comprised of the following components: ordering cost, purchasing cost, holding cost, selling price of fresh item, selling price of for damageable item and interest charged for pre-payment, transportation cost. Evaluation of components is done in the following way:

(a) Total ordering cost for the cycle is: $OC = O_c$
(b) Total purchasing cost for the replenishment cycle is: $A_c = P_c q$.
(c) Total holding cost for the cycle length is:

$$H_c = h_c\left(\int_0^{t_2} Q(t)dt + \int_{t_2}^{T} Q(t)dt\right)$$
$$= h_c\left(qt_2 - \frac{D(S_p)t_2^2}{2} - \frac{(D(S_p)+B(q))(T-t_2)^2}{2}\right)$$

(d) Total selling price:

$$SP = S_p\left(\int_0^{t_2} Q(t)dt + \int_{t_2}^{T} Q(t)dt\right)$$
$$= S_p\left(qt_2 - \frac{D(S_p)t_2^2}{2} - \frac{(D(S_p)+B(q))(T-t_2)^2}{2}\right).$$

(e) Total number of damageable units is:
$\int_{t_2}^{T} B(q)dt = B(q)(T - t_2)$.
Selling price for each damaged item is: $n_0 P_c$, $0 \le n_0 \le 1$.
Total selling price for damageable item is: $D_c = B(q)(T - t_2)n_0 P_c$.

(f) Total transportation cost for the cycle length: $TC = t_c q$.

(g) Fig. 3 shows the interest that is assessed for pre-payment for the cycle as follows:

$$IC = I_e\left(\frac{\alpha A_c}{n}\left(\frac{t_1}{n}\right)(1+2+3+\ldots+n)\right) = \frac{I_e\alpha t_1(1+n)}{2n}A_c = \frac{P_c q I_e \alpha t_1(1+n)}{2n}$$

From the above arguments, the annual total average profit incurred to the retailer is:

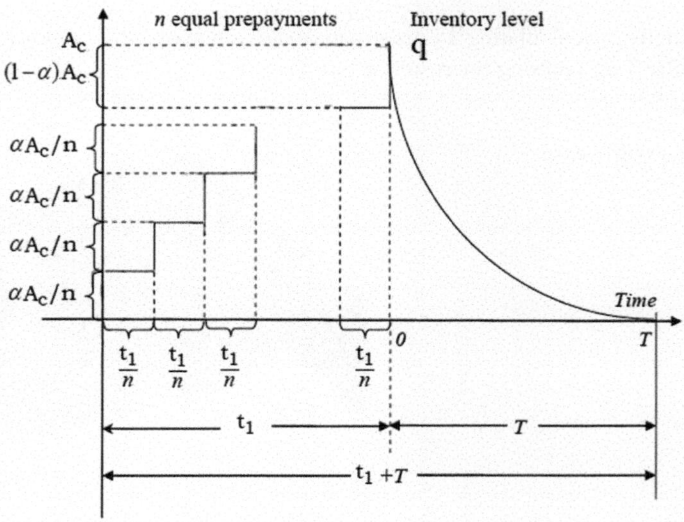

Fig. 3 Diagrammatic representation of the model

$$TP(T,S_p)=\frac{SP+D_c-OC-A_c-H_c-TC-IC}{T}$$
$$=\frac{S_p}{T}\left(qt_2-\frac{D(S_p)t_2^2}{2}-\frac{(D(S_p)+B(q))(T-t_2)^2}{2}\right)+\frac{B(q)(T-t_2)n_0P_c}{T}-\frac{O_c}{T}-\frac{P_cq}{T}- \quad (4)$$
$$\frac{h_c}{T}\left(qt_2-\frac{D(S_p)t_2^2}{2}-\frac{(D(S_p)+B(q))(T-t_2)^2}{2}\right)-\frac{t_cq}{T}-\frac{I_e\alpha t_1(1+n)}{2n}\frac{P_cq}{T}.$$

5.2 Concavity Analysis

From the Fig. 4 it is clearly visible that the objectives function, i.e. total average profit is a concave function of selling price and cycle length. Hence, we can have unique global optimum of total average profit.

To find T^*, for any particular $S_p > 0$, calculating the 1st order partial derivative of $TP(T, S_p)$, w.r.t T, putting the result at zero gives us,

$$\frac{\partial TP}{\partial T}=-\frac{1}{T^2}\left[\begin{array}{l}(S_p-h_c)\left(qt_2-\frac{D(S_p)t_2^2}{2}-\frac{(D(S_p)+B(q))(T-t_2)^2}{2}\right)+\\ B(q)(T-t_2)n_0P_c-O_c-P_cq-t_cq-\frac{I_e\alpha t_1(1+n)P_cq}{2n}\end{array}\right]+$$
$$\frac{1}{T}\left[\begin{array}{l}(S_p-h_c)\left(qt_2'-D(S_p)t_2t_2'-(D(S_p)+B(q))(T-t_2)(1-t_2')-B'(q)(T-t_2)^2\right)\\ +B(q)(1-t_2')n_0P_c+B'(q)(T-t_2)n_0P_c\end{array}\right]=0$$

Additionally, by calculating 1st order partial derivative of $TP(T, S_p)$ w.r.t. S_p, for any particular T and putting the result at zero gives us,

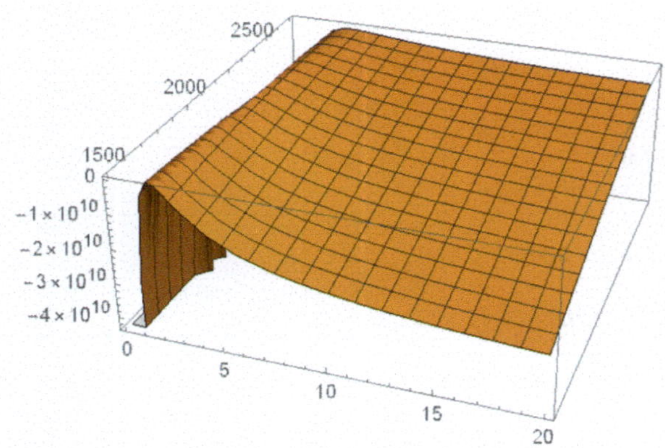

Fig. 4 Profit graph of crisp model

$$\frac{\partial TP}{\partial S_p} = \frac{1}{T}\left[\begin{array}{l}\left(qt_2 - \frac{D(S_p)t_2^2}{2} - \frac{(D(S_p)+B(q))(T-t_2)^2}{2}\right) + S_p\left(-\frac{D'(S_p)t_2^2}{2} - \frac{D'(S_p)(T-t_2)^2}{2}\right) + \\ h_c\left(-\frac{D'(S_p)t_2^2}{2} - \frac{D'(S_p)(T-t_2)^2}{2}\right)\end{array}\right] = 0$$

$$f_1(T) = S_p\left(qt_2 - \frac{D(S_p)t_2^2}{2} - \frac{(D(S_p)+B(q))(T-t_2)^2}{2}\right) + B(q)(T-t_2)n_0P_c - O_c - P_cq - \\ h_c\left(qt_2 - \frac{D(S_p)t_2^2}{2} - \frac{(D(S_p)+B(q))(T-t_2)^2}{2}\right) - t_cq - \frac{I_e\alpha t_1(1+n)P_cq}{2n} \tag{5}$$

$$g_1(T) = T \tag{6}$$

Considering the 1st order and 2nd order derivatives of $f_1(T)$ w.r.t T,

$$f_1'(T) = (S_p - h_c)\left(qt_2' - D(S_p)t_2t_2' - (D(S_p)+B(q))(T-t_2)(1-t_2') - B'(q)(T-t_2)^2\right) \\ +B(q)(1-t_2')n_0P_c + B'(q)(T-t_2)n_0P_c \tag{7}$$

$$f\prime_2(T) = (S_p - h_c)\left(\begin{array}{l} qt\prime_2 - D(S_p)(t_2'^2 + Tt\prime_2) + B(q)(T-t_2)t\prime_2 \\ +(B(q)+B'(q))(T-t_2)(1-t'_2)\end{array}\right) - \\ B(q)t\prime_2n_0P_c + 2B'(q)(1-t'_2)n_0P_c \tag{8}$$

$$= (S_p - h_c)\left(qt\prime_2 - D(S_p)(t_2'^2 + Tt\prime_2) + (B(q)+B'(q))(T-t_2)(1-t'_2)\right) + \\ B(q)\left((S_p-h_c)(T-t_2) - n_0P_c\right)t\prime_2 + 2B'(q)(1-t'_2)n_0P_c$$

Here,

$$t_2 = \frac{(D(S_p)+B(q))T - q}{B(q)} = \frac{D(S_p)}{a_1q^r} + T - \frac{q}{a_1q^rT} \tag{9}$$

$$1 - t'_2 = -\frac{q}{a_1q^rT^2} < 0,\ t\prime\prime_2 = -\frac{2q}{a_1q^rT^3} < 0,\ (t_2'^2 + Tt\prime\prime_2) = 1 + \frac{q^2}{a_1q^rT^4} > 0,\ T > t_2 \tag{10}$$

It can be shown after rearranging the terms that $f_1''(T) < 0$ for suitable choice of parameters, which has been experimented numerically. Therefore, $TP(T, S_p) = \frac{f_1(T)}{g_1(T)}$ is a strictly pseudo-concave in nature w.r.t T.

5.3 *Optimality Analysis*

For any fixed T, taking the 2nd order partial derivatives of $TP(S_p, T)$ w.r.t. S_p, we get

$$\frac{\partial^2 TP}{\partial S_p^2} = \frac{1}{T}\left[\left(\frac{bt_2^2}{2} - \frac{(-b + B(q))(T - t_2)^2}{2}\right) + (S_p - h_c)\left(\frac{bt_2^2}{2} + \frac{b(T - t_2)^2}{2}\right)\right] \tag{11}$$

For any fixed S_p, taking the 2nd order partial derivatives of $TP(S_p, T)$ w.r.t. T, we get

$$\begin{aligned}\frac{\partial^2 TP}{\partial T^2} &= \frac{2}{T^3}\left[\begin{array}{l}(S_p - h_c)\left(qt_2 - \frac{D(S_p)t_2^2}{2} - \frac{(D(S_p) + B(q))(T - t_2)^2}{2}\right) + \\ B(q)(T - t_2)n_0P_c - O_c - P_cq - t_cq - \frac{I_e\alpha t_1(1 + n)P_cq}{2n}\end{array}\right] + \\ &\frac{1}{T}\left[\begin{array}{l}(S_p - h_c)\left(qt''_2 - D(S_p)(t'^2_2 + Tt''_2) + (B(q) + B'(q))(T - t_2)(1 - t'_2)\right) + \\ B(q)\left((S_p - h_c)(T - t_2) - n_0P_c\right)t''_2 + 2B'(q)(1 - t'_2)n_0P_c\end{array}\right] - \\ &\frac{1}{T^2}\left[\begin{array}{l}(S_p - h_c)\left(qt'_2 - D(S_p)t_2t'_2 - (D(S_p) + B(q))(T - t_2)(1 - t'_2) - B'(q)(T - t_2)^2\right) \\ +B(q)(1 - t'_2)n_0P_c + B'(q)(T - t_2)n_0P_c\end{array}\right]\end{aligned} \tag{12}$$

Again,

$$\begin{aligned}\frac{\partial^2 TP}{dS_pdT} &= -\frac{1}{T^2}\left[\begin{array}{l}\left(qt_2 - \frac{D(S_p)t_2^2}{2} - \frac{(D(S_p) + B(q))(T - t_2)^2}{2}\right) + \\ S_p\left(-\frac{D'(S_p)t_2^2}{2} - \frac{D'(S_p)(T - t_2)^2}{2}\right) - \\ h_c\left(-\frac{D'(S_p)t_2^2}{2} - \frac{D'(S_p)(T - t_2)^2}{2}\right)\end{array}\right] + \\ &\frac{1}{T}\left[\begin{array}{l}\left(qt'_2 - D(S_p)t_2t'_2 - (D(S_p) + B(q))(T - t_2)(1 - t'_2) - \frac{B'(q)(T - t_2)^2}{2}\right) \\ +(S_p - h_c)\left(-D'(S_p)t_2t'_2 - D'(S_p)(T - t_2)(1 - t'_2)\right)\end{array}\right]\end{aligned} \tag{13}$$

Now, computing the value below we have,

$$\frac{\partial^2 TP}{\partial S_p^2}\frac{\partial^2 TP}{\partial T^2} - \left(\frac{\partial^2 TP}{\partial T \partial S_p}\right)^2 = 9.44387 \times 10^9 > 0, \ \frac{\partial^2 TP}{\partial T^2} = -2.91676 \times 10^9 < 0. \tag{14}$$

The condition of maximum holds for the objective function.

5.4 *Intuitionistic Fuzzy Model*

Considering the parameters $\tilde{P}_c$, $\tilde{t}_c$, $\tilde{h}_c$, $\tilde{O}_c$, $\tilde{I}_e$ as the triangular intuitionistic fuzzy numbers, the annual total intuitionistic fuzzy average profit incurred to the retailer is:

$$\tilde{T}P(T, S_p) = \frac{S_p}{T}\left(qt_2 - \frac{D(S_p)t_2^2}{2} - \frac{(D(S_p)+B(q))(T-t_2)^2}{2}\right) + \frac{B(q)(T-t_2)n_0\tilde{P}_c}{T} - \frac{\tilde{O}_c}{T} - \frac{\tilde{P}_c q}{T} - \frac{\tilde{h}_c}{T}\left(qt_2 - \frac{D(S_p)t_2^2}{2} - \frac{(D(S_p)+B(q))(T-t_2)^2}{2}\right) - \tilde{t}_c q - \frac{\tilde{I}_e \alpha t_1(1+n)}{2n}\frac{\tilde{P}_c q}{T} \tag{15}$$

5.5 *Crispified Model*

Now, in order to crispify the intuitionistic fuzzy model given in (15), we used the method described in Sect. 4. The annual total crispified average profit incurred at the retailer is:

$$d_{avg}(\tilde{T}P(T, S_p), 0) = \frac{S_p}{T}\left(qt_2 - \frac{D(S_p)t_2^2}{2} - \frac{(D(S_p)+B(q))(T-t_2)^2}{2}\right) + \frac{B(q)(T-t_2)n_0 d_{avg}(\tilde{P}_c, 0)}{T} - \frac{d_{avg}(\tilde{O}_c, 0)}{T} - \frac{\tilde{P}_c q}{T} - \frac{d_{avg}(\tilde{h}_c, 0)}{T}\left(qt_2 - \frac{D(S_p)t_2^2}{2} - \frac{(D(S_p)+B(q))(T-t_2)^2}{2}\right) - d_{avg}(\tilde{t}_c, 0)q - \frac{d_{avg}(\tilde{I}_e, 0)\alpha t_1(1+n)}{2n}\frac{d_{avg}(\tilde{P}_c, 0)q}{T} \tag{16}$$

6 Numerical Examples and Result Analysis

6.1 *Numerical Examples*

Our aim in this section is to demonstrate theoretical results numerically using Mathematica and obtain the global optimal solution. We have formed two numerical examples corresponding to the crisp and intuitionistic fuzzy environments and are given in Example 1 and Example 2 (Table 1).

Table 1 Intuitionistic fuzzy input data

Parameters	Values	Crispified values
$\tilde{P}_c$	(1200,1400,1600; 1300,1400,1700)	1425
$\tilde{t}_c$	(0.05,0.25,0.30; 0.10,0.25,0.4)	0.2300
$\tilde{h}_c$	(0.10,0.12,0.14; 0.11,0.12,0.15)	0.1225
$\tilde{o}_c$	(11,13,14; 12,13,15)	13.000
$\tilde{I}_e$	(0.01,0.025,0.035; 0.015,0.025,0.04)	0.02500

Table 2 Output data

Model	T^*	S_p^*	TP^*
Crisp	2.24419	2218.10	43,536.30
Intuitionistic Fuzzy	2.29722	2259.77	50,179.40

Example 1: Crisp model

$O_c = \$10/\text{order},\ P_c = \$1500/\text{unit},\ h_c = \$0.1/\text{unit/year},\ I_e = 0.02,\ a = 100,\ b = 0.025,\ \alpha = 0.04,$
$n = 10,\ q = 100\ \text{unit},\ n_0 = 0.02\ \text{year},\ r = 0.02,\ a_1 = 0.10,\ t_1 = 0.5\ \text{year},\ t_c = \$0.2/\text{unit/year}.$

Example 2: Intuitionistic Fuzzy Model and Crispified Model

6.2 *Result Analysis*

From Table 3 we can see that:

i. TP^* is a decreasing function of $\tilde{P}_c$ and α but increasing function of $\tilde{t}_c$, $\tilde{o}_c$ and n_0.
ii. T^* is an increasing function of n_0, $\tilde{t}_c$ and $\tilde{o}_c$.
iii. S_p^* is an increasing function of $\tilde{t}_c$, $\tilde{o}_c$.
iv. When $\tilde{h}_c$ changes from 0.10 to 0.12, T^*, S_p^* and TP^* all are increasing, whereas, when $\tilde{h}_c$ changes from 0.14 to 0.16 they are decreasing.
v. When t_1 changes from 0.3 to 0.4 and 0.6 to 0.7, T^*, S_p^* and TP^* all are decreasing, whereas, when t_1 changes from 0.4 to 0.6 they are increasing.
vi. When $\tilde{P}_c$ changes from 1350 to 1450, T^* and S_p^* are decreasing but when it changes from 1450 to 1500 they are increasing.
vii. When α changes from 0.01 to 0.03, T^* increases but when it changes from 0.03 to 0.07, T^* decreases.

Table 3 Sensitivity analysis on parameters of crispified model

Parameters	Parameters change	T^*	S_p^*	TP^*
$\tilde{P}_c$	1350	2.39447	2329.97	59,617.70
	1400	2.21551	2194.84	46,327.20
	1450	2.20382	2185.42	43,000.60
	1500	2.20556	2186.85	40,834.60
$\tilde{t}_c$	0.21	2.19428	2177.38	43,699.60
	0.22	2.22948	2206.12	46,214.80
	0.24	2.24263	2216.90	46,659.10
	0.25	2.34362	2294.20	51,905.10
$\tilde{h}_c$	0.10	2.25689	2228.17	47,650.00
	0.12	2.28051	2246.62	49,004.90
	0.14	2.24817	2220.95	47,566.30
	0.16	2.23487	2210.50	46,508.80
$\tilde{o}_c$	11	2.19428	2177.38	43,699.60
	12	2.22948	2206.12	46,514.10
	14	2.24263	2216.90	46,659.10
	15	2.34362	2294.20	51,905.10
α	0.01	2.25819	2228.85	48,253.10
	0.03	2.27256	2210.50	48,139.50
	0.05	2.24238	2216.78	46,456.10
	0.07	2.2298	2206.60	45,869.20
n_0	6	2.18896	2172.73	43,485.60
	7	2.21585	2195.35	44,811.60
	8	2.26192	2231.94	48,301.10
	9	2.35121	2299.30	53,679.30
t_1	0.3	2.29436	2257.37	496,430.80
	0.4	2.20405	2185.49	44,346.80
	0.6	2.38139	2321.31	54,110.10
	0.7	2.20550	2187.04	43,630.03

viii. When α changes from 0.01 to 0.03 and 0.05 to 0.07, S_p^* decreases but when it changes from 0.03 to 0.05, S_p^* increases.

ix. When n_0 changes from 6 to 7 and 8 to 9, S_p^* increases but when it changes from 7 to 8, S_p^* decreases.

7 Managerial Insights

In this intuitionistic fuzzy EOQ model, we have incorporated a pre-payment financing scheme with equal instalments based on the selling price-dependent demand rate and initial stock dependent damageable units. Cycle length and selling price play a significant role in decision-making. Here, we are primarily focused on maximizing the average profit. By observing the several characteristics stated below, the average profit can be increased in various companies. Hence, a manager can easily include some essential components for optimizing his investments to earn maximum profit.

(i) The demand rate is considered as selling price-dependent, it is more realistic rather than taking it as constant. The manager will identify the actual profit by taking the selling price-dependent demand rate.
(ii) Here, we considered that the products break or damage after some time period. In real-life situation large stocks of the products displayed in the mall, supermarket, or street market to attract the customers. This type of storage can reduce the actual utility, glamour, and brightness of the products after some time period.
(iii) The manager can commence the pre-payment policy to recover any financial crisis.
(iv) To handle the impreciseness of the parameters, the organization can adopt the parameters as fuzzy, intuitionistic fuzzy, type-2 fuzzy, neutrosophic fuzzy, stochastic or combination of these.
(v) The manager can reduce the parameters $\tilde{P}_c$ and α and increase the parameters $\tilde{t}_c$, $\tilde{o}_c$ and n_0 to reach the goal to achieve maximum profit of the inventory system keeping other parameters same.

8 Conclusions and Future Scopes

The purpose of this chapter is to examine an inventory model that includes initial stock levels dependent damageable units with the policy of pre-payment with n equal instalments before receiving the order. Demand rates are based on selling prices, and damageability rates are based on the initial stock level to go through real-life scenarios. We first design the crisp model, followed by formulating its intuitionistic fuzzy counterpart with the intuitionistic parameter values. Our crispified model then allowed us to reduce computation time by converting intuitionistic fuzzy parameters to crisp values using the signed distance method. After solving the crisp and crispified model, the sensitivities of the parameters are tabulated. Additionally, we have produced managerial insights of our work so that the business managers can easily recognize the business benefits from our studies. Researchers can apply the concepts discussed here to other inventory models and can see many more prospects for future work in inventory modelling. During the analysis and to find out the solutions, we have used Mathematica 11.2. Nevertheless, one can also use some soft computing algorithms, such as the GA, PSO, ACO, ABC, or an extension of these algorithms, or MATLAB, etc. By using real data, researchers can manage the impreciseness of parameters in stochastic, neutrosophic fuzzy, type-2 fuzzy or any other fuzzy number.

Acknowledgements The authors are thankful to the anonymous Reviewers, Editor, and Editorial Board members for considering our work for publication.

References

Chakrabortty S, Pal M, Nayak PK (2011) Intuitionistic fuzzy optimization technique for the solution of an EOQ model. In: Fifteenth Int. Conf. on IFSs, Burgas, 2011

Chiang J, Yao JS, Lee HM (2005) Fuzzy inventory with backorder defuzzification by signed distance method. J Inf Sci Eng 21:673–694

Gupta RK, Bhunia AK, Goyal SK (2009) An application of genetic algorithm in solving an inventory model with advance payment and interval valued inventory cost. Math Comput Modell 49:893–905

Khan A Md, Shaikh AA, Panda GC, Bhunia AK, Konstantaras I (2020) Non-instantaneous deterioration effect in ordering decisions for a two-warehouse inventory system under advance payment and backlogging. Ann Oper Res 289:243–275.https://doi.org/10.1007/s10479-020-03568-x

Khan A Md, Shaikh AA, Cárdenas-Barrón LE (2021) An inventory model under linked-to-order hybrid partial advance payment, partial credit policy, all-units discount and partial backlogging with capacity constraint. Omega. https://doi.org/10.1016/j.omega.2021.102418

Kundu A, Guchhait P, Panigrahi G, Maiti M (2018) An EPQ model for Breakable Item under Imperfect Production Process. Ann Pure Appl Math 16:487–499. https://doi.org/10.22457/apam.v16n2a27

Majumder P, Bera UK, Maiti M (2000) A multi-item EPQ model with imperfect production process for time varying demand with shortages. AMS Classif

Maragatham M, Gnanvel G (2017) A Purchasing Inventory Model for Breakable items with Permissible Delay in Payments and Price Discount. Ann Pure Appl Math 15:305–314. https://doi.org/10.22457/apam.v15n2a17

Priyan S, Palanivel M, Uthayankumar R (2014) Mathematical modeling for EOQ inventory system with advance payment and fuzzy parameters. Int J Oper Manage 1:260–278

Rahman MS, Khan MA, Halim MA, Nofal TA, Shaikh AA, Mahmoud EE (2021) Hybrid price and stock dependent inventory model for perishable goods with advance payment related discount facilities under preservation technology. Alexandria Eng J 60:3455–3465

Saha A, Roy A, Kar S, Maiti M (2010) Inventory models for breakable items with stock dependent demand and imprecise constraints. Math Comput Model 52:1771–1782

Saha A, Roy A, Kar S, Maiti M (2012) Multi-item two storage inventory models for breakable items with fuzzy cost and resources based on different defuzzification techniques. Opsearch. https://doi.org/10.1007/s12597-012-0072-7

Supakar P, Mahato S K (2018) Fuzzy-stochastic advance payment inventory model having no shortage and with uniform demand using ABC algorithm. Int J Appl Comput Math 4.https://doi.org/10.1007/s40819-018-0539-1

Supakar P, Mahato S K (2020) An EPQ model with time proportion deterioration and ramp type demand under different payment schemes with fuzzy uncertainties. Int. J. Syst. Sci.: Oper. Logist. https://doi.org/10.1080/23302674.2020.1826594

Teng J, Cárdenas-Barrón LE, Chang H, Wu J, Hu Y (2016) Inventory lot-size policies for deteriorating items with expiration dates and advance payments. Appl Math Modell. https://doi.org/10.1016/j.apm.2016.05.022

Tsao YC, Fitriana Restu Putri RK, Zhang C, Linh V (2019) Optimal pricing and ordering policies for perishable products under advance-cash-credit payment scheme. J Ind Eng Int 15:131–146. https://doi.org/10.1007/s40092-019-00325-z
Zhang Q, Tsao YC, Chen TH (2014) Economic order quantity under advance payment. Appl Math Modell 38:5910–5921

Chapter 12
Impact of Blockchain on Retail Supply Chain

Piyusha Nayyar and Pratibha Garg

Abstract Initially blockchain technology was viewed as a synonym for crypto currency 'Bitcoin'. But blockchain has unique qualities which make it a potential disruptive technology that can replace many existing technologies and services in the industry. Blockchain has drawn attention of various industries and has potential to be utilized in manufacturing, healthcare, consumer goods, Government and Public Sector, food industry, and many more, apart from financial space. One of the game-changing applications of blockchain can be in the retail supply chain. Today's retail consumers are more cautious and concerned about the authenticity of the information provided or the claims made by the retailer, about the merchandise. On the other hand, if retailers are able to provide the information or claims in such a way that it cannot dispute, it will help the retailer to win the trust of the consumer. This is where the application of blockchain will be really handy. This paper brings out how the use of blockchain technology in retail supply chain can help retailers and consumers to exchange authentic and un-disputable information. The paper also highlights how operational and cost efficiencies can be achieved in retail supply chain using blockchain.

1 Introduction

Supply chains are a ecosystem of networked individuals, resources, organizations, activities, and technologies which are involved from the start of manufacturing a product or service to the sale of the product to the end user (Zhao and Feng 2020). In today's competitive world, businesses are trying to improve efficiencies, reduce costs, and increase customer satisfaction in every area of business. Supply Chain Management is one key area that is continuously under focus for improving efficiencies. The expanding size of businesses, increasing geographical markets, shortening of the life cycle of a product, and rapid variations in consumer demand make SCM very crucial

P. Nayyar · P. Garg (✉)
Amity School of Business, Amity University, Noida, India
e-mail: Pgarg1@amity.edu

I. Ali et al. (eds.), *Computational Modelling in Industry 4.0*,
https://doi.org/10.1007/978-981-16-7723-6_12

for success of an organization (Butner 2010). Therefore, the improvements in SCM have become one of the key strategies of business in any industry.

Blockchain as one of the disruptive technologies allows the exercise of peer-to-peer network involving all the nodes in validating the transaction and therefore eliminate the need for central entity to validate the transaction (Christidis and Devetsikiotis 2016). Blockchain has found applications in various sectors of Industry like Healthcare and Life Sciences, Manufacturing, Food Industry, Retail, Oil and Gas, Gaming, supply chain, and logistics etc. (Gupta 2017). Blockchain advocates traceability, immutability, transparency, authenticity, reduced cost, speed of transactions, accurate and faster decisions, improves supply chain performance, and customer satisfaction (Babich and Hilary 2019). Blockchain keeps track of every single transfer of goods along the supply chain and it cuts the transaction settlement time as the transactions are authenticated more efficiently all along the supply chains.

However, many studies have examined the impact of blockchain technology on supply chains but still understanding of its potential remains narrow. This paper starts with a discussion on the architecture of blockchain technology, blockchain in retail supply chain including supply chain challenges and role of blockchain to handle these challenges, conclusion and lastly references.

2 Blockchain Technology

The Blockchain is a Distributive Ledger Technology (DLT) which was introduced in 2008 by Satoshi Nakamota (Nakamoto and Bitcoin 2008) in a paper titled 'Bitcoin' and was successfully used for crypto currency 'Bitcoin' in 2009. In 2010, the first time a product purchase happened using bitcoin and in 2011, the exchange rate of bitcoin reached parity with US$ (Pilkington 2016). In 2016, Japan's cabinet recognized the bitcoin as having a function similar to real money. By 2021, the market capitalization of Bitcoin and other cryptocurrencies is expected to reach over US$ 1 Trillion (Economic Times 2021).

Crypto currency in the Financial sector initially was the only application of Blockchain. But the technology has potential for far-reaching applications where it can disrupt the existing technologies and industries to their core. The technology has the potential to be used in settlement applications, smart contracts, patient record management, real estate record management, and so on.

A blockchain is defined as a decentralized, continuously growing list of records which are called blocks, across a peer-to-peer network that are linked and secured using cryptography (Attaran and Attaran 2007). In other words, Blockchain is a time-stamped series of records of data that is immutable and is managed by a cluster of computers that are not owned by any single entity.

Blockchain is a distributed database or a ledger that maintains a permanent, chronologically recorded set of transactions. It is a database and a network that can store and transmit data. Blockchain is the type of distributed ledger that stores a chain of data 'blocks' comprised of all historic and unchangeable data (Bashir

2017). Every time a new block is created and added to the linear chain, transactions, participant addresses and unique identifiers of the previous block in the chain are stored in the new block. Every subsequent block contains a unique identifier of the group derived out of the previous block in the chain ensuring that no historic data has been tampered with and remained unchanged.

Blockchain is of broadly three types i.e. Public blockchain, private blockchain, and Hybrid blockchain. Public blockchain such as Bitcoin, IOTA, Ethereumare those which are accessible to anyone to participate, read, or write Itis decentralized, does not have any central authority and the participants may be anonymous. Private Blockchains are those, where all participants are known and trusted. It is managed by a network administrator and participants need permission to participate in the network. These are useful for traditional businesses and governance Hybrid Blockchains are those which are a combination of a private and public blockchain. This blockchain offers benefits of both Public and private Blockchains. Figure 1 gives a comparison between Public and Private Blockchain (Wang et al. 2017).

The important characteristics that make blockchain unique, useful, and can disrupt existing technologies are ‘decentralization’, ‘immutability’, ‘transparency’, and speed of transactions. Blockchain is a peer-to-peer network and all the nodes are involved in validating the transaction and therefore eliminate the need for central entity to validate the transaction. Table 1.2 lists outs the various characteristics of blockchain technology (Fig. 2).

In the blockchain process, when a user initiates a transaction, a block of information is created and broadcasted to all the users on the network for validation. Anyone on the network can validate the transaction. The user or node that validates the transaction first (Called Miner) wins the ‘proof of work’ and adds the block in the distributed ledger which is broadcasted to all the users (Fig. 3).

	PERMISSIONLESS	PERMISSIONED
PUBLIC	•Anyone can participate, read and write. •Anonymous •True decentralization •Scalability is low	•Anyone can participate, read and write. •Only authorized and trusted participants can write. •Scalability is generally medium.
PRIVATE	•Only authorized and trusted participants are allowed. They can read or write. •Identity of the participants are known. •Scalability is high	•Only authorized and trusted participants are allowed. They can read. •Only network administrator or permissioned participants can write. •Identity of the participants are known. •Scalability is very high

Fig. 1 Types of blockchain technology

CHARACTERISTICS	DESCRIPTION
DE-CENTRALIZATION	There is no requirement of central co-ordinator / Service provider in the systems uses Blockchain.
IMMUTABILITY OF TRANSACTION RECORDS:	Once the transaction is recorded in the database , it cannot be reversed or altered as it is maintained in multiple copies of database over the network.
TRANSPARENCY	There is complete transparency as any transaction which is recorded in the database is visible to everyone over the network.
SPEED OF TRANSACTION	Transactions on the Blockchain Network are completed extremely fast as there are no central co-ordinator.
TRUST	Enforces trust through crypographic security, hard coded rules and behavioural economics. Provides one verified record of the truth which is trusted by all parties.
CONSENSUS	Network consenses is the mechanism used to decide the truth. No single point of control and validation of all network actions.
AUTOMATION	It removes the need of human involvement, significantly reducing manual processing and reconciliation efforts

Fig. 2 Characteristics of blockchain technology

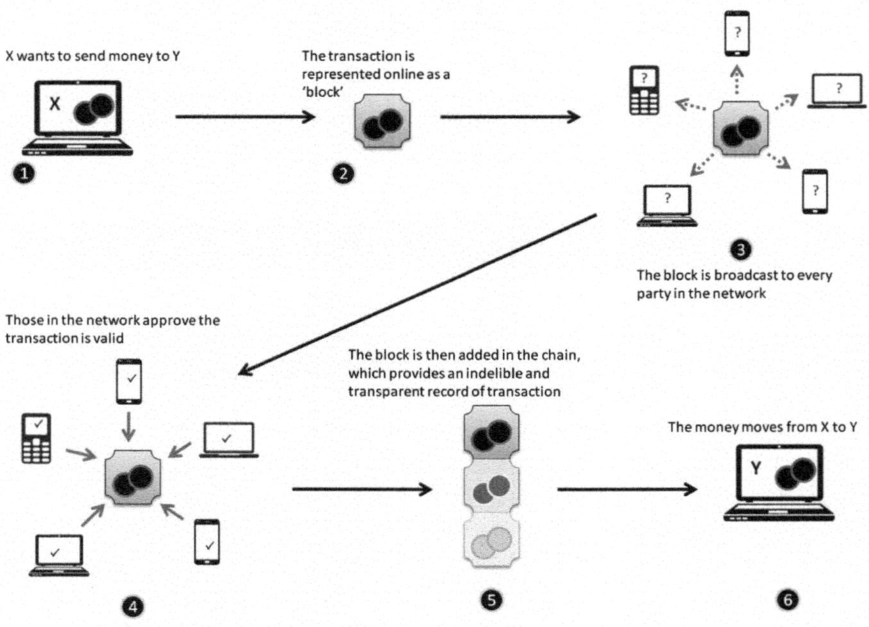

Fig. 3 The blockchain process

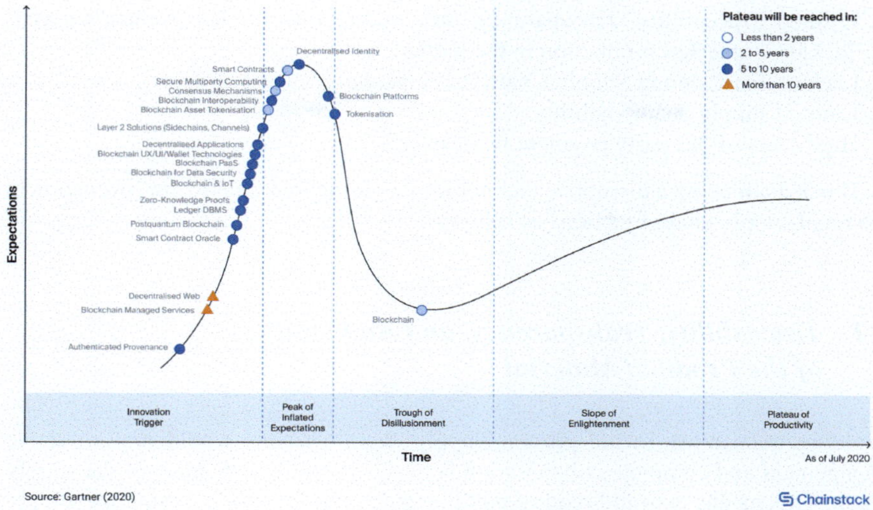

Fig. 4 Gartner's report: hype cycle for blockchain technologies, 2020

One of the most useful applications of blockchain is 'smart contracts'. A smart contract is a set of promises created in the digital form which includes the terms and protocols under which the parties involved perform the promises. These smart contracts are created and stored in the blockchain as a distributed ledger which gets self-executed without any human intervention when the predefined terms and conditions between the two parties are met. Smart contract reduces the risk of fraud and executes quickly. An example of smart contract in blockchain may be in the music or film industry. The royalty to the singer or music composer gets automatically distributed when the terms of the agreement are satisfied.

According to Gartner report 2020, blockchain is one of the top 10 emerging technology trends. Innovative solutions supported by blockchain are mainly in the experimentation phase or limited scale production. Still, early adopters are using blockchains to digitally transform their businesses especially in supply chain related and payments related use (Gartner 2020). The below depicted Gartner Hype cycle indicates that some of the applications/uses like 'smart contracts' and 'Blockchain Asset Tokenisation' may reach the plateau in 2 to 5 years (Fig. 4).

3 Blockchain in Retail Supply Chain

The present supply chain function is facing several challenges as follows (Kadia 2020; Scott 2020):

- Limited Traceability, Transparency, and Auditability of the product/material, Tackle Counterfeit production in the market.
- Lack of Trust among Supply Chain Participants.
- Limited supply chain visibility.
- High costs of 3rd party payment processors.

Blockchain being a disruptive technology can bring in dramatic improvements in the retail supply chain functions as follows.

3.1 Traceability, Transparency, and Auditability of the Product/ Material

Building traceability in the present Supply chain Systems would require active engagement and complete collaboration between all the stakeholders of the supply chain. This means it would require building a centralised system across the supply chain participants, which is practically not possible (Kim and Laskowski 2018).

Consumers today are demanding more authentic information about the product they want to buy. For example, a consumer buying an organic vegetable, wants to have authentic proof that the vegetables she is buying have actually grown in the organic farm not just a label of 'organic' is pasted on the vegetable pack. Consumers are willing to pay a premium on the product if they get authentic information about the purity, origin, and content/ingredients of the product.

Blockchain can help the retailer and consumer by providing this authentic product information (Subramanian and Chaudhuri 2020). This can be achieved by recording the information about a product in a blockchain that follows the entire journey of a product from farm, warehousing, distribution to the retailer (Paliwal et al. 2020). This information may also contain data like soil test reports, sensors readings, storage temperature information etc. (Fig. 5). At each stage, information is recorded in the blockchain as immutable records. This information can be made available to the consumer using a digital ID on the product packaging (Shah 2021; Treiblmaier 2018).

Another challenge for adopting traceability in the present supply chain systems is the entry barrier. Few centralized legacy ERP systems like SAP, Oracle has been designed to implement traceability but these are very expensive systems. SMEs, individuals, farmers etc. which are part of the supply chain, cannot afford these. Using blockchain, only a smartphone will enable them to join the supply chain with complete traceability (Kim and Laskowski 2018).

Tackling counterfeit products in the market

Counterfeit products have been a challenge globally for businesses. Counterfeit products sales estimated over US$ 450 Billion globally which accounts for approx. 2.5% of the global trade. This means that a significant number of sales is lost for the brands and the retailers. On the other hand, the consumer does not get the desired product

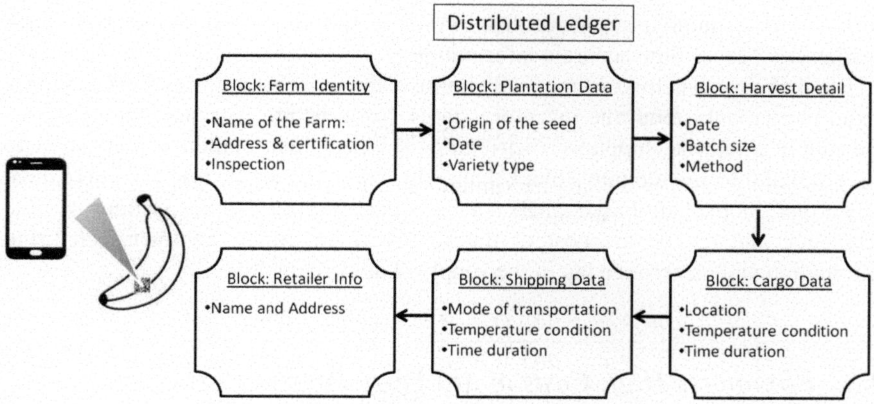

Fig. 5 With using the Blockchain technology, the organic bananas may carry traceable and immutable information of the entire history from farm to market place

and feels cheated as he overpays the price of the product and does not get the value for money paid. Even more damaging aspect of counterfeit products is that consumer trust in the brand is eroded. The process of proving authenticity amounts to significant recurring costs for the manufacturer, retailer, and even for law-enforcing agencies. Amazon has started filing suits in court, against counterfeit product sellers.

Blockchain can help retailers and consumers overcome this issue by providing authentic information about the product for every stage of the supply chain right from raw material, manufacturing, warehousing, distribution, and retail. This information will be indisputable as it would come as an immutable record. This information can be recorded on a smart tag printed on 3D printer. This smart tag can be attached to the product packaging which consumer/retailer can scan using a smartphone and can get complete information about the product right from the raw material stage to retailer (Min 2019).

3.2 *Enhancing Supply Chain Visibility*

Visibility across supply chain participants is very important. Lack of visibility may result in over-ordering in the upstream which may cause bull-whip effect whereas a shortage in inventory may cause lost sales for the retailer and price hike or dissatisfaction for the consumer (Scott 2020). To make the right decision on supply chain operations, the availability of accurate and on time/real-time information regarding various stages of supply chain like warehouse inventory, in Transit, manufacturing lead time etc. is extremely crucial. Many legacy ICT systems have been designed to make this information available. As there are many supply chain participants with varied technologies and localized systems, this remains a challenge to have availability of real-time accurate information.

Blockchain can help to dramatically improve the visibility among complex supply chain participants about the inventory status, location, etc. Various participants of the supply chain like suppliers, distributors, shippers, and retailers can consolidate their digital information into a blockchain which will be a single true version of data.

Further, blockchain-based smart contracts can improve drastically improves the efficiency and reduce the cost of the supply chain. Smart contracts can bring automation and can reduce lag in the supply chain (Zhang 2019).

3.3 *Eliminates High Costs of 3rd Party Payment Processors*

In the present retail supply chain, all participants are dependent upon 3rdparty for transaction settlement. A significant price is paid by the SC participants for these services for validation of exchange of goods/services, money, and data. Further, this 3rd party transaction settlement adds to delay which further increases the finance cost of the supply chain participants.

Using blockchain, this need of 3rd party services, for transaction settlement will be eliminated. This will reduce the cost for retailers and other SC stakeholders which is charged by 3rd parties (banks and other similar agencies). Also, the use of blockchain will reduce the risk of fraud and thereby cut the cost of auditors and accounting staff and lawyers, for fraud detection and recovering damages.

3.4 *Building Trust Among Supply Chain Participants*

Today's supply chains are complex and some of these have a global footprint. Supply chain participants might be from different geographies, with different languages and cultures. For the smooth functioning of the supply chain, trust among the participants is extremely essential.

As the blockchain is a technology that uses the concept of consensus, all the participants are involved in authenticating a transaction and each one can verify the history of transactions. The data cannot be tempered by anyone. This brings in more transparency among participants and as the transactions are secured and thereby enhance trust among supply chain participants (Min 2019; Reyna 2018).

4 Conclusion

Blockchain technology will transform and generate value across the entire business ecosystem. Its implementation would increase transparency across the industry ecosystem, eliminate manual reconciliation, automate contract execution, audits, and reporting, increase data accuracy and enable real-time analytics, leverage one single, immutable version of the truth accessible at the same time to all the authorized stack holders, and integrate the various processes into a single environment.

The use of blockchain can revolutionize the retail supply chain as it can bring transparency and enhance trust among retailers and consumers. There are tremendous opportunities to utilize blockchain technology to address the issue of counterfeit products and to gain operational and cost efficiencies in the retail supply chain.

References

Attaran M, Attaran S (2007) Collaborative supply chain management. Bus Proc Manag J

Babich V, Hilary G (2018) Blockchain and other distributed ledger technologies in operations. Found Trends Technol Infor Operat Manag Now Publ 12:1–21

Bashir I (2017) Mastering blockchain. Packt Publishing Ltd

Butner K (2010) The smarter supply chain of the future. Strategy & Leadership

Christidis K, Devetsikiotis M (2016) Blockchains and smart contracts for the internet of things. IEEE Access 4:2292–2303

Economic Times (2021) Bitcoin hits $1 trillion market cap, surges above $56,000. https://economictimes.indiatimes.com/markets/stocks/news/bitcoin-hits-1-trillion-market-cap-surges-above-55000/articleshow/81121700.cms. Accessed 15 Feb 2021

Gartner (2020) Gartner's report: hype cycle for blockchain technologies

Gupta M (2017) Blockchain for dummies. John Wiley & Sons

Kadia P (2020) The success story of tradelens. https://businessblockchainhqcom/business-blockchain-news/the-success-story-of-tradelens/. Accessed 10 Dec 2020

Kim HM, Laskowski M (2018) Toward an ontology-driven blockchain design for supply-chain provenance. Intel Syst Acc Fin Manag 25(1):18–27

Min H (2019) Blockchain technology for enhancing supply chain resilience. Bus Horiz 62(1):35–45

Nakamoto S, Bitcoin A (2008) A peer-to-peer electronic cash system. https://bitcoin.org/bitcoin. Accessed 15 Feb 2021

Paliwal V, Chandra S, Sharma S (2020) Blockchain technology for sustainable supply chain management: a systematic literature review and a classification framework. Sustainability 12(18):7638

Pilkington M (eds) (2016) Blockchain technology: principles and applications. research handbook on digital transformations. SSRN 2662660

Reyna A, Martín C, Chen J, Soler E, Díaz M (2018) On blockchain and its integration with IoT Challenges and opportunities. Futur Gener Comput Syst 88:173–190

Scott T (2020) Tradelens: how IBM and Maersk are sharing blockchain to build a global trade platform. https://www.ibm.com/blogs/think/2018/11/tradelens-how-ibm-and-maersk-are-sharing-blockchain-to-build-a-global-trade-platform/. Accessed 10 Feb 2021

Shah D (2017) Unblocking the retail supply chain with blockchain. https://www.tcs.com/content/dam/tcs/pdf/Industries/Retaillogistics/Abstract/Unblocking-retail-supply-chain-with-blockchain-1017-1. Accessed 10 Jan 2021

Subramanian N, Chaudhuri A, Kayıkcı Y (2020) Blockchain and supply chain logistics: evolutionary case studies, Springer Nature

Treiblmaier H (2018) The impact of the blockchain on the supply chain: a theory-based research framework and a call for action. Supply Chain Manag Int J

Wang J, Wu P, Wang X, Shou W (2017) The outlook of blockchain technology for construction engineering management. Front Eng Manag 12:67–75

Zhao J, Ji M, Feng B (2020) Smarter supply chain: a literature review and practices. J Data Inf Manag 19:1–6

Zhang J (2019) Deploying blockchain technology in the supply chain. In Computer Security Threats Intech Open

Chapter 13
Optimization of System Reliability with Time-Dependent Reliability Components in Imprecise Environment Using Hybridized QPSO

Nabaranjan Bhattacharyee, Nirmal Kumar, Sanat Kumar Mahato, and Asoke Kumar Bhunia

Abstract To design reliable systems, the optimization of system reliability (SR) is a highly concerned topic in the engineering design and industry. The reliability optimization aims to develop a reliable system with higher reliability to perform satisfactorily under certain conditions and up to a specified period. This chapter considers the redundancy allocation problem as a highly non-linear and integer programming constrained optimization problem. To cope up with reality with unpredictability, we desire to consider the reliabilities of the time-dependent components that lead to a reliable time-dependent system. Further, to incorporate the fluctuating behaviour of the system's controlling parameters and uncertainty of the situations of the environment in which the system is operated, we developed the fuzzy model. As the problem is combinatorial and highly non-linear, we developed and implemented the hybridized metaheuristic technique derived by combining QPSO, a variant of particle swarm optimization, and the Big-M penalty technique to find the solution. The crisp and fuzzy (triangular and pentagonal) models are solved, and the comparative studies are presented. The statistical computations and the sensitivity studies of the HQPSO parameters are also presented corresponding to the numerical experiments.

Keywords Mission Design Life (MDL) · Redundancy Allocation Problem (RAP) · Time-varying reliability · Triangular Fuzzy Numbers (TFN) · Pentagonal Fuzzy Numbers (PtFN) · Hybridized Quantum behaved PSO (HQPSO)

N. Bhattacharyee · S. K. Mahato (✉)
Department of Mathematics, Sidho-Kanho-Birsha University, Purulia, WB 723104, India
e-mail: sanatkr_mahato.math@skbu.ac.in

N. Kumar · A. K. Bhunia
Department of Mathematics, The University of Burdwan, Bardhaman, WB 713104, India
e-mail: akbhunia@math.buruniv.ac.in

I. Ali et al. (eds.), *Computational Modelling in Industry 4.0*,
https://doi.org/10.1007/978-981-16-7723-6_13

1 Introduction

During the last few decades, many researchers have shown their keen interest in the study of reliability optimization. A wide area of applications of reliability design is observed such as engineering and industry, machine design and productions, including networking communications and transportations, etc. Also, industrialists and machine designers have been showing their interest in reliability theory and practice as it has many practical applications. Reliability analysis is an important part of many developmental works in system designing, communication systems, infrastructure development, etc. System reliability is practically the probability of successful performance of a system up to a given period under some predetermined conditions. The reliability components are taken with fixed values in most of the works reported in the literature. However, it is more realistic to consider the reliability of a system as a function of time since it undoubtedly decreases with time. In this research area, some researchers have presented such a genuine attempt in the literature review of reliability optimization. The reliability practitioners always desire to maximize the system reliability and the considered system's lifetime under certain constraints.

Several types of reliability optimization problems have been designed and solved in the literature such as Redundancy Allocation Problem (RAP), Reliability Redundancy Allocation Problem (RRAP), etc. Most attempts are found in redundancy allocation problems. In this work, our main target is to consider the RAP type of problem in which redundant components are allocated with regard to some constraints to optimize the system reliability. The renowned researchers like, Tillman et al. (1980), Sun and Li (2002), Mahapatra and Roy (2011), Mahato et al. (2020), Garg et al. (2014), Gupta et al. (2009), Mahato et al. (2013), Sahoo et al. (2013), etc. have reported important contributions in the literature. The heuristic technique in optimal reliability allocation reported by Nakagawa and Nakashima (1977), the fuzzy environment is used reliability optimization by Chen (1977), the reduced gradient method utilized by Hwang et al. (1979), a detailed study of the optimization of the reliability of a system is done by Tillman et al. (1980), the surrogate constrained algorithm is used by Nakagawa and Miyazaki (1981). The remarkable researchers such as Chern and Jan (1986), Misra (1986), Park (1987), Misra and Sharma (1991), Huang (1996), Sung and Cho (2000), Kuo et al. (2001) Sun and Li (2002), Mahapatra and Roy (2006, 2009, 2011), Gupta et al. (2009), Bhunia et al. (2010), Sahoo et al. (2010), Bhattacharyee et al. (2021), etc. are notable in the field of reliability optimization.

The uncertainty concepts such as interval, fuzzy, and intuitionistic fuzzy are introduced and studied by many researchers in the reliability analysis. These studies gave a new direction in the study of reliability models. These models are proved to be more realistic in terms of real-life phenomena. The concepts of generalized fuzzy number (Mahapatra and Roy 2011, 2012; Garg 2013; Mahato et al. 2013; Sahoo et al. 2013, 2014; Mahato et al. 2020), interval number (Bhunia and Sahoo 2011; Mahapatra and Roy 2012; Sahoo et al. 2012; Mahato et al. 2012), intuitionistic fuzzy (Garg 2013; Garg and Rani 2013; Garg et al. 2014; Garg 2015; Jamkhaneh 2017, Bhattacharyee et al. 2021) are introduced and studied in reliability theory.

The use of soft computing techniques is observed to have a great impact on reliabity optimization. Usually, the designed problems are found to be highly non-linear combinatorial problems and the analytical solutions are very difficult. So to handle such problems several soft computing algorithms are designed and implemented to solve reliability optimization problems. Soft computing techniques like, GA, PSO, and ABC, etc. are proved to be highly effective in finding the optimal reliability for any type of reliability optimization problems. The works of several researchers like, Garg (2013), Garg and Rani (2013), Khalili-Damghani et al. (2013), Garg et al. (2014a, b), Sahoo et al. (2012, 2013, 2014), Garg (2015, 2016, 2017), Gupta et al. (2009), Mahato et al. (2013, 2020), Bhattacharyee et al. (2021) are worth mentioning. The list is not exhaustive but there are lots of researchers who have developed and utilized several algorithms to solve the problems of reliability maximization.

The time-dependent reliability models are also very much relevant in reliability theory and practice. To make the models more realistic, the reliabilities of the components should be considered to be a function of time. Only a few researchers have designed the reliability models with time-dependent reliability. The works, in this regard, of Mori and Ellingwood (1993), Hamadani and Khorshidi (2013), Ganzalezet al. (2015), Hu and Mahadevan (2015), Mourelatos et al. (2015), Wang et al. (2015), Zhu and Zhifu (2016), Mostafa (2017), Ahmadivala et al. (2019), Zafar and Wang (2020), Bhattacharyee et al. (2021), etc. are noteworthy.

This chapter's main goal is to consider the reliable system having time-varying component reliabilities and the impreciseness of the environments. We have considered here the component's reliabilities to follow exponentially decreasing function of time. So, the reliability model developed here becomes time-dependent. This work has included the impreciseness in terms of triangular fuzzy and pentagonal fuzzy numbers to handle the fluctuating situations, which certainly looks to be more realistic. Hence, we have three models including the two imprecise models, viz., the crisp model, the triangular fuzzy model, and the pentagonal fuzzy model. We developed a new soft computing algorithm to solve the problems. The newly proposed algorithm is named Hybridized Quantum-behaved PSO (HQPSO) which is a variant of PSO involving the Big-M penalty technique.

Organization of the Chapter

Section 1	Introduction
Section 2	Research Gaps
Section 3	Notation and Assumptions
Section 4	Mathematical Foundations
Subsection 4.1	Relevent Definitions
Subsection 4.2	Method of Defuzzification of Fuzzy Numbers
Section 5	Problem Formulation
Subsection 5.1	The Crisp Model
Subsection 5.2	The Fuzzy Models

(continued)

(continued)

Section 6	Solution Procedure
Subsection 6.2	Particle Swarm Optimization
Subsection 6.3	Quantum behaved Particle Swarm Optimization (QPSO)
Subsection 6.4	Proposed Hybridized QPSO
Section 7	Numerical Experiments
Section 8	Result Discussions
Section 9	Acknowledgements

2 Research Gaps

It is clearly to be noted from the existing literature that most of the reliability optimization problems focused on precise environments. Some researchers have recently presented their research on imprecise environments that include interval, fuzzy, intuitionistic fuzzy, stochastic, and a mixture of these. Few works in this area are found to attempt the problems on reliability optimization using GA, PSO, hybridized PSO, ABC algorithms, Cuckoo search algorithm, and other heuristic algorithms. Moreover, in most of the works related to our paper, the reliability components are of constant values and only a few are observed to consider these as time-varying functions. A few works are also found to consider the machine design life as the objective function, and the others have taken system reliability or the cost function as the objective function.

Again, the problem's constraints are handled in several ways; only our research group in this field has incorporated the Big-M penalty method. Thus, we have been motivated eagerly to formulate a problem in reliability studies which has the machine design life as the objective function, the reliability component as exponentially decreasing functions of time, utilize the Big-M penalty technique to tackle the constraints. We use Simpson's 1/3 rule to handle the integration to get the machine design life from system reliability function and develop a hybridized Quantum-behaved Particle Swarm Optimization due to Big-M penalty method.

3 Notation and Assumptions

Throughout the chapter, we use the symbols described below. Also, the necessary assumptions to formulate the problem under consideration are given below.

3.1 Notation

Symbols	Meanings
$\tilde{P}$, $\widehat{P}$	Triangular and Pentagonal fuzzy number respectively
$\mu_{\tilde{P}}(x)$, $\mu_{\widehat{P}}(x)$	Membership function of x $\in X$ w.r.t. $\tilde{P}$, $\widehat{P}$ respectively
$Cr1\left(\tilde{P}\right), Cr2(\widehat{P})$	Defuzzified value of the fuzzy number $\tilde{P}$, $\widehat{P}$ respectively
$u = (u_1, u_2, \ldots, u)$	Redundancy vector (decision variable)
$R_1(u, \lambda, t)$,$\tilde{R}_2\left(u, \tilde{\lambda}, t\right)$, $\widehat{R}_3\left(u, \hat{\lambda}, t\right)$	System reliability in crisp, triangular fuzzy and pentagonal fuzzy forms respectively
$M_1(u, \lambda, M_T)$, $\tilde{M}_2\left(u, \tilde{\lambda}, M_T\right)$, $\widehat{M}_3(u, \hat{\lambda}, M_T)$	MDL in crisp, triangular fuzzy and pentagonal fuzzy forms respectively
$g_{1j}(u)$, $\tilde{g}_{2j}(u)$, $\widehat{g}_{3j}(u)$	Constraints usability functions in crisp, triangular fuzzy and pentagonal fuzzy environments
b_{1j},$\tilde{b}_{2j}$, $\widehat{b}_{3j}$	Availability of resources of j-th constraint in crisp, triangular fuzzy and pentagonal fuzzy environments
l_{1i}, l_{2i}	Lower bound and upper bound of u_i
S_{size}	Swarm size in QPSO
$f(p_i)$	Value of fitness function of i-th particle in its best position
$m_i^{(z)}$	Mean best position of j-thcomponent at z-th iteration
$m^{(z)}$	Mean best position vector at z-th iteration
$x_{ij}^{(z)}$:	The position of i-thparticle in the j-th swarm at z-th iteration
$A_{ij}^{(z)}$	Local attractor of the j-th component of the i-th particle at z-th iteration
Mg	Maximum number of generations
n	Dimension of the variables
F_R	Feasible region

3.2 Assumptions

To develop our proposed model, we have careful into consideration of the assumptions given here:

(i) The proposed system is a series–parallel system.
(ii) At a particular stage, the subsystem contains identical components.
(iii) Without any repair, the redundancies are always active.
(iv) Reliability of each component is an exponentially decreasing function of time.
(v) The MDL function is defined as the integral of system reliability.
(vi) Component failure in each subsystem might not be tantamount to the system to its failure.
(vii) Throughout all environments, the control parameters are well known, viz. crisp and fuzzy (triangular and pentagonal).
(viii) Fuzzy numbers (triangular and pentagonal) are of linear type.

4 Mathematical Foundations

4.1 Relevant Definitions

Definition 4.1: The fuzzy set is the pair $(x,\mu_{\tilde{P}}(x))$, where $x \in X$ and $\mu_{\tilde{P}}(x)$:X $\rightarrow$ [0, 1], X being the universe of discourse and it is represented as $\tilde{P}=\{(x,\mu_{\tilde{P}}(x)): x \in$ X}, where $\mu_{\tilde{P}}(x)$ denotes the membership function of $x \in X$ w.r.t. $\tilde{P}$.

Definition 4.2: The fuzzy set $\tilde{P}$ becomes convex iff $\mu_{\tilde{P}}(\lambda x_1 + (1-\lambda)x_2) \geq \min\{\mu_{\tilde{P}}(x_1), \mu_{\tilde{P}}(x_2)\}$, for all $x_1, x_2 \in X$, where $\lambda \in [0, 1]$.

Definition 4.3: The fuzzy set $\tilde{P}$ becomes normal if $\mu_{\tilde{P}}(x) = 1$, for some $x \in X$.

Definition 4.4: A fuzzy set becomes a fuzzy number provided it is (Fig. 1).

(i) Normal
(ii) convex

The membership function of a fuzzy number $\tilde{P}$ can be described as

$$\mu_{\tilde{P}}(x) = \begin{cases} l(x), p_1 \leq x < p_2 \\ 1, \ p_2 \leq x \leq p_3 \\ u(x), p_3 < x \leq p_4 \\ 0, \ \text{otherwise.} \end{cases}$$

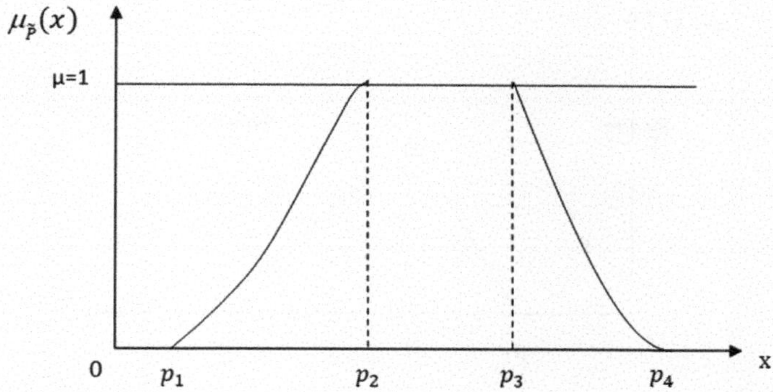

Fig. 1 General fuzzy number

l(x) and *u(x)* being the left and right shape functions, respectively.

Definition 4.5: Linear Triangular Fuzzy Number (LTFN)
An LTFN $\tilde{P}$ is represented by the triplet (p_1, p_2, p_3) and can be defined by the continuous membership function $\mu_{\tilde{P}}(x) : X \to [0, 1]$ as follows:

$$\mu_{\tilde{P}}(x) = \begin{cases} \dfrac{x - p_1}{p_2 - p_1} & \text{if } p_1 \le x \le p_2 \\ 1 & \text{if } x = p_2 \\ \dfrac{p_3 - x}{p_3 - p_2} & \text{if } p_2 \le x \le p_3 \\ 0 & \text{otherwise.} \end{cases}$$

Definition 4.6: Linear Pentagonal Fuzzy Numbe (LPtFN)
A fuzzy pentagonal number $\widehat{P}= (p_1, p_2, p_3, p_4, p_5)$ satisfies the conditions given below:

(1) it has the continuous membership function $\mu_{\widehat{P}}(x)$ in [0,1]
(2) the membership function $\mu_{\widehat{P}}(x)$ is strictly non-decreasing in $[p_1, p_2]$ and $[p_2, p_3]$
(3) the membership function $\mu_{\widehat{P}}(x)$ is strictly non-increasing in $[p_3, p_4]$ and $[p_4, p_5]$ (Figs. 2 and 3)

4.2 *Method of Defuzzification of Fuzzy Number*

There are several methods of defuzzification available in the literature. The most commonly used technique for defuzzification of a fuzzy number is the centre of area (COA) method.

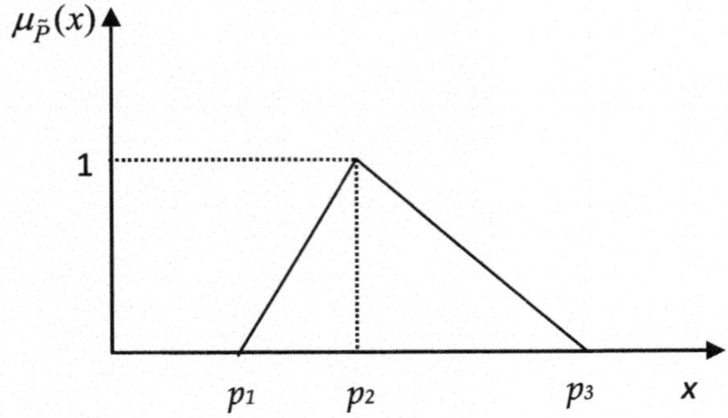

Fig. 2 Linear triangular fuzzy number

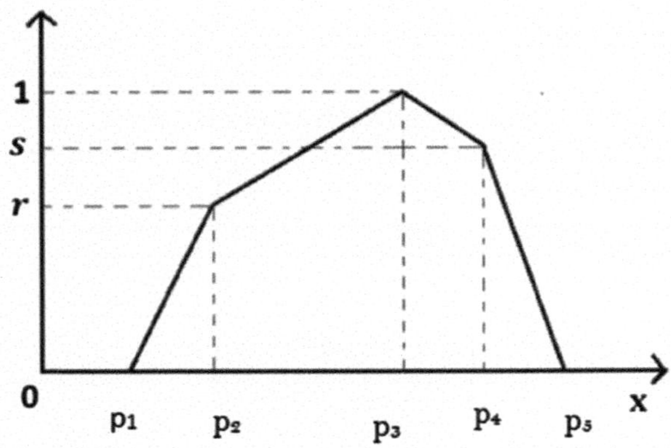

Fig. 3 Linear pentagonal fuzzy number

Let the fuzzy number $\tilde{P}$ has a continuous membership function $\mu_{\tilde{P}}(x)$ then the COA formula for crispification is defined as follows (Mahato and Bhunia 2016).

$Cr1\left(\tilde{P}\right) = \frac{\int \mu_{\tilde{P}}(x) x dx}{\int \mu_{\tilde{P}}(x) dx}.$

4.2.1 Crispification Formula for Linear Triangular Fuzzy Number

The crispification formula for Linear Triangular Fuzzy Number $\tilde{P} = (p_1, p_2, p_3)$ can be defined as (Mahato and Bhunia 2016)

$Cr1(\tilde{P}) = (p_1 + p_2 + p_3)/3.$

Example 4.1: For $\tilde{P} = (2, 3, 4)$, $p_1 = 2$, $p_2 = 3$, $p_3 = 4$, so

$$
\begin{aligned}
Cr1(\tilde{P}) &= \tfrac{1}{3}(p_1 + p_2 + p_3) \\
&= \tfrac{1}{3}(2 + 3 + 4) \\
&= 3
\end{aligned}
$$

Example 4.2: For $\tilde{P} = (1.6, 2.9, 3.8)p_1 = 1.6$, $p_2 = 2.9$, $p_3 = 3.8$ and so

$$Cr1(\tilde{P}) = \tfrac{1}{3}(p_1 + p_2 + p_3)$$
$$= \tfrac{1}{3}(1.6 + 2.9 + 3.8)$$
$$= 2.7666666667$$

4.2.2 Crispification Formula for Linear Pentagonal Fuzzy Number

The crispification formula for Linear Pentagonal Fuzzy Number $\widehat{P}= (p_1, p_2, p_3, p_4, p_5)$ is defined as (Mahato and Bhunia 2016)

$$Cr2(\widehat{P}) = \frac{p_5^2+p_4^2+p_5 p_4-p_1 p_2-p_2^2-p_1^2}{3(p_5+p_4-p_2-p_1)}$$

Example 4.3: For $\widehat{P} = (1, 2, 3, 4, 6)$, $p_1 = 1$, $p_2 = 2$, $p_3 = 3$, $p_4 = 4$, $p_5 = 6$ and so

$$Cr2(\widehat{P}) = \frac{p_5^2+p_4^2+p_5 p_4-p_1 p_2-p_2^2-p_1^2}{3(p_5+p_4-p_2-p_1)}$$
$$= 3.2857142857$$

Example 4.4: For $\widehat{P} = (2.5, 3.3, 4.4, 5.8, 6.4)$, $p_1 = 2.5$, $p_2 = 3.3$, $p_3 = 4.4$, $p_4 = 5.8$, $p_5 = 6.4$ and so

$$Cr2(\widehat{P}) = \frac{p_5^2+p_4^2+p_5 p_4-p_1 p_2-p_2^2-p_1^2}{3(p_5+p_4-p_2-p_1)} = 4.496354$$

5 Problem Formulation

This section covers the formulation of a series–parallel reliability redundancy allocation problem using the fact that the components have time-dependent reliabilities (Bhattacharyee et al. 2021). It is supposed that the reliability components obey exponential distributions, leading to the reliability of the system being time-dependent.

Moreover, we are inspired for considering the mission design life and desire to get the maximum value of the system reliability with a proper choice of the redundancy allocation vector. Evidently, it is better not to take the controlling parameters as fixed numbers by some deterministic rule but to consider these as imprecise numbers to retain the reliable system's unpredictable nature. The parameters' estimated values cannot be predicted precisely due to the reliability system's fluctuating character. This unpredictable situation can be handled by considering the impreciseness in terms of fuzzy, intuitionistic fuzzy, interval, stochastic, or combination. In the fuzzy approach, we need to know the membership function for a given fuzzy number, while for an intuitionistic fuzzy approach, we should know both the membership function and the non-membership function. For the interval method, the parameters are taken as closed intervals. Some known probability distributions are taken in the stochastic approach.

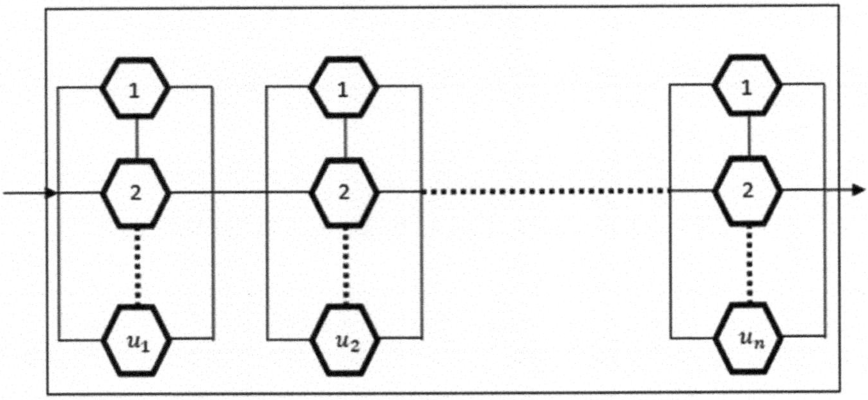

Fig. 4 Series–parallel system with n-stages

In this work, we assume the impreciseness/vagueness in the form of fuzzy numbers (TFN and PtFN). Thus, depending upon the nature of the controlling parameters, we develop three models corresponding to the series-parallelsystem (Fig. 4).

5.1 The Crisp Model

To solve the redundancy allocation problem (RAP) with time-varying reliability, we considered the reliability component following the exponential failure rate and time-dependent component reliability function. We have considered a system with n subsystems connected in series, and each subsystem consists of $u_i(i = 1, 2, \ldots, n)$ the number of active redundant components which are identical.

Let the failure density function be $f_i(t) = \lambda_i e^{-\lambda_i t}$ and the reliability function of each component of the i-th subsystem be $r_i(t) = e^{-\lambda_i t}, t > 0, \lambda_i$ is constant, $i = 1, 2, \ldots, n$.

Then using the combinatorial theory of probability, the reliability of the series–parallel system (Fig. 4) becomes $R_1(u, t; \lambda) = \prod_{i=1}^{n} \left[1 - (1 - e^{-\lambda_i t})^{u_i}\right]$.

To fulfil the aims like mission time, i.e. the system's non-stop successful functioning, cost-effectiveness, etc. the design is to be done suitably. The Mission Design Life (MDL) function is defined as (Mostafa et al. 2017; Bhattacharyee et al. 2021) $M_1(u, \lambda, M_T) = \int_0^{M_T} R_1(u, \lambda, t)dt$.

It is easily understood that the optimization problem of maximizing the system reliability $R_1(u, \lambda, t)$ is equivalent to maximizing $M_1(u, \lambda, M_T)$. For the known parameters λ and M_T, the optimization problem can be stated as,

$$Maximize\ M_1(u, \lambda, M_T) = \int_0^{M_T} R_1(u, \lambda, t)dt$$

$$= \int_0^{M_T} \left(\prod_{i=1}^{n} [1 - (1 - e^{-\lambda_i t})^{u_i}] \right) dt$$

$$\text{subject to} \tag{1}$$

$$g_{1j}(u_1, u_2, \ldots, u_n) \leq b_{1j}, j = 1, 2, \ldots, m$$

$$l_{1i} \leq y_i \leq l_{2i}, i = 1, 2, \ldots, n$$

$u = (u_1, u_2, \ldots, u_n)$ being the redundancy vector, u_i is a non-negative integer representing the redundancy level of the i-th component.

5.2 *The Fuzzy Models*

The two fuzzy models in a fuzzy environment are developed in which the control parameters are taken as linear triangular fuzzy numbers (LTFN) and linear pentagonal fuzzy numbers (LPtFN) with linear membership functions. Thus, the fuzzy models can be stated as:

5.2.1 Triangular Fuzzy Model

$$Maximize\ \underset{2}{\tilde{M}}(u, \tilde{\lambda}, M_T) = \int_0^{M_T} R_2(u, \tilde{\lambda}, t) dt$$

$$= \int_0^{M_T} \left(\prod_{i=1}^{n} \left[1 - (1 - e^{-\tilde{\lambda}_i t})^{u_i} \right] \right) dt$$

$$\text{subject to} \tag{2}$$

$$\tilde{g}_{2j}(u_1, u_2, \ldots, u_n) \leq \tilde{b}_{2j}, j = 1, 2, \ldots, m$$

$$l_{1i} \leq u_i \leq l_{2i}, i = 1, 2, \ldots, n$$

$u = (u_1, u_2, \ldots, u_n)$ being the redundancy vector, u_i is a nonnegative integer representing the redundancy level of the i-th component.

5.2.2 Pentagonal Fuzzy Model

$$\begin{aligned} Maximize\ \hat{M}_2(u,\ \hat{\lambda},\ M_T) &= \int_0^{M_T} \hat{R}_3(u,\ \hat{\lambda},\ t)dt \\ &= \int_0^{M_T} \left(\prod_{i=1}^{n}\left[1-(1-e^{-\hat{\lambda}_i t})^{u_i}\right]\right)dt \end{aligned}$$

subject to (3)

$$\hat{g}_{3j}(u_1, u_2, \ldots, u_n) \le \hat{b}_{3j},\ j = 1, 2, \ldots, m$$

$$l_{1i} \le u_i \le l_{2i},\ i = 1, 2, \ldots, n$$

$u = (u_1, u_2, \ldots, u_n)$ being the redundancy vector, u_i is a non-negative integer representing the redundancy level of the i-th component.

6 Solution Procedure

The objective functions in problems (1), (2) and (3) all are highly non-linear and the problems are combinatorial optimization problems. The objective functions are to be maximized that involve the integration of the system reliability of a series–parallel system. The integration is quite difficult to evaluate analytically so the Simpson's 1/3 rule is utilized to evaluate the approximate integral value.

6.1 Particle Swarm Optimization (PSO)

Kennedy and Eberhart (1995) reported the new algorithm as being inspired by the social behaviours of fish schooling and birds flocking. This is known as the PSO algorithms and proved to be efficient enough in solving global optimization problems. In this algorithm, every solution of the swarm is represented as bird/fish like particles and they have the liberty to fly throughout the solution space with the common goal to land on or near of the optimal position. The position of each particle is updated by the combined knowledge of the individual and the group of the swarm. Each particle remembers its personal best (*pbest*) along with the group best position or global best (*gbest*).

Let us take,

$x_i^{(z)} = \left(x_{i1}^{(z)}, x_{i2}^{(z)}, \ldots, x_{in}^{(z)}\right)$, as the current position

$v_i^z = \left(v_{i1}^{(z)}, v_{i2}^{(z)}, \ldots, v_{in}^{(z)}\right)$, as the current velocity

$p_i^z = \left(p_{i1}^{(z)}, p_{i2}^{(z)}, \ldots, p_{in}^{(z)}\right)$, as the *pbest* position

$p_g^{(z)} = \left(p_{g1}^{(z)}, p_{g2}^{(z)}, \ldots, p_{gn}^{(z)}\right)$, as the *gbest* position respectively at the z-th iteration of the i-th swarm.

Then the updation formulae for the velocity and position of the i-th particle in the j-th direction at the z-th iteration are given by

$$v_{ij}^{(z+1)} = v_{ij}^{(z)} + c_1 r_{1j}^{(z)}\left(p_{ij}^{(z)} - x_{ij}^{(z)}\right) + c_2 r_{2j}^{(z)}\left(p_{gj}^{(z)} - x_{ij}^{(z)}\right) \tag{5}$$

$$x_{ij}^{(z+1)} = x_{ij}^{(z)} + v_{ij}^{(z+1)} \tag{6}$$

where $i = 1, 2, \ldots, S_{size}; j = 1, 2, \ldots, n; z = 1, 2, \ldots, Mg; c_1(> 0), c_2(> 0)$ are the acceleration coefficients and $r_{1j}^{(z)}, r_{2j}^{(z)} \sim U(0, 1)$.

6.2 *Quantum Behaved Particle Swarm Optimization (QPSO)*

The strategies done by traditional PSO completely fail in quantum space because the velocity and position cannot be specified concurrently according to '*Heisenberg's Uncertainty Principle*'. So it needs to describe the particles in terms of the wave function. While moving in the quantum space, the wave function $\psi(x, t)$ must satisfy the Schrödinger wave equation and by solving the equation, we get the density function $|\psi|^2$. Utilizing the Monte Carlo technique, the updating formula is obtained as stated below

$$x_{ij}^{(z+1)} = A_{ij}^{(z)} + \beta\left|x_{ij}^{(z)} - m_j^{(z)}\right| \log\left(\frac{1}{u_{ij}^{(z)}}\right) \quad if \;\; r \geq 0.5 \tag{7}$$

$$x_{ij}^{(z+1)} = A_{ij}^{(z)} - \beta\left|x_{ij}^{(z)} - m_j^{(z)}\right| \log\left(\frac{1}{u_{ij}^{(z)}}\right) \quad if \;\; r < 0.5 \tag{8}$$

where $A_{ij}^{(z)} = \phi_j p_{ij}^{(z)} + (1 - \phi_j) p_{gj}^{(z)}$,

$m^{(z)}$ = averages of all *pbest* positions

$$= \left(m_1^{(z)}, m_2^{(z)}, \ldots, m_n^{(z)}\right)$$

$$= \left(\frac{1}{S_{size}}\sum_{i=1}^{S_{size}} p_{i1}^{(z)}, \frac{1}{S_{size}}\sum_{i=1}^{S_{size}} p_{i2}^{(z)}, \ldots, \frac{1}{S_{size}}\sum_{i=1}^{S_{size}} p_{in}^{(z)}\right) \tag{9}$$

β = expansion contraction parameter

$u_{ij}^{(z)}$, r are random numbers in (0,1).

6.3 Proposed Hybridized QPSO (HQPSO)

In order to use the QPSO after combing with the Big-M penalty, we have modified the QPSO algorithm and developed the hybrid form of it. This hybrid algorithm combines the features of the QPSO and the Big-M penalty technique. The Big-M penalty function techniques has the capability to wipe out the infeasible solutions from the search region reducing the constrained optimization into unconstrained one. This is similar to the notion that the particles will never search for food in the points once observed not to contain any food. We have developed the HQPSO especially to solve the pure integer programming problems of combinatorial type involving the integration of highly non-linear integrand. The constrained optimization problems can easily be solved by implementing this new HQPSO. In the Big-M penalty method, a very big/small value is assigned as the fitness value corresponding to the infeasible points/positions according to the problem (minimization/maximization). In this method, the infeasible points/positions are never revisited and so the efficiency of the algorithm increases with quick convergence in the feasible region

$$F_R = \left\{u = (u_1, u_2, \ldots, u_n) : g_{1j}(u_1, u_2, \ldots, u_n) \le b_{1j};\ j = 1, 2, \ldots, n\right\}.$$

The iterative steps of HQPSO are given.

Step 1: Start
Step 2: Initialize QPSO parameters and also the bounds of the variables
Step 3: Create a random particles' swarm, i.e. randomly generate
$X_{ij}(i = 1(1)S_{size}; j = 1(1)n)$
Step 4: Set this initial positions as **pbest** *position i.e.* $p_i = x_i$ *for* $i = 1,2,\ldots, S_{size}$
Step 5: Determine **gbest** *position,* $g = arg(max\ (f(p_i)))$ *for* $i = 1(1)\ S_{size}$
Step 6: Set $z = 1$
Step 7: Calculate mean best position m using Eq. (9)
Step 8: Generate $\phi = rand(0, 1)$
Step 9: Compute local attractor $A_{ij} = \phi p_{ij} + (1 - \phi)p_{gj}$
Step 10: Generate $r = rand(0, 1)$
Step 11: If $r > 0.5$, $x_{ij} = A_{ij} + \beta|m_j - x_{ij}|ln(\frac{1}{u_{ij}})$
Step 12: Otherwise, $x_{ij} = A_{ij} - \beta|m_j - x_{ij}|ln(\frac{1}{u_{ij}})$
Step 13: If $f(x_i) \in F_R$ *assign* $f(x_i) = -M$
Step 14: Iff $(p_i) < f(x_i)$, *set* $p_i = x_i$
Step 15: Otherwise, $g = arg(max(f(p_i)))$
Step 16: if $z < Mg$, $z = z + 1$ *and follow Step 7*
Step 17: Otherwise, print the result
Step 18: Stop.

---

7 Numerical Experiments

For the illustration of the methodology, we have considered three numerical examples given below (Bhattacharyee et al. 2021). These examples are provided with crisp data. The input data for the triangular and pentagonal fuzzy models can be found in Tables 1, 2, 3, 4, 5 and 6, respectively. The defuzzified data are computed by the formulae described in Sect. 4.2 (Tables 7, 8, 9, 10, 11 and 12).

Example 1: Crisp Form

$$Maximize M_1(u, \lambda, M_T) = \int_0^{M_T} R_1(u, \lambda, t)dt$$

$$where\ R_1(u, \lambda t) = \left(\prod_{i=1}^{4}[1 - (1 - e^{-\lambda_i t})^{u_i}]\right)$$

subject to

$$C_s = \sum_{i=1}^{4} c_i u_i \leq C$$

$$W_s = \sum_{i=1}^{4} w_i u_i \leq W$$

Table 1 Data for Example 1 (LTFN)

i	$\tilde{\lambda}_i\ (\times 10^{-4})$	$\tilde{c}_i$	$\tilde{w}_i$
1	(9.415,9.431,9.446)	(0.2,1.2,2.4)	(3,5,6)
2	(5.118,5.129,5.138)	(2.0,2.3,2.8)	(2,4,5)
3	(8.323,8.338,8.349)	(3.0,3.4,3.9)	(6,8,9)
4	(16.151,16.252,16.354)	(4.0,4.5,4.8)	(5,7,8)

$\tilde{C}$= (50,56,60); $\tilde{W}$= (113,120,125)

Table 2 Data for Example 1 (LPtFN)

i	$\hat{\lambda}_i(\times 10^{-4})$	$\hat{c}_i$	$\hat{w}_i$
1	(9.102,9.415,9.431,9.446,9.521)	(0.1,0.2,1.2,2.4,2.6)	(2,3,5,6,8)
2	(5.108,5.118,5.129,5.138,5.141)	(1.8,2.0,2.3,2.8,2.9)	(1,2,4,5,6)
3	(8.320,8.323,8.338,8.349,8.356)	(2.7,3.0,3.4,3.9,4.1)	(4,6,8,9,10)
4	(16.143,16.151,16.252,16.354,16.361)	(3.7,4.0,4.5,4.8,4.9)	(3,5,7,8,10)

$\widehat{C}$= (48,50,56,60,63); $\widehat{W}$= (111,113,120,125,129)

Table 3 Data for Example 2 (LTFN)

i	$\tilde{\lambda}_i$ ($\times 10^{-4}$)	$\tilde{v}_i$	$\tilde{c}_i$	$\tilde{w}_i$
1	(9.415,9.431,9.446)	(0.15,1.0,1.48)	(6.4,7.0,7.7)	(6.7,7.0,7.6)
2	(5.118,5.129,5.138)	(1.4,2.0,2.7)	(6.5,7.0,7.8)	(7.8,8.0,8.4)
3	(8.323,8.338,8.349)	(2.4,3.0,3.8)	(4.8,5.0,5.6)	(7.5,8.0,8.3)
4	(16.151,16.252,16.354)	(3.6,4.0,4.8)	(8.3,9.0,9.5)	(5.1,6.0,6.7)
5	(7.147,7.257,7.361)	(1.6,2.0,2.7)	(3.4,4.0,4.7)	(8.7,9.0,9.7)
$\tilde{V}$= (103,110,123); $\tilde{C}$ = (167, 175, 188); $\tilde{W}$= (178,200,227)				

$$u_i \in Z^+, r_i(t) = e^{-\lambda_i t}, i = 1, 2, 3, 4;\ M_T = 100hrs,$$

$$C = 56, c = (c_1, c_2, c_3, c_4) = (1.2, 2.3, 3.4, 4.5),$$

$$W = 120, w = (w_1, w_2, w_3, w_4) = (5, 4, 8, 7)$$

$$\lambda = (\lambda_1, \lambda_2, \lambda_3, \lambda_4) = (0.0009431, 0.0005129, 0.0008338, 0.0016252).$$

Example 2: Crisp Form

$$Maximize M_1(u, \lambda, M_T) = \int_0^{M_T} R_1(u, \lambda, t)dt$$

$$where\ R_1(u, \lambda t) = \left(\prod_{i=1}^{5}[1 - (1 - e^{-\lambda_i t})^{u_i}]\right)$$

subject to

$$V_s = \sum_{i=1}^{5} v_i u_i^2 \leq V$$

$$C_s = \sum_{i=1}^{5} c_i[u_i + e^{\frac{u_i}{4}}] \leq C$$

$$W_s = \sum_{i=1}^{5} w_i[u_i e^{\frac{u_i}{4}}] \leq W$$

$$u_i \in Z^+, r_i(t) = e^{-\lambda_i t}, i = 1, 2, 3, 4, 5;\ M_T = 100hrs,$$

Table 4 Data for Example 2 (LPtFN)

i	$\widehat{\lambda}_i(\times 10^{-4})$	$\widehat{v}_i$	$\widehat{c}_i$	$\widehat{w}_i$
1	(9.408,9.415,9.431,9.446,9.449)	(0.13,0.15,1.0,1.48,1.50)	(6.3,6.4,7.0,7.7,7.9)	(6.5,6.7,7.0,7.6,7.9)
2	(5.109,5.118,5.129,5.138,5.140)	(1.2,1.4,2.0,2.7,2.9)	(6.3,6.5,7.0,7.8,7.9)	(7.7,7.8,8.0,8.4,8.6)
3	(8.321,8.323,8.338,8.349,8.350)	(2.2,2.4,3.0,3.8,3.9)	(4.5,4.8,5.0,5.6,5.8)	(7.3,7.5,8.0,8.3,8.6)
4	(16.149,16.151,16.252,16.354,16.357)	(3.5,3.6,4.0,4.8,4.9)	(8.1,8.3,9.0,9.5,9.8)	(5.0,5.1,6.0,6.7,6.9)
5	(7.141,7.147,7.257,7.361,7.364)	(1.3,1.6,2.0,2.7,2.9)	(3.1,3.4,4.0,4.7,4.9)	(8.4,8.7,9.0,9.7,9.9)

$\widehat{V}$= (101,103,110,123,125); $\widehat{C}$ = (160, 167, 175, 188, 194); $\widehat{W}$=(171,178,200,227,234)

Table 5 Data for Example 3 (LTFN)

i	$\tilde{\lambda}_i$ $(\times 10^{-4})$	$\tilde{c}_i$	$\tilde{w}_i$
1	(9.415,9.431,9.446)	(3,5,6)	(2,5,6)
2	(5.118,5.129,5.138)	(2,4,7)	(3,4,7)
3	(8.323,8.338,8.349)	(8,9,11)	(7,9,10)
4	(16.151,16.252,16.354)	(5,7,8)	(6,7,8)
5	(7.147,7.257,7.361)	(4,7,9)	(5,7,10)
6	(2.015,2.020,2.027)	(4,5,8)	(4,5,7)
7	(9.422,9.431,9.448)	(5,6,9)	(3,6,8)
8	(21.054,21.072,21.083)	(7,9,10)	(6,9,11)
9	(4.071,4.082,4.093)	(2,4,5)	(2,4,5)
10	(16.241,16.252,16.354)	(4,5,7)	(4,5,7)
11	(6.173,6.188,6.197)	(5,6,8)	(5,6,8)
12	(23.562,23.572,23.579)	(4,7,8)	(5,7,8)
13	(1.001,1.005,1.016)	(7,9,10)	(6,9,10)
14	(5.109,5.129,5.137)	(6,8,11)	(7,8,11)
15	(23.561,23.572,23.583)	(3,6,8)	(4,5,7)
$\tilde{C}$= (376,400,460); $\tilde{W}$= (375,414,475)			

Table 6 Data for Example 3 (LPtFN)

i	$\hat{\lambda}_i(\times 10^{-4})$	$\hat{c}_i$	$\widehat{w}_i$
1	(9.414,9.415,9.431,9.446,9.449)	(1,3,5,6,7)	(1,2,5,6,7)
2	(5.108,5.118,5.129,5.138,5.140)	(1,2,4,7,8)	(2,3,4,7,8)
3	(8.320,8.323,8.338,8.349,8.354)	(7,8,9,11,12)	(5,7,9,10,12)
4	(16.142,16.151,16.252,16.354,16.359)	(3,5,7,8,9)	(4,6,7,9,10)
5	(7.144,7.147,7.257,7.361,7.363)	(3,4,7,9,11)	(4,5,7,10,12)
6	(2.008,2.015,2.020,2.027,2.029)	(2,4,5,8,11)	(3,4,5,7,9)
7	(9.415,9.422,9.431,9.448,9.451)	(1,5,6,9,10)	(1,3,6,8,10)
8	(21.049,21.054,21.072,21.083,21.085)	(5,7,9,10,12)	(4,6,9,11,12)
9	(4.068,4.071,4.082,4.093,4.095)	(1,2,4,5,7)	(1,2,4,5,7)
10	(16.235,16.241,16.252,16.354,16.359)	(1,4,5,7,9)	(3,4,5,7,8)
11	(6.171,6.173,6.188,6.197,6.199)	(4,5,6,8,9)	(3,5,6,8,9)
12	(23.555,23.562,23.572,23.579,23.582)	(3,4,7,8,10)	(4,5,7,8,9)
13	(1.000,1.001,1.005,1.016,1.018)	(5,7,9,10,11)	(4,6,9,10,11)
14	(5.107,5.109,5.129,5.137,5.139)	(4,6,8,11,12)	(4,7,8,11,13)
15	(23.559,23.561,23.572,23.583, 23.585)	(1,3,6,8,11)	(3,4,5,7,9)
$\widehat{C}$= (371,376,400,460,465); $\widehat{W}$= (368,375,414,475,482)			

Table 7 Crispified data for Example 1 (LTFN)

i	$\tilde{\lambda}_i$ ($\times 10^{-4}$)	$\tilde{c}_i$	$\tilde{w}_i$
1	9.430667	1.266667	4.666667
2	5.128333	2.366667	3.666667
3	8.336667	3.433333	7.666667
4	16.25233	4.433333	6.666667
$\tilde{C}$= 55.33333; $\tilde{W}$= 119.3333			

Table 8 Crispified data for Example 1 (LPtFN)

i	$\hat{\lambda}_i(\times 10^{-4})$	$\hat{c}_i$	$\hat{w}_i$
1	9.353899	1.325532	4.777778
2	5.126107	2.373684	3.50000
3	8.337054	3.423188	7.222222
4	16.25225	4.346667	6.50000
$\hat{C}$= 55.26667; $\hat{W}$= 119.533			

Table 9 Crispified data for Example 2 (LTFN)

i	$\tilde{\lambda}_i$ ($\times 10^{-4}$)	$\tilde{v}_i$	$\tilde{c}_i$	$\tilde{w}_i$
1	9.430667	0.876667	7.033333	7.10000
2	5.128333	2.033333	7.10000	8.066667
3	8.336667	3.066667	5.133333	7.933333
4	16.25233	4.133333	8.933333	5.933333
5	7.25500	2.10000	4.033333	9.133333
$\tilde{V}$= 112; $\tilde{C}$ = 176.6667; $\tilde{W}$= 201.6667				

Table 10 Crispified data for Example 2 (LPtFN)

i	$\hat{\lambda}_i(\times 10^{-4})$	$\hat{v}_i$	$\hat{c}_i$	$\hat{w}_i$
1	9.429454	0.81500	7.075862	7.176812
2	5.126124	2.05000	7.124138	8.126667
3	8.335745	3.074194	5.173016	7.926984
4	16.25275	4.2	8.926437	5.925714
5	7.253245	2.123457	4.023656	9.173333
$\hat{V}$=113; $\hat{C}$ = 177.2303; $\hat{W}$= 202.5				

Table 11 Crispified data for Example 3 (LTFN)

i	$\tilde{\lambda}_i$ ($\times 10^{-4}$)	$\tilde{c}_i$	$\tilde{w}_i$
1	9.430667	4.666667	4.333333
2	5.128333	4.333333	4.666667
3	8.336667	9.333333	8.666667
4	16.25233	6.666667	7.333333
5	7.25500	6.666667	7.333333
6	2.020667	5.666667	5.333333
7	9.433667	6.666667	5.666667
8	21.06967	8.666667	8.666667
9	4.08200	3.666667	3.666667
10	16.28233	5.333333	5.333333
11	6.18600	6.333333	6.333333
12	23.5710	6.333333	6.666667
13	1.007333	8.666667	8.333333
14	5.12500	8.333333	8.666667
15	23.57200	5.666667	5.333333
$\tilde{C}$= 412; $\tilde{W}$= 421.3333			

Table 12 Crispified data for Example 3 (LPtFN)

i	$\hat{\lambda}_i(\times 10^{-4})$	$\hat{c}_i$	$\widehat{w}_i$
1	9.43101	2.222222	4.000000
2	5.125846	4.500000	5.000000
3	8.336522	9.500000	8.500000
4	16.25149	6.222222	7.222222
5	7.253749	6.769231	7.769231
6	2.019636	6.282051	5.777778
7	9.433946	6.153846	5.500000
8	21.06772	8.500000	8.230769
9	4.081741	3.777778	3.777778
10	16.29725	4.666667	5.500000
11	6.185000	6.500000	6.222222
12	23.56942	6.272727	6.500000
13	1.008758	8.222222	7.727273
14	5.123000	8.230769	8.717949
15	23.57200	5.777778	5.777778
$\widehat{C}$ = 418; $\widehat{W}$ = 425			

$$V = 110, v = (v_1, v_2, v_3, v_4, v_5) = (1, 2, 3, 4, 2),$$

$$C = 175, c = (c_1, c_2, c_3, c_4, c_5) = (7, 7, 5, 9, 4),$$

$$W = 200,\ w = (w_1, w_2, w_3, w_4, w_5) = (7, 8, 8, 6, 9)$$

$$\lambda = (\lambda_1, \lambda_2, \lambda_3, \lambda_4, \lambda_5) = (0.0009431, 0.0005129, 0.0008338, 0.0016252, 0.0007257).$$

Example 3: Crisp Form

$$Maximize M_1(u, \lambda, M_T) = \int_0^{M_T} R_1(u, \lambda, t)dt$$

$$where\ R_1(u,\ \lambda t)\ =\ \left(\prod_{i=1}^{15}[1 - (1 - e^{-\lambda_i t})^{u_i}]\right)$$

subject to

$$C_s\ =\ \sum_{i=1}^{15} c_i u_i\ \le\ C$$

$$W_s = \sum_{i=1}^{15} w_i u_i \le W$$

$$u_i \in Z^+, r_i(t) = e^{-\lambda_i t}, i = 1, 2, \ldots, 15, M_T\ =\ 100hrs, C = 400,$$
$$c = (c_1, c_2, \ldots, c_{15}) = (5, 4, 9, 7, 7, 5, 6, 9, 4, 5, 6, 7, 9, 8, 6),$$

$$W = 414, w = (w_1, w_2, \ldots, w_{15}) = (5, 4, 9, 7, 7, 5, 6, 9, 4, 5, 6, 7, 9, 8, 5)$$

$$\lambda = (\lambda_1, \lambda_2, \ldots, \lambda_{15}) = \begin{pmatrix} 0.0009431, 0.0005129, 0.0008338, 0.0016252, 0.0007257, \\ 0.0002020, 0.0009431, 0.0021072, 0.0004082, 0.0016252, \\ 0.0006188, 0.0023572, 0.0001005, 0.0005129, 0.0023572 \end{pmatrix}.$$

8 Result Discussions

Here, we have formulated and solved three numerical experiments for testing the efficiency of our proposed method to maximize the MDL as well as to maximize reliability of the system under optimal redundancies. In this work, we have executed 30 independent runs for each numerical problem with the help of HQPSO algorithm coded in C++ in a notebook with Intel i3 processor, 4 GB RAM in Linux operating system. To study the robustness, we have procured the results to identify the best and worst values of MDL, its average value, standard deviation and the execution time along with the corresponding system reliability. In this HWQPSO, the population size and the maximum number of generations for the three experiments are taken respectively as 70, 100; 80, 150 and 300, 700.

From Tables 13, 14, and 15, we can see the results of problems 1, 2, and 3, respectively. It is evident that the respective standard deviations are 0, 0, and 0.0023557. Figures 5, 6 and 7, respectively, show convergence history of the objective functions (MDL) in crisp form as stable w.r.t. the number of generations in wide range. Table 16 presents the comparative results of Example 1 in crisp, triangular fuzzy, and pentagonal fuzzy cases indicating that it achieved the best result in PtFN case. From Table 17, the comparative results of Example 2 can be seen and it is noticed that the best result corresponds to PtFN case. The comparative results of Example 3 are given in Table 18 indicating that the best output is obtained in PtFN case.

Table 13 Statistical results for a crisp form of Example 1 (S_{size}-70, Maxgen-100)

Best	Worst	Average	Standard deviation	Average running time
99.9987559819	99.9987559819	99.9987559819	0	0.068688

Table 14 Statistical results for a crisp form of Example 2 (S_{size}-80, Maxgen-150)

Best	Worst	Average	Standard deviation	Average running time
98.4880171489	98.4880171489	98.4880171489	0	0.0729564

Table 15 Statistical results for a crisp form of Example 3 (S_{size}-100, Maxgen-200)

Best	Worst	Average	Standard deviation	Average running time
99.9774303072	99.9684707239	99.97656270762	0.0023557	5.623954

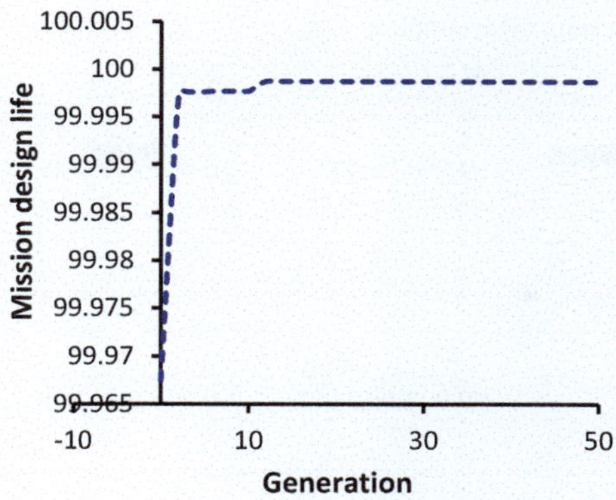

Fig. 5 Convergence history of MDL of Example 1 (crisp form) using HQPSO

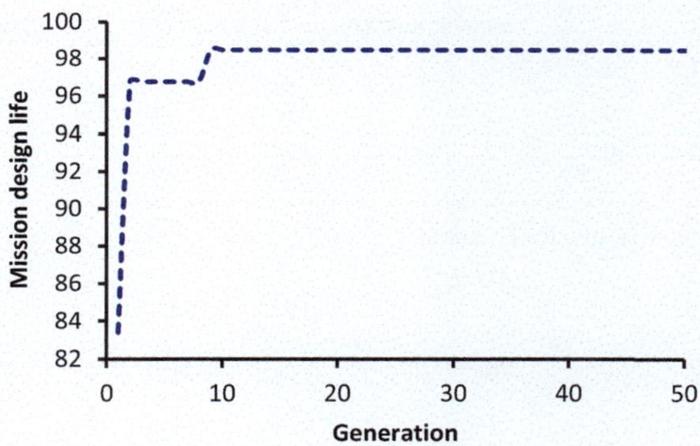

Fig. 6 Convergence history of MDL of Example 2 (crisp form) using HQPSO

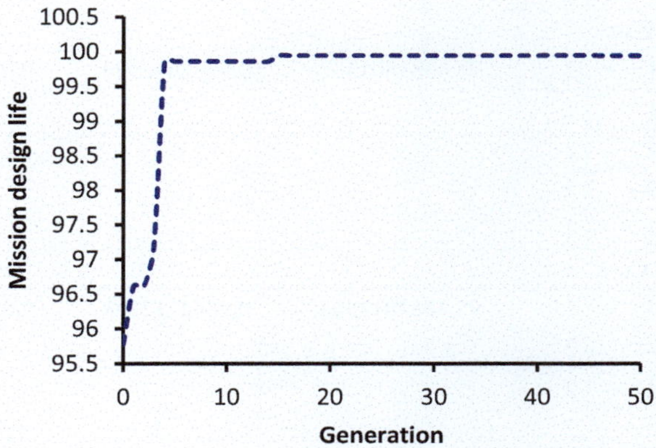

Fig. 7 Convergence history of MDL of Example 3(crisp form) using HQPSO

Table 16 Comparative results for Example 1

i	R_i	u_i	MDL^*	SR^*	CPU Time (s)
Crisp Case					
1	0.999994	5	99.9987559819	0.9999355000	0.069200
2	0.999994	4			
3	0.999959	4			
4	0.999989	6			
Triangular Fuzzy Case					
1	0.999994	5	99.9987565642	0.9999355280	0.054976
2	0.999994	4			
3	0.999959	4			
4	0.999989	6			
Pentagonal Fuzzy Case					
1	0.999994	5	**99.9987605507**	**0.999935757**	**0.051477**
2	0.999994	4			
3	0.999959	4			
4	0.999989	6			

Table 17 Comparative results for Example 2

i	R_i	u_i	MDL^*	SR^*	CPU Time (s)
Crisp Case					
1	0.991900	2	98.4880171489	0.9562601906	0.068281
2	0.997500	2			
3	0.993600	2			
4	0.977500	2			
5	0.995100	2			
Triangular Fuzzy Case					
1	0.991901	2	98.4881801367	0.9562648909	0.071150
2	0.997501	2			
3	0.993602	2			
4	0.977499	2			
5	0.995103	2			
Pentagonal Fuzzy Case					
1	0.991903	2	**98.4884012663**	**0.9562712705**	**0.072423**
2	0.997503	2			
3	0.993604	2			
4	0.977498	2			
5	0.995105	2			

Table 18 Comparative results for Example 3

i	R_i	u_i	MDL^*	SR^*	CPU Time (s)
Crisp Case					
1	0.999934	4	99.9774303072	0.9989171631	6.184960
2	0.999994	4			
3	0.999959	4			
4	0.999924	5			
5	0.999976	4			
6	0.999992	3			
7	0.999934	4			
8	0.999752	5			
9	0.999936	3			
10	0.999924	5			
11	0.999987	4			
12	0.999914	6			
13	0.999900	2			
14	0.999875	3			
15	0.999914	6			
Triangular Fuzzy Case					
1	0.999934	4	99.9811300948	0.9991167640	4.947174
2	0.999994	4			
3	0.999959	4			
4	0.999924	5			
5	0.999976	4			
6	0.999992	3			
7	0.999934	4			
8	0.999753	6			
9	0.999936	3			
10	0.999923	5			
11	0.999987	4			
12	0.999914	6			
13	0.999900	2			
14	0.999875	3			
15	0.999914	6			
Pentagonal Fuzzy Case					
1	0.999934	4	**99.9890618878**	**0.9993957023**	**4.844208**
2	0.999994	4			
3	0.999959	4			

(continued)

Table 18 (continued)

i	R_i	u_i	MDL^*	SR^*	CPU Time (s)
4	0.999924	5			
5	0.999976	4			
6	0.999992	3			
7	0.999934	4			
8	0.999953	6			
9	0.999997	4			
10	0.999923	5			
11	0.999987	4			
12	0.999914	6			
13	0.999999	3			
14	0.999994	4			
15	0.999914	6			

9 Conclusions and Future Directions

This work explores a more realistic and practical form of redundancy allocation problem where time-dependent reliabilities for the components in decreasing exponential function are considered. We use the mission design life (MDL) as the objective function rather than the traditional system reliability. Integrating the system reliability between zero (0) and mission time (M_T), the MDL is obtained. The objective function, MDL is then maximized along with the system reliability under optimal redundancy allocations (Fig. 8 and Tables 19, 20, 21).

Table 19 Best found solution of Example 1 (LPtFN case). [Popsize-70; Maximum generations-100]

x_1	x_2	x_3	x_4	R_1	R_2	R_3	R_4	Mission Design Life	System Reliability
5	4	4	6	0.999994	0.999994	0.999959	0.999989	99.9987605507	0.999935757

Table 20 Worst found a solution of Example 1 (LPtFN case). [Popsize-70; Maximum generations-100]

x_1	x_2	x_3	x_4	R_1	R_2	R_3	R_4	Mission Design Life	System Reliability
5	4	4	6	0.999994	0.999994	0.999959	0.999989	99.9987605507	0.999935757

Table 21 Statistical data for Example 1 (LPtFN case)

Best	Worst	Average	Standard deviation	Average running time
99.9987605507	99.9987605507	99.9987605507	0	0.064034

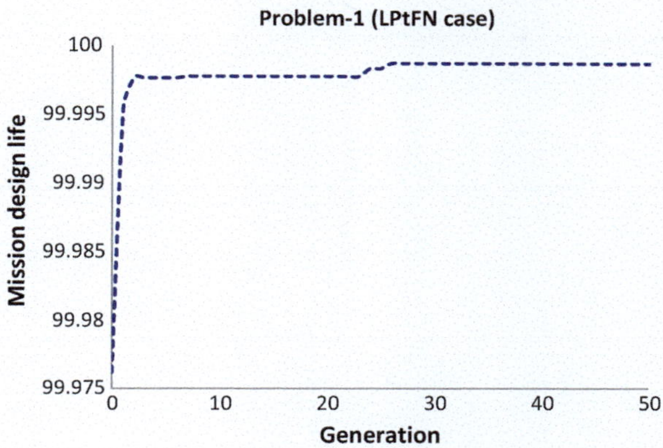

Fig. 8 Convergence history of MDL of Example 1 (LPtFN case) using HQPSO

The two fuzzy models (triangular fuzzy and pentagonal fuzzy) are developed to show the effects of uncertainty along with the crisp one of the reliability system. To evaluate the MDL as an integral of the quite complex integral, Simpson's 1/3 rule is utilized. A new PSO algorithm is developed by combining the characteristics of QPSO and the Big-M penalty technique. This hybridized algorithm HQPSO is implemented for solving the three numerical examples under consideration in three different forms, namely crisp, triangular fuzzy, and pentagonal fuzzy. The performance of the proposed algorithm is well established in these experiments (Fig. 9 and Tables 22, 23, 24).

Table 22 Best found solution of Example 2 (LPtFN case). [Popsize-80; Maximum generations-150]

x_1	x_2	x_3	x_4	x_5	R_1	R_2	R_3	R_4	R_5	Mission Design Life	System Reliability
2	2	2	2	2	0.991903	0.997503	0.993604	0.977498	0.995105	98.4884012663	0.9562712705

Table 23 Worst found a solution of Example 2 (LPtFN case). [Popsize-80; Maximum generations-150]

x_1	x_2	x_3	x_4	x_5	R_1	R_2	R_3	R_4	R_5	Mission Design Life	System Reliability
2	2	2	2	2	0.991903	0.997503	0.993604	0.977498	0.995105	98.4884012663	0.9562712705

Table 24 Statistical data for Example 2 (LPtFN case)

Best	Worst	Average	Standard deviation	Average running time
98.4884012663	98.4884012663	98.4884012663	0	0.084681

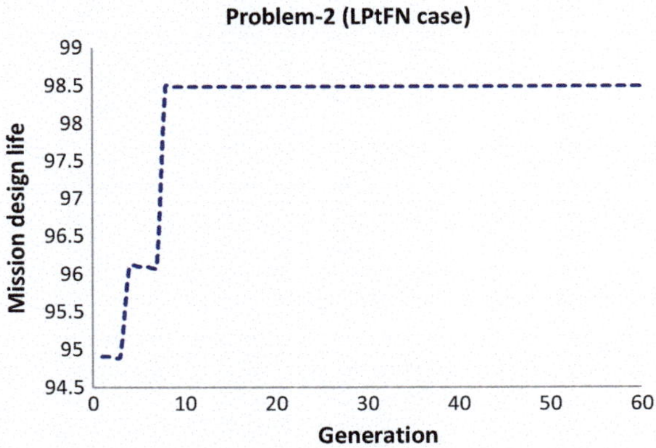

Fig. 9 Convergence history of MDL of Example 2 (LPtFN case) using HQPSO

The proposed methodology for maximizing MDL can be applied in the field of reliability optimization, system design, engineering design, industrial problems, etc. Soft computing techniques like GA, ABC algorithm, DE, Taboo search, Cuckoo search, Neural Network, Tournament-based PSO, etc. can be employed to solve this kind of problem. To consider the uncertainty, several other imprecise environments can be considered (Fig. 10 and Tables 25, 26, 27).

Table 25 Best found solution of Example 3 (LPtFN case). [Popsize-300; Maximum generations-700]

x_1	x_2	x_3	x_4	x_5	x_6	x_7	x_8	x_9	x_{10}	x_{11}	x_{12}	x_{13}	x_{14}	x_{15}
4	4	4	5	4	3	4	6	4	5	4	6	3	4	6

R_1	R_2	R_3	R_4	R_5	R_6	R_7	R_8	R_9	R_{10}
0.999934	0.999994	0.999959	0.999924	0.999976	0.999992	0.999934	0.999953	0.999997	0.999923

R_{11}	R_{12}	R_{13}	R_{14}	R_{15}	Mission Design Life	System Reliability
0.999987	0.999914	0.999999	0.999994	0.999914	99.9890618878	0.9993957023

Table 26 Worst found solution of Example 3 (LPtFN case). [Popsize-300; Maximum generations-700]

x_1	x_2	x_3	x_4	x_5	x_6	x_7	x_8	x_9	x_{10}	x_{11}	x_{12}	x_{13}	x_{14}	x_{15}
5	4	4	5	4	3	5	5	4	6	4	6	3	3	6

R_1	R_2	R_3	R_4	R_5	R_6	R_7	R_8	R_9	R_{10}
0.999994	0.999994	0.999959	0.999924	0.999976	0.999992	0.999994	0.999753	0.999997	0.999988

R_{11}	R_{12}	R_{13}	R_{14}	R_{15}	Mission Design Life	System Reliability
0.999987	0.999914	0.999999	0.999875	0.999914	99.9859984351	0.9992618745

Table 27 Statistical data for Example 3 (LPtFN case)

Best	Worst	Average	Standard deviation	Average running time
99.9890618878	99.9859984351	99.9885610794	0.000616189	5.91237

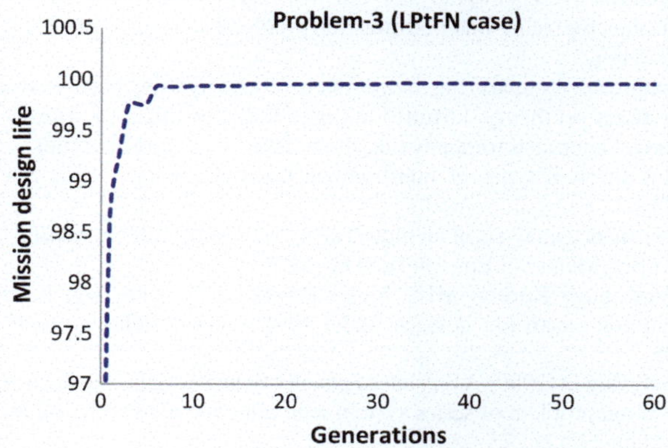

Fig. 10 Convergence history of MDL of Example 3 (LPtFN case) using HQPSO

Acknowledgements The authors are thankful to the anonymous referees for their constructive comments and suggestions. Also, the authors would like to express their hearty thanks to the Editorial Board Members for their cooperation. The third author would like to acknowledge the financial support provided by the DSTBT, Govt. of West Bengal, India vide Memo No. 30 (Sanc)/ST/P/S&T/16G-43/2017 Dated 12/06/2018.

References

Ahmadivala M, Mattrand C, Gayton N, Dumasa A, Yalamas T, Orcesi A (2019) Application of AK-SYS method for time-dependent reliability analysis. 24ème CongrèsFrançais de Mécanique Brest, 26 au 30 Août 2019

Ardakan M, Zahra M, Hamadani A, Elsayed EA (2017) Reliability optimization by considering time dependent reliability for components. Qual Reliab Eng Int 33(8):1641–1654

Bhattacharyee N, Kumar N, Mahato SK, Bhunia AK (2021) Development of a blended particle swarm optimization to optimize mission design life of a series-parallel reliable system with time dependent component reliabilities in imprecise environments. Soft Computing, Accepted

Bhunia AK, Sahoo L, Roy D (2010) Reliability stochastic optimization for a series system with interval component reliability via genetic algorithm. Appl Math Comput 216:929–939

Bhunia AK, Sahoo L (2011) Genetic algorithm based reliability optimization in interval environment. In: Nedjah N (ed) Innovative computing methods and their applications to engineering problems, SCI 357. Springer-Verlag, Berlin Heidelberg, pp 13–36

Chen DX (1977) Fuzzy reliability of the bridge circuit system. Syst Eng Theory Pract 11:109–112

Chern MS, Jan RH (1986) Reliability optimization problems with multiple constraints. IEEE Trans Reliab 35(4):431–436

Garg H (2015) Multi-objective optimization problem of system reliability under intuitionistic fuzzy set environment using cuckoo Search algorithm J. Intell Fuzzy Syst 29(4):1653–1669
Garg H, Rani M (2013) An approach for reliability analysis of industrial systems using PSO and IFS technique. ISA Trans Elsevier 52(6):701–710
Garg H, Rani M, Sharma SP (2014a) An approach for analyzing the reliability of industrial systems using soft-computing based technique. Expert Syst Appl 41(2):489–501
Garg H, Rani M, Sharma SP, Vishwakarma Y (2014b) Intuitionistic fuzzy optimization technique for solving multi-objective reliability optimization problems in interval environment. Expert Syst Appl 41(7):3157–3167
Garg H (2013) An approach for analyzing fuzzy system reliability using particle swarm optimization and intuitionistic fuzzy set theory. J Multi Val Logic Soft Comput 21(3–4):335–354
Garg H (2016) A novel approach for analyzing the reliability of series-parallel system using credibility theory and different types of intuitionistic fuzzy numbers. J Braz Soc Mech Sci Eng 38(3):1021–1035
Garg H (2017) Performance analysis of an industrial system using soft computing based hybridized technique. J Braz Soc Mech Sci Eng 39(4):1441–1451
González LCM, Rodríguez Borbón MIR, Valles-Rosales DJ, Valle AD, Rodriguez A (2015) Reliability model for electronic devices under time varying voltage. Qual Reliab Eng Int 32(4):1295–1306
Gupta R, Bhunia AK, Roy D (2009) A GA based penalty function technique for solving constrained redundancy allocation problem of series system with interval valued reliabilities of components. J Comput Appl Math 232(2):275–284
Hamadani AZ, Khorshidi HA (2013) System reliability optimization using time value of money. Int J Adv Manuf Technol 66(14):97–106
Hu and Mahadevan (2015) Time-dependent system reliability analysis using random field discretization. J mech design 137(10). https://doi.org/10.1115/1.4031337
Huang H (1996) Fuzzy multi-objective optimization decision-making of reliability of series system. Microele Reliab 37(3):447–449
Hwang CL, Tillman FA, Kuo W (1979) Reliability optimization by generalized Lagrangian-function based and reduced-gradient methods. IEEE Trans Reliab 28:316–319
Jamkhaneh E (2017) System reliability using generalized intuitionistic fuzzy exponential lifetime distribution. Int J Soft Comput Eng (IJSCE) 7:2231–2307
Kennedy J, Eberhart R (1995) Particle swarm optimization. In Proceedings of ICNN'95-International Conference on Neural Networks 4:1942–1948
Khalili-Damghani K, Abtahi AR, Tavana M (2013) A new multi objective particle swarm optimization method for solving reliability redundancy allocation problems. Reliab Eng Syst Saf 111:58–75
Kuo W, Prasad V, Tillman FA, Hwang CL (2001) Optimal reliability design fundamentals and applications. Cambridge University Press, Cambridge LINGO User Guide (2013) Lindo Systems Inc, Chicago, IL
Mahapatra GS, Roy TK (2012) Reliability evaluation of bridge system with fuzzy reliability of components using interval nonlinear programming. Electron J Appl Stat Anal 5(2):151–163
Mahapatra GS, Roy TK (2006) Fuzzy multi-objective mathematical programming on reliability optimization model. Appl Math Comput 174(1):643–659
Mahapatra GS, Roy TK (2009) Reliability evaluation using triangular intuitionistic fuzzy numbers arithmetic operations. World Acad Sci Eng Technol 50:574–581
Mahapatra GS, Roy TK (2011) Optimal redundancy allocation in series-parallel system using generalized fuzzy number. Tamsui Oxford J Inf Math Sci 27(1):1–20
Mahato SK, Bhattacharyee N, Paramanik R (2020) Fuzzy reliability redundancy optimization with signed distance method for defuzzification using genetic algorithm. Int J Oper Res (IJOR) 37(3):307–323
Mahato SK, Bhunia AK (2016) Reliability optimization in fuzzy and interval environments. Lap Lambert Academic Publishing (Book)

Mahato SK, Sahoo L, Bhunia AK (2012) Reliability-redundancy optimization problem with interval valued reliabilities of components via genetic algorithm. J Inf Comput Sci (UK) 7(4):284–295
Mahato SK, Sahoo L, Bhunia AK (2013) Effects of defuzzification methods in redundancy allocation problem with fuzzy valued reliabilities via genetic algorithm. Int J Inf Comput Sci 2(6):106–115
Misra K (1986) On optimal reliability design: a review. Syst Sci 12(4):5–30
Misra B, Sharma U (1991) An efficient algorithm to solve integer-programming problems arising in system-reliability design. IEEE Trans Reliab 40(1):81–91
Mori Y, Ellingwood BR (1993) Time-dependent system reliability analysis by adaptive importance sampling, vol 12, issue 1, pp 59–73
Mourelatos ZP, Majcher M, Pandey V, Baseski I (2015) Time-dependent reliability analysis using the total probability theorem. ASME J Mech Des 137(3). https://doi.org/10.1115/1.4029326
Nakagawa Y, Miyazaki S (1981) Surrogate constraints algorithm for reliability optimization problems with two constraints. IEEE Trans Reliab 30(2):181–184
Nakagawa Y, Nakashima K (1977) A heuristic method for determining optimal reliability allocation. IEEE Trans Reliab 26(3):156–161
Park K (1987) Fuzzy apportionment of system reliability. IEEE Trans Reliab 36(1):129–132
Sahoo L, Bhunia AK, Kapur PK (2012) Genetic algorithm based multi-objective reliability optimization in interval environment. Comput Ind Eng 62(1):152–160
Sahoo L, Bhunia AK, Mahato SK (2014) Optimization of system reliability for series system with fuzzy component reliabilities by genetic algorithm J. Uncertain Syst 8(2):136–148
Sahoo L, Bhunia AK, Roy D (2010) A genetic algorithm based reliability redundancy optimization for interval valued reliabilities of components. J Appl Quant Methods 5(2):270–287
Sahoo L, Bhunia AK, Roy D (2013) Reliability optimization with high and low level redundancies in interval environment via genetic algorithm. Int J Syst Assur Eng Manage. https://doi.org/10.1007/s13198-013-0199-9
Sun XL, Li D (2002) Optimization condition and branch and bound algorithm for constrained redundancy optimization in series system. Optim Eng 3(1):53–65
Sung CS, Cho YK (2000) Reliability optimization of a series system with multiple choice and budget constraints. Eur J Oper Res 127(1):159–171
Tillman FA, Hwang CL, Kuo W (1980) System effectiveness models: an annotated bibliography. IEEE Trans Reliab 29(4):295–304
Wang L, Wang X, Wang R, Chen X (2015) Time-dependent reliability modeling and analysis method for mechanics based on convex process, mathematical problems in engineering, vol 2015, Article ID 914893, p 16. http://dx.doi.org/https://doi.org/10.1155/2015/914893
Zafar T, Wang Z (2020) Time-dependent reliability prediction using transfer learning. Struct Multidiscip Optim 62:147–158. https://doi.org/10.1007/s00158-019-02475-5
Zhu Z (2016) Efficient time-dependent system reliability analysis Doctoral Dissertations, 2552. https://scholarsmine.mst.edu/doctoral_dissertations/2552

Chapter 14
Mediative Fuzzy Pythagorean Algorithm to Multi-criteria Decision-Making and Its Application in Medical Diagnostic

M. K. Sharma, Nitesh Dhiman, Vandana, and Vishnu Narayan Mishra

Abstract Soft computing approaches have been used for confronting the fuzzy embedded process in various decision-making problems. Traditional fuzzy sets (FS), intuitionistic fuzzy sets (IFSs), and Pythagorean fuzzy sets (PFS) frequently applied in the measurement of correlation coefficient techniques based on similarity and distance for various parameters. But in the present era of conflicting factors, it is quite difficult to handle these problems with the existing techniques. Correlation coefficient plays a leading role between the existing computing techniques due to its nature to estimate the interrelationship and their interdependency between factors of different FS. This article proposes a mediative Pythagorean fuzzy technique with a novel mediative correlation coefficient for multi-criteria in decision-making and its implementation in diagnostic process. The accuracy, efficacy, and the dominance of proposed approach over the previous approaches are explained with measurement of correlation coefficient between mediative PF Samong and looks more promising than existing techniques. In this work we also proved some theoretical results, theorems, and described the numerical validation of this work. We corroborate the mediative fuzzy Pythagorean fuzzy approach with some properties with loss of generality and numerical validation of our technique.

Keywords Mediative fuzzy logic · Mediative Pythagorean fuzzy set · Medical knowledge · Correlation coefficient · Mediative fuzzy relation

---

M. K. Sharma (✉) · N. Dhiman
Department of Mathematics, Chaudhary Charan Singh University, Meerut 250004, India

Vandana
Department of Computer Application SCRIET, Chaudhary Charan Singh University, Meerut 250004, India

V. N. Mishra
Department of Mathematics, Indira Gandhi National Tribal, University, Lalpur, Amarkantak, Anuppur 484887, Madhya Pradesh, India

I. Ali et al. (eds.), *Computational Modelling in Industry 4.0*,
https://doi.org/10.1007/978-981-16-7723-6_14

1 Introduction

Ambiguity is a big barricade to acknowledge many decision-making problems in many real-life situations are embroil with indecisions. An incredible vision of alleviation to the decision-makers has been brought by the discovery of fuzzy set (FS) theory (Zadeh 1965), due to the potential of modelling the models to restrict the embedded unpredictability in decision-making. But FS theory could not resolve the entire decision-making problem accurately, due to its consideration of membership grade of the concerned variable only. For the better modelling of real-life arguments (Atanassov in 1986) introduced intuitionistic fuzzy sets (IFSs). Intuitionistic fuzzy sets constituent membership grade together with non-membership grade with the tendency of hesitation margin of the concerned information. Intuitionistic fuzzy set theory has been used in many decision-making problems with addition to several tools including distance measures, similarity measures by (Atanassov 1999; Boran and Akay 2014; Chen and Chang 2016; Chen et al. 2016a; Zeng et al.2019; Zhang et al.2020). In 1895, Karl Pearson introduced the correlation coefficient. Correlation coefficient plays a major role in the measurement of similarity, interdependency, and interrelationship among the data. Chiang and Lin (1999) and Murthy et al. (1985) generalized the correlation coefficient analysis into fuzzy domain to handle the uncertain fuzzy data. Ejegwa (2020), Mitchell (2004), Xuan Thao (2018), Xu et al. (2008), Hong and Hwang (1995), and Zeng and Li (2007) also described the correlation coefficients in the surroundings of intuitionistic fuzzy data and also used to handle many 'multi-criteria decision-making problems (MCDMP)'. Gerstenkorn and Manko (1991) inaugurated the work on correlation coefficient between IFSs by using correlation and informational energies. Garg (2016), introduced Pythagorean fuzzy set (PFS) to describe the correlation coefficient and also used this concept over MCDMP. Thao (2019) and Ejegwa (2020), also used the concept of PFS in the measurement of correlation coefficient and also applied this concept over medical diagnosis process. So, there are many literatures that exist to enhance the applications of traditional fuzzy set, IFS and PFS in many research fields. But some real-life-based problem exists in which, contradictory, non-contradictory, and doubtful cannot be handled by the previous logic studies. To overcome these types of situations (Montiel et al. 2005a) introduced the mediative fuzzy logic. There are so many existing studies that can elaborate the applications of mediative fuzzy logic in many contradictory information management problems (Montiel et al. 2008; Dhiman and Sharma 2019a, b, c). Our basic objectives of this research work are

(a) To define a mediative PFS that can handle the uncertainty, contradictory, and non-contradictory information.
(b) To develop a mathematical model for 'mediative Pythagorean fuzzy sets' with the help of correlation coefficient based on mediative fuzzy set (Dhiman and Sharma 2019a, b, c) called mediative fuzzy correlation coefficient, designate over [0, 1].
(c) To determine a multi-criteria decision-making medical diagnostic process in the context of mediative fuzzy correlation coefficient.

We will develop a mediative Pythagorean correlation coefficient and it will indicate a value closer to one, when two mediative fuzzy data sets are strongly related, and a value closer to zero, when two mediative fuzzy data sets are weakly related. But whenever the mediative correlation coefficient will zero, it will indicate that there is no relationship between the data. The main purpose of this work.

(i) Study the performance of methods given by Gerstenkorn and Manko (1991, Zeng and Li (2007), Xu et al. (2008), and Garg (2016) for calculating the correlation coefficients and Ejegwa et al. (2020) calculate correlation coefficient based on PFS and thus got an idea to introduced mediative Pythagorean fuzzy correlation coefficient.
(ii) Will prove some mathematical properties for the consistency mediative Pythagorean fuzzy correlation coefficient.
(iii) Will authenticate the efficacy of the proposed techniques over the Pythagorean fuzzy correlation method.
(iv) Will determine the superior method for multi-criteria decision-making given by Sanchez (1979, Szmidt and Kacprzyk (2001, 2004, 2010), and by Yager (2013, 2014, 2016; Yager and Abbasov 2013) by using this PFS.

For an easy elaboration, we have divided the present work into eight sections. The second section represents some basic concepts of mediative fuzzy logic, Pythagorean mediative fuzzy set, and mediative fuzzy compositional relations. Section three consists of the subsisting approaches for qualification of correlation coefficients designate over [0, 1]. In section four we proposed mediative fuzzy correlation coefficient defined within [0, 1]. Some theorems on mediative fuzzy correlation coefficient have also been proved in section five. In section six, we compared our introduced approach with the previous techniques and showed the supremacy of our technique over the existing techniques. Section seven shows the medical illustration of proposed work in MCDMP for diagnosis process and also simulation for mediative correlation coefficients is defined in this section. In section eight we conclude the discussion of the entire work.

2 Basic Notions

This segment, contains some basic concepts allied with mediative fuzzy sets (MFSs), mediative fuzzy relations, and mediative Pythagorean fuzzy sets.

2.1 *Mediative Fuzzy Logic (MFL)*

Let U^* denote a universal set and a collection of triplets (membership and non-membership and hesitation function) A is called an IFS if.

(1) Membership grade, i.e. $\mu_A(\sigma)$ and non-membership grade, i.e. $\nu_A(\sigma)$ lies between [0, 1].
(2) $\mu_A(\sigma) + \nu_A(\sigma) + \pi_A(\sigma) = 1$..

where $\mu_A(\sigma) : U^* \rightarrow [0, 1]$., $\nu_A(\sigma) : U^* \rightarrow [0, 1]$ and clearly hesitation grade $\pi_A(\sigma)$ also lies between closed interval [0, 1].

Total outcome of an IFL-based system is estimated with the linear relationship between Y_μ and Y_ν as

$$Y = (1 - \pi)Y_\mu + \pi Y_\nu \tag{1}$$

This output of intuitionistic fuzzy logic (IFL)-based system is traditional and if hesitation grade is 0 then it will turnout as the outcome of fuzzy system, but when $\pi \neq 0$ then we will get the unlike output values for the IFL-based systems. The output of IFL is more valuable due to its inclusion of non-favourable cases. But there are some real-life situations in which, contradictory, non-contradictory, and doubtful situations arise, then we cannot deal with such kind of problems by the intuitionistic and classical fuzzy logic, in such conditions MFL comes under the conducted study. MFL was initially given by Montiel et al. (2005a), This MFL can control contradictory and non-contradictory knowledge provided by the experts. By mediative fuzzy logic, we can easily handle such real-life problems. In MFL, we have the contradiction FS $\overline{Cn}$ in X defined as

$$\zeta_{\overline{Cn}}(\sigma) = min\{\mu_{\overline{Cn}}(\sigma), \nu_{\overline{Cn}}(\sigma)\} \tag{2}$$

Montiel et al. (2008) introduced the following expressions for calculating the output of an inference system based on MFL;

$$Y_1 = \left(1 - \pi - \frac{\zeta}{2}\right)Y_\mu + \left(\pi + \frac{\zeta}{2}\right)Y_\nu \tag{3}$$

$$Y_2 = \min\left\{Y, \left(1 - \frac{\zeta}{2}\right)\right\} \tag{4}$$

$$Y_3 = Y\left(1 - \frac{\zeta}{2}\right) \tag{5}$$

where, $Y = (1 - \pi)Y_\mu + \pi Y_\nu$

Now, there are two cases;

(1) If $\zeta = 0$, the output of the system can be converted into the output of IFL-based system
(2) Or, if $\pi = 0$ & $\zeta = 0$, the output can be reduced into the traditional fuzzy logic-based system output.

2.2 Compositional Mediative Fuzzy Medical Knowledge Representation Among (Symptoms, Patient, and Disease)

With the increment in the information available to medical experts, the process of classifying disease for diagnosis becomes difficult. So, for modelling the diagnostic processes, the fuzzy logic approaches applied to different mathematical models. Sanchez (1979) expressed the medical information in the form of fuzzy relation between patient's symptoms and disease. Consider a set of four patients P, MFS S of symptoms of the patient and D denotes the MFS of disease. Let R_1 on a set $P \times S$ representing a relation to represent a medical knowledge called mediative fuzzy relation, that relates A with D expressed as

$$R_1 = P \odot S \tag{6}$$

Let R_2 on a set $S \times D$ representing a relation, that relates S with D expressed as

$$R_2 = S \odot D \tag{7}$$

Let R be a mediative fuzzy relation described on a set $P \times D$ representing a compositional relation that relates set of patients P to the set of disease D (as shown in Fig. 1) as

$$R = R_1 \times R_2 = P \odot D \tag{8}$$

2.3 Mediative Pythagorean Fuzzy Set

Let $\overline{I} =< (\sigma, \mu_{\overline{I}}(\sigma), \nu_{\overline{I}}(\sigma)) >$ be an IFS defined on$X$. Then it is said to be a PFS if

$$0 \leq \mu_{\overline{I}}^2(\sigma) + \nu_{\overline{I}}^2(\sigma) \leq 1 \tag{9}$$

With some hesitation margin

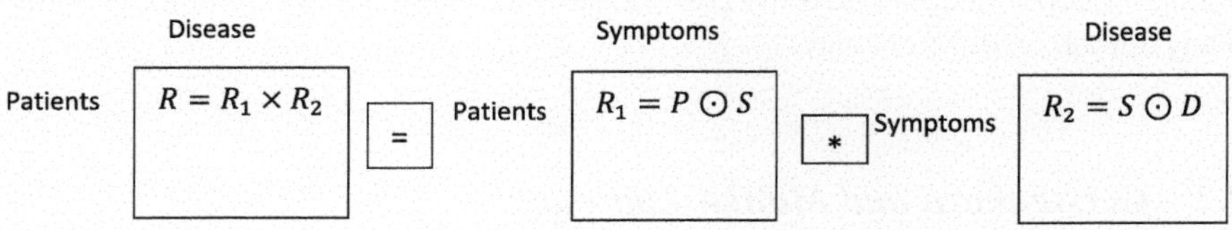

Fig. 1 Mediative fuzzy sets and compositional fuzzy relation involved in medical diagnostic process

$$\pi_{\overline{I}}(x) = \sqrt{1 - \left[\mu_{\overline{I}}^2(\sigma) + \nu_{\overline{I}}^2(\sigma)\right]} \tag{10}$$

MFS $\overline{I}$ on a universal set X is defined (Dhiman and Sharma 2019a, b, c) in the following manner

$$\bar{I} = \left\{\left(\sigma, \mu_{\bar{I}}(\sigma),\ \nu_{\bar{I}}(\sigma)\right) : \sigma \in \mathrm{X}\right\} \tag{11}$$

with $\zeta_{\overline{I}}(\sigma) = \min(\mu_{\overline{I}}(\sigma), \nu_{\overline{I}}(\sigma)))$ called contradiction factor of I in X and

$$\mu_{\overline{I}}(\sigma) : \mathrm{X} \rightarrow [0, 1]$$

$$\nu_{\overline{I}}(\sigma) : \mathrm{X} \rightarrow [0, 1]$$

With $(\gamma_{\overline{I}}(\sigma) + \frac{\zeta_{\overline{I}}(\sigma)}{2}) = 1 - \mu_{\overline{I}}(\sigma) - \nu_{\overline{I}}(\sigma)$ and $0 \leq (\gamma_{\overline{I}}(\sigma) + \frac{\zeta_{\overline{I}}(\sigma)}{2}) \leq 1$ with $- 0.25 \leq \gamma_{\overline{I}} \leq 1$,

Where $\gamma_{\overline{I}}(\sigma) + \frac{\zeta_{\overline{I}}(\sigma)}{2}$ is called the index of uncertainty with contradiction about 'x'.

from this definition we observe that $0 \leq \mu_{\overline{I}}^2(\sigma) + \nu_{\overline{I}}^2(\sigma) \leq 1$ with hesitation margin

$$(\gamma_{\overline{I}}(\sigma) + \frac{\zeta_{\overline{I}}(\sigma)}{2}) = \sqrt{1 - \left[\mu_{\overline{I}}^2(\sigma) + \nu_{\overline{I}}^2(\sigma)\right]} \tag{12}$$

Then it turns into PFS and forms a novel set called mediative Pythagorean fuzzy set.

3 Subsisting Approaches to Quantify Correlation Coefficients Designated Over [0, 1]

Many subsiting approaches have been introduced in the existing literature to evaluate the correlation coefficient. Before introducing our proposed mediative correlational approach, we studied the correlation coefficient techniques introduced by many authors in different ways.

3.1 Gerstenkorn and Manko Approach

Gerstenkorn and Manko (1991) initiated the intuitionistic fuzzy approach-based correlation coefficient, which was given as

$$\gamma_1(\overline{A}, B) = \frac{\theta(\overline{A}, \overline{B})}{\sqrt{\phi(\overline{A})\phi(\overline{B})}} \tag{13}$$

where $\theta(\overline{A}, \overline{B})$ denotes the correlation between IFS $\overline{A} =< (\sigma, \mu_{\overline{A}}(\sigma), \nu_{\overline{A}}(\sigma)) >$ and $\overline{B} =< (\sigma, \mu_{\overline{B}}(\sigma), \nu_{\overline{B}}(\sigma)) >$, with informational energies$\phi(\overline{A}) and \phi(\overline{B})$, respectively, as follows;

$$\theta(\overline{A}, \overline{B}) = \sum_{i=1}^{n}[\mu_{\overline{A}}(\sigma_i)\mu_{\overline{B}}(\sigma_i) + \nu_{\overline{A}}(\sigma_i)\nu_{\overline{B}}(\sigma_i)] \tag{14}$$

$$\phi(\overline{A}) = \sum_{i=1}^{n}[\mu_{\overline{A}}^2(\sigma_i) + \nu_{\overline{A}}^2(\sigma_i)] \tag{15}$$

$$\phi(\overline{B}) = \sum_{i=1}^{n}[\mu_{\overline{B}}^2(\sigma_i) + \nu_{\overline{B}}^2(\sigma_i)] \tag{16}$$

3.2 *Xu et al. Approach*

In 2008, Xu et al. gave the modification of correlation coefficient defined by Gerstenkorn and Manko as

$$\gamma_2(\overline{A}, \overline{B}) = \frac{\theta(\overline{A}, \overline{B})}{\sqrt{\phi(\overline{A})\phi(\overline{B})}} \tag{17}$$

where $\theta(\overline{A}, \overline{B})$ denotes the correlation between IFS $\overline{A} =< (\sigma, \mu_{\overline{A}}(\sigma), \nu_{\overline{A}}(\sigma)) >$ and $\overline{B} =< (\sigma, \mu_{\overline{B}}(\sigma), \nu_{\overline{B}}(\sigma)) >$, with informational energies $\phi(\overline{A}) and \phi(\overline{B})$ respectively, as follows;

$$\theta(\overline{A}, \overline{B}) = \sum_{i=1}^{n}[\mu_{\overline{A}}(\sigma_i)\mu_{\overline{B}}(\sigma_i) + \nu_{\overline{A}}(\sigma_i)\nu_{\overline{B}}(\sigma_i) + \pi_{\overline{A}}(\sigma_i)\pi_{\overline{B}}(\sigma_i)] \tag{18}$$

$$\phi(\overline{A}) = \sum_{i=1}^{n}[\mu_{\overline{A}}^2(\sigma_i) + \nu_{\overline{A}}^2(\sigma_i) + \pi_{\overline{A}}^2(\sigma_i)] \tag{19}$$

$$\phi(\overline{B}) = \sum_{i=1}^{n}[\mu_{\overline{B}}^2(\sigma_i) + \nu_{\overline{B}}^2(\sigma_i) + \pi_{\overline{B}}^2(\sigma_i)] \tag{20}$$

3.3 Zeng and Li Approach

In 2007, Zeng and Li has given the correlation coefficient given as

$$\gamma_3(\overline{A},\overline{B}) = \frac{\theta(\overline{A},\overline{B})}{\sqrt{\phi(\overline{A})\phi(\overline{B})}} \tag{21}$$

where $\theta(\overline{A},\overline{B})$ denotes the correlation between IFS $\overline{A} =< (\sigma, \mu_{\overline{B}}(\sigma), \nu_{\overline{B}}(\sigma)) >$ and $\overline{B} =< (\sigma, \mu_{\overline{B}}(\sigma), \nu_{\overline{B}}(\sigma)) >$, with informational energies$\phi(\overline{A}) and \phi(\overline{B})$, respectively, as follows;

$$\theta(A,B) = \frac{\sum_{i=1}^{n}\left[\mu_A(\sigma_i)\mu_{\overline{B}}(\sigma_i) + \nu_A(\sigma_i)\nu_{\overline{B}}(\sigma_i) + \pi_{\overline{A}}(\sigma_i)\pi_{\overline{B}}(\sigma_i)\right]}{n} \tag{22}$$

$$\phi(\overline{A}) = \frac{\sum_{i=1}^{n}\left[\mu_{\overline{A}}^2(\sigma_i) + \nu_{\overline{A}}^2(\sigma_i) + \pi_{\overline{A}}^2(\sigma_i)\right]}{n} \tag{23}$$

$$\phi(\overline{B}) = \frac{\sum_{i=1}^{n}\left[\mu_{\overline{B}}^2(\sigma_i) + \nu_{\overline{B}}^2(\sigma_i) + \pi_{\overline{B}}^2(\sigma_i)\right]}{n} \tag{24}$$

3.4 Garg Approach

In 2016, Garg modified the (Gerstenkorn and Manko and Xu et al.) approach

$$\gamma_4(\overline{A},\overline{B}) = \frac{\theta(\overline{A},\overline{B})}{\sqrt{\phi(\overline{A})\phi(\overline{B})}} \tag{25}$$

where $\theta(\overline{A},\overline{B})$ denotes the correlation between IFS $\overline{A} =< (\sigma, \mu_A(\sigma), \nu_A(\sigma)) >$ and $\overline{B} =< (\sigma, \mu_{\overline{B}}(\sigma), \nu_{\overline{B}}(\sigma)) >$, with informational energies$\phi(\overline{A}) \& \phi(\overline{B})$, respectively, as follows;

$$\theta(\overline{A},\overline{B}) = \sum_{i=1}^{n}\left[\mu_{\overline{A}}(\sigma_i)\mu_B(\sigma_i) + \nu_{\overline{A}}(\sigma_i)\nu_B(\sigma_i) + \pi_{\overline{A}}(\sigma_i)\pi_B(\sigma_i)\right] \tag{26}$$

$$\phi(\overline{A}) = \sum_{i=1}^{n}\left[\mu_{\overline{A}}^4(\sigma_i) + \nu_{\overline{A}}^4(\sigma_i) + \pi_{\overline{A}}^4(\sigma_i)\right] \tag{27}$$

$$\phi(\overline{B}) = \sum_{i=1}^{n}[\mu_{\overline{B}}^{4}(\sigma_i) + \nu_{\overline{B}}^{4}(\sigma_i) + \pi_{\overline{B}}^{4}(\sigma_i)] \quad (28)$$

3.5 *Ejegwa et al. Approach*

Ejegwa et al. (2020) have also expressed the correlation coefficient in the following manner

$$\gamma_5(\overline{A}, \overline{B}) = \frac{\theta(\overline{A}, \overline{B})}{\sqrt{\phi(\overline{A})\phi(\overline{B})}} \quad (29)$$

where $\theta(\overline{A}, \overline{B})$ denotes the correlation between IFS $\overline{A} =< (\sigma, \mu_{\overline{A}}(\sigma), \nu_{\overline{A}}(\sigma)) >$ and $\overline{B} =< (\sigma, \mu_{\overline{B}}(\sigma), \nu_{\overline{B}}(\sigma)) >$, with informational energies$\phi(\overline{A})\&\phi(\overline{B})$, respectively, defined as;

$$\theta(\overline{A}, \overline{B}) = \frac{\sum_{i=1}^{n}[\mu_{\overline{A}}(\sigma_i)\mu_{\overline{B}}(\sigma_i) + \nu_{\overline{A}}(\sigma_i)\nu_{\overline{B}}(\sigma_i) + \pi_{\overline{A}}(\sigma_i)\pi_{\overline{B}}(\sigma_i)]}{n-1} \quad (30)$$

$$\phi(\overline{A}) = \frac{\sum_{i=1}^{n}\left[\mu_{\overline{A}}^{2}(\sigma_i) + \nu_{\overline{A}}^{2}(\sigma_i) + \pi_{\overline{A}}^{2}(\sigma_i)\right]}{n-1} \quad (31)$$

$$\phi(\overline{B}) = \frac{\sum_{i=1}^{n}\left[\mu_{\overline{B}}^{2}(\sigma_i) + \nu_{\overline{B}}^{2}(\sigma_i) + \pi_{\overline{B}}^{2}(\sigma_i)\right]}{n-1} \quad (32)$$

4 Mathematical Formulation of Proposed Method (Correlation Measurement Based on a Mediative Fuzzy Algorithm)

Let $\overline{A} =< (\sigma, \mu_A(\sigma), \nu_A(\sigma)) >$ Let $\overline{B} =< \left(\sigma, \mu_{\overline{B}}(\sigma), \nu_{\overline{B}}(\sigma)\right) >$ be two MFS defined on universal set$X$. The mediative fuzzy logic-based correlation coefficient is given as

$$\gamma(\overline{A}, \overline{B}) = \frac{\theta(A, \overline{B})}{\sqrt{\phi(A)\phi(\overline{B})}} \quad (33)$$

where, the correlation between $\overline{A}\&\overline{B}$ denoted by $\theta(\overline{A}, \overline{B})$, with informational energies $\phi(\overline{A})\&\phi(\overline{B})$, respectively, defined as

Where,

$$\phi\left(\overline{A}\right) = \sum_{i=1}^{n} \alpha_i^2 \tag{34}$$

With

$$\alpha_i = \left(1 - \pi_i - \frac{\zeta_i}{2}\right)\mu_i + \left(\pi_i + \frac{\zeta_i}{2}\right)\nu_i \tag{35}$$

And

$$\phi\left(\overline{B}\right) = \sum_{i=1}^{n} \beta_i^2 \tag{36}$$

with

$$\beta_i = \left(1 - \pi_i - \frac{\zeta_i}{2}\right)\mu_i + \left(\pi_i + \frac{\zeta_i}{2}\right)\nu_i \tag{37}$$

and

$$\theta(\overline{A}, \overline{B}) = \sum_{i=1}^{n} \alpha_i \beta_i$$

5 Theorem and Properties

(i) *If* $\overline{A} = \overline{B}$

Then $\phi(\overline{A}) = \phi(\overline{B}) = \theta(\overline{A}, \overline{A})$

It is easy to show, so we omit the solution.

(ii) $\gamma(\overline{A}, \overline{B}) = \gamma(\overline{B}, A)$

$$\gamma\left(\overline{A}, \overline{B}\right) = \frac{\theta\left(\overline{A}, \overline{B}\right)}{\sqrt{\phi\left(\overline{A}\right)\phi\left(\overline{B}\right)}} = \frac{\sum_{i=1}^{n} \alpha_i \beta_i}{\sqrt{\sum_{i=1}^{n} \alpha_i^2 \sum_{i=1}^{n} \beta_i^2}}$$

$$= \frac{\sum_{i=1}^{n} \beta_i \alpha_i}{\sqrt{\sum_{i=1}^{n} \beta_i^2 \sum_{i=1}^{n} \alpha_i^2}} = \gamma\left(\overline{B}, \overline{A}\right)$$

(iii) $\gamma(\overline{A}, \overline{B}) = 1, if\,\overline{A} = \overline{B}$

$$\gamma(\overline{A}, \overline{B}) = \frac{\theta(\overline{A}, \overline{B})}{\sqrt{\phi(\overline{A})\phi(\overline{B})}} = \frac{\theta(\overline{A}, \overline{A})}{\sqrt{\phi(\overline{A}, \overline{A})\phi(\overline{A}, \overline{A})}} = 1$$

(iv) $0 \le \gamma(\overline{A}, \overline{B}) \le 1$

Hint: *as* $0 \le \alpha_i \le 1$, $0 \le \beta_i \le 1$, *so* $0 \le \alpha_i \beta_i \le 1$ *and* $0 \le \alpha_i^2, \beta_i^2 \le 1$

also, $\sum_{i=1}^{n} \alpha_i \beta_i \leqslant \sqrt{\sum_{i=1}^{n} \alpha_i^2 \sum_{i=1}^{n} \beta_i^2}$

$$\Rightarrow \frac{\sum_{i=1}^{n} \alpha_i \beta_i}{\sqrt{\sum_{i=1}^{n} \alpha_i^2 \sum_{i=1}^{n} \beta_i^2}} \le 1$$

Since, $\sum_{i=1}^{n} (\alpha_i x - \beta_i)^2 \ge 0$, $\forall x$ and $i = 1, 2....n$.

$$\Rightarrow \sum_{i=1}^{n} (\alpha_i x)^2 - 2\sum_{i=1}^{n} (\alpha_i \beta_i)x + \sum_{i=1}^{n} (\beta_i)^2 \ge 0$$

$$\Rightarrow (2\sum_{i=1}^{n} (\alpha_i \beta_i))^2 - 4\sum_{i=1}^{n} (\alpha_i)^2 \sum_{i=1}^{n} (\beta_i)^2 \le 0$$

$$\Rightarrow 4\left(\sum_{i=1}^{n} (\alpha_i \beta_i)\right)^2 \le 4\sum_{i=1}^{n} (\alpha_i)^2 \sum_{i=1}^{n} (\beta_i)^2 0$$

$$\Rightarrow \sum_{i=1}^{n} \alpha_i \beta_i \le \sqrt{\sum_{i=1}^{n} \alpha_i^2 \sum_{i=1}^{n} \beta_i^2}$$

$$\Rightarrow \frac{\sum_{i=1}^{n} \alpha_i \beta_i}{\sqrt{\sum_{i=1}^{n} \alpha_i^2 \sum_{i=1}^{n} \beta_i^2}} \le 1$$

6 Comparison of Our Approach With Previous Approaches

The proposed mediative fuzzy correlational technique gives a comprehensive as well as optimal performance index in comparison with previous existing studies. In this section, we have given a comparison between our study with Gerstenkorn and Manko; Xu et al.; Zeng and Li; Garg and Ejegwa et al. as shown in Table 2. We have taken two examples (Ejegwa et al. 2020) to trail the accomplishment of our developed technique than the performance of existing techniques.

Example 1: Let $\overline{A} =< (x, 0.3, 0.6), (y, 0.5.0.3), (z, 0.4, 0.5) >$

$$\overline{B} =< (x, 0.3, 0.55), (y, 0.5.0.3162), (z, 0.3873, 0.5) >$$

We have, $\alpha_1 = \left(1 - \pi_1 - \frac{\zeta_1}{2}\right)\mu_1 + \left(\pi_1 + \frac{\zeta_1}{2}\right)\nu_1 = 0.375$

$$\alpha_2 = (1 - \pi_2 - \frac{\zeta_2}{2})\mu_2 + (\pi_2 + \frac{\zeta_2}{2})\nu_2 = 0.5$$

$$\alpha_3 = (1 - \pi_3 - \frac{\zeta_3}{2})\mu_3 + (\pi_3 + \frac{\zeta_3}{2})\nu_3 = 0.43$$

So that $\psi(\overline{A}) = 0.63$
Also $\beta_1 = \left(1 - \pi_1 - \frac{\zeta_1}{2}\right)\mu_1 + \left(\pi_1 + \frac{\zeta_1}{2}\right)\nu_1 = 0.375$

$$\beta_2 = (1 - \pi_2 - \frac{\zeta_2}{2})\mu_2 + (\pi_2 + \frac{\zeta_2}{2})\nu_2 = 0.437$$

$$\beta_3 = (1 - \pi_3 - \frac{\zeta_3}{2})\mu_3 + (\pi_3 + \frac{\zeta_3}{2})\nu_3 = 0.5405$$

So that $\psi(\overline{B}) = 0.5067$ and

$$\theta(\overline{A}, \overline{B}) = \alpha_1\beta_1 + \alpha_2\beta_2 + \alpha_3\beta_3 = 0.5658$$

So , $\sigma(\overline{A}, \overline{B}) = \frac{\varnothing(A,B)}{\sqrt{\psi(A)\psi(B)}} = \frac{0.5405}{\sqrt{0.5746*0509}} = 0.9996$

Example 2: Let $\overline{A} =< (x, 0.5, 0.3), (y, 0.3.0.7), (z, 0.9, 0.1), (t, 0.6, 0.3) >$

$$\overline{B} =< (x, 0.5, 0.5), (y, 0.7.0.2), (z, 0.8, 0.1), (t, 0.7, 0.1) >$$

We have, $\alpha_1 = \left(1 - \pi_1 - \frac{\zeta_1}{2}\right)\mu_1 + \left(\pi_1 + \frac{\zeta_1}{2}\right)\nu_1 = 0.43$

$$\alpha_2 = (1 - \pi_2 - \frac{\zeta_2}{2})\mu_2 + (\pi_2 + \frac{\zeta_2}{2})\nu_2 = 0.36$$

$$\alpha_3 = (1 - \pi_3 - \frac{\zeta_3}{2})\mu_3 + (\pi_3 + \frac{\zeta_3}{2})\nu_3 = 0.86$$

$$\alpha_4 = (1 - \pi_4 - \frac{\zeta_4}{2})\mu_4 + (\pi_4 + \frac{\zeta_4}{2})\nu_4 = 0.525$$

So that $\psi(\overline{A}) = 1.32972$
Also $\beta_1 = \left(1 - \pi_1 - \frac{\zeta_1}{2}\right)\mu_1 + \left(\pi_1 + \frac{\zeta_1}{2}\right)\nu_1 = 0.5$

$$\beta_2 = (1 - \pi_2 - \frac{\zeta_2}{2})\mu_2 + (\pi_2 + \frac{\zeta_2}{2})\nu_2 = 0.58$$

$$\beta_3 = (1 - \pi_3 - \frac{\zeta_3}{2})\mu_3 + (\pi_3 + \frac{\zeta_3}{2})\nu_3 = 0.695$$

$$\beta_4 = (1 - \pi_4 - \frac{\zeta_4}{2})\mu_4 + (\pi_4 + \frac{\zeta_4}{2})\nu_4 = 0.55$$

So that $\boldsymbol{\phi}(\overline{B}) = 1.371925$ and

$$\boldsymbol{\theta}(\overline{A}, \overline{B}) = \alpha_1\beta_1 + \alpha_2\beta_2 + \alpha_3\beta_3 = 1.31025$$

So $\boldsymbol{\gamma}(\overline{A}, \overline{B}) = \frac{\boldsymbol{\theta}(A,B)}{\sqrt{\psi(A)\psi(B)}} = \frac{1.31025}{\sqrt{1.329725 * 1.371925}} = 0.97008$

7 Illustration of the Proposed Technique in Multi-criteria Decision-Making of Medical Diagnostic

Consider 4 patients namely; Patient-i, Patient-ii, Patient-iii, and Patient-iv visited a medical hospital for their disease diagnostic and we represented by the set of patient P as

$$P = \{P_1, P_2, P_3, P_4\}$$

After critical analysis of samples collected from patients the set of major symptoms can be represented as

$$S = \{S_1, S_2, S_3, S_4\}$$

Let us assume a group of likely diseases namely; 'Typhoid', 'Pneumonia', 'Dengue', and 'Malaria' based on the set of symptoms and denote set of diseases as follows

$$D = \{D_1, D_2, D_3, D_4\}$$

Namely D_1= Typhoid, D_2= pneumonia, D_3= dengue, D_4= malaria. The medical experts select the four parameters of the symptoms to measure the four diseases.

Based on this hypothetically medical information of the patient diseases, medical intelligence based on the computed mediative fuzzy logic is shown in Table 6. On behalf of the medical investigation for the set P, the diseases' medical information based on mediative fuzzy logic is given in Table 7. In the proposed study we will find out the most likely disease from which the patient is agonizing by evaluating the correlational value between the diseases and the patients.

7.1 Algorithm

The architecture of proposed method is shown in Fig. 2. Further, we also give the simulation-based algorithm which describes the process of computing the mediative correlation coefficient between patients and diseases as shown in Table 5.

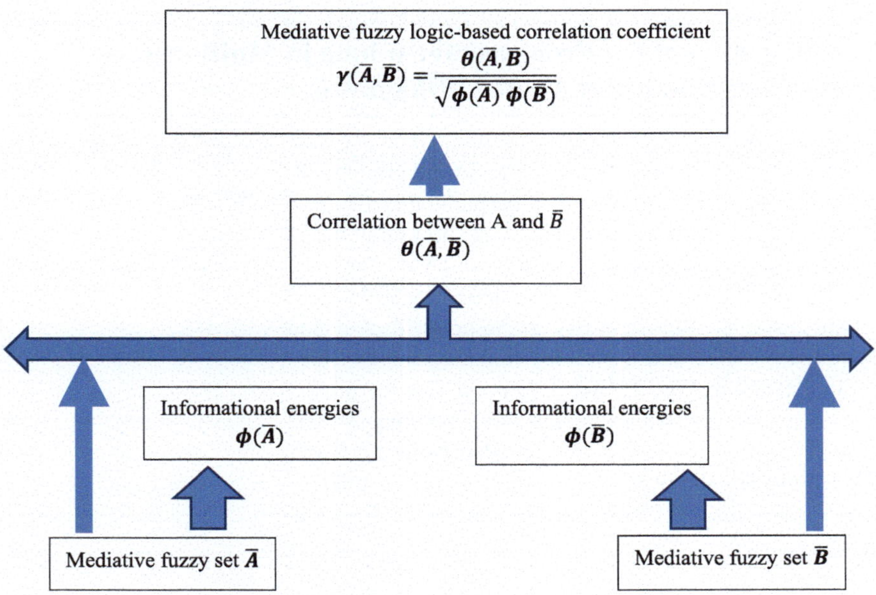

Fig. 2 Architecture of proposed method

From the computational Table 8, we have the following observations.

(1) Patient $\boldsymbol{P_1}$ is diagnosed highly with disease $\boldsymbol{D_1}$, followed by diseases $\boldsymbol{D_3}$,$\boldsymbol{D_2}$, $\boldsymbol{D_4}$, respectively.
(2) In the same manner, **patient** $\boldsymbol{P_2}$ is suffering from disease $\boldsymbol{D_1}$, followed by diseases $\boldsymbol{D_2}$, $\boldsymbol{D_4}$, and $\boldsymbol{D_3}$.
(3) **Patient** P_3 should also highly be severe with disease $\boldsymbol{D_3}$, followed by diseases $\boldsymbol{D_1}$, $\boldsymbol{D_4}$, and $\boldsymbol{D_2}$.
(4) **Patient** $\boldsymbol{P_4}$ is also diagnosed with disease $\boldsymbol{D_1}$, followed by diseases $\boldsymbol{D_3}$, $\boldsymbol{D_2}$, $\boldsymbol{D_4}$, respectively.

8 Conclusion

In this present era of statistical research, the role of correlation coefficient in the relationship measurement of two statistical data (may be crisp or fuzzy) is very crucial. In this article we have developed a mediative-based correlational coefficient formula within the closed interval [0, 1]. We also described the superiority of our proposed method over the previously existing methodologies. For the application prospective, the developed algorithm has also been used in the medical diagnosis process. Proposed technique is more reliable and optimal in the context of correlational value over the MFSs. The numerical commutation illustrates the utility of our proposed technique. Some theorems and results have also given on the basis of proposed mediative correlation coefficient. The entire work in this work illustrates the following points as follows.

1. We have developed the Mediative Pythagorean fuzzy set that can handle the uncertainty, contradictory, and non-contradictory information.
2. Mathematical model for the measurement of correlation coefficient called mediative fuzzy correlation coefficient, designate over [0, 1] has been defined in mediative fuzzy environment.
3. Present work also describes a diagnosis process for the MCDMP with the help of developed mediative fuzzy correlation coefficient.
4. The efficacy of proposed approach over the previously existing approaches has also been described in Tables 1, 2, 3, and 4.
5. Statistical validation and some theorems based on the proposed mediative fuzzy correlational technique have also been discussed in this work.

Table 1 Correlation coefficient for the first data

α_i	β_i	$\phi(\overline{A})$	$\phi(\overline{B})$	$\theta(\overline{A}, \overline{B})$	$\gamma(\overline{A}, \overline{B})$
0.525	0.43	0.63	0.5067	0.5658	**0.9996**
0.37	0.37				
0.47	0.43				

Table 2 Computational study of correlation coefficients

Approaches	$\phi(\overline{A})$	$\phi(\overline{B})$	$\theta(\overline{A}, \overline{B})$	$\gamma(\overline{A}, \overline{B})$
Gerstenkorn and Manko (1991)	1.2000	1.1425	1.1698	0.9991
Xu et al. (2008)	1.2600	1.2115	1.9828	>1
Zeng and Li (2007)	0.4200	0.4038	0.6609	>1
Garg (2016)	1.3825	1.4086	1.3920	0.9975
Ejegwa et al. (2020)	1.5000	1.5000	1.4989	0.9993
New approach	0.63	0.5067	0.5645	**0.9996**

Table 3 Correlation coefficient for the second data

α_i	β_i	$\phi(\overline{A})$	$\phi(\overline{B})$	$\theta(\overline{A}, \overline{B})$	$\gamma(\overline{A}, \overline{B})$
0.525	0.43	1.32972	1.371925	1.31025	**0.97008**
0.37	0.37				
0.47	0.43				

Table 4 Computation of correlation coefficients

Approaches	$\phi(\overline{A})$	$\phi(\overline{B})$	$\theta(\overline{A}, \overline{B})$	$\gamma(\overline{A}, \overline{B})$
Gerstenkorn and Manko (1991)	*2.19*	*2.18*	*1.93*	0.8833
Xu et al. (2008)	1.24	2.24	1.95	0.8705
Zeng and Li (2007)	0.56	0.56	0.4875	0.8705
Garg (2016)	2.0596	1.8600	1.7099	0.8736
Ejegwa et al. (2020)	1.333	1.3333	1.2414	0.9311
New approach	1.32972	1.371925	1.31025	**0.97008**

6. Simulation for the proposed algorithm is defined in Table 5. That is understandable and quite easy in computation.
7. Table 6 represents the mediative fuzzy medical information in between patients and symptoms, Table 7 describes the mediative relation between the symptoms and diseases. From computational Table 8, we can observe the severity of the patient, so that doctors or physicians can recommend the dose accordingly.

Table 5 Simulator for mediative correlation coefficients

```
#include<stdio.h>
#include<math.h>
#include<conio.h>
main()
{float x[4][2];
 float y[4][2];
float val,alfa[4],beta[4];
float hza[4][2], hzb[4][2], up,down,down2;
int i, size=4;
up=down=down2=0.0;
clrscr();
fflush(stdin);
printf("enter the data membership and non-membership values for x");
for(i=0;i<4;i++)
{ scanf("%f %f",&x[i][0],&x[i][1]);
}
printf("enter the data membership and non-membership values for y");

for(i=0;i<4;i++)
{ scanf("%f %f",&y[i][0],&y[i][1]);
}


for (i=0;i<size;i++)
{ hza[i][0]=1-(x[i][0]+x[i][1]);
hza[i][1]=(x[i][0]<x[i][1]?x[i][0]:x[i][1]);
hzb[i][0]= 1-(y[i][0]+y[i][1]);
hzb[i][1]=(y[i][0]<y[i][1]?y[i][0]:y[i][1]);
printf(" \n hesitation%d %f zeta%d %f", i+1,hzb[i][0],i+1, hzb[i][1]);
 alfa[i]=((1-hza[i][0]-hza[i][1]/2)*x[i][0])+((hza[i][0]+hza[i][1]/2)*x[i][1]);
```

(continued)

Table 5 (continued)

```
beta[i]=((1-hzb[i][0]-hzb[i][1]/2)*y[i][0])+((hzb[i][0]+hzb[i][1]/2)*y[i][1]);
printf(" \n alfa%d %fbeta%d %f",i+1,alfa[i],i+1,beta[i]);

up=up+alfa[i]*beta[i];
down=down+(alfa[i]*alfa[i]);
down2=down2+(beta[i]*beta[i]);

}
down=down*down2;

val=up/sqrt(down);
printf(" \n value(A,B)=%f",val);
getch();
}
```

Table 6 Mediative fuzzy medical information between patients and symptoms

Patients\Symptoms	S_1	S_2	S_3	S_4
P_1	<0.7, 0.2>	<0.8, 0.1>	<0.6, 0.3>	<0.4, 0.2>
P_2	<0.9, 0.1>	<0.5, 0.4>	<0.5, 0.2>	<0.5, 0.2>
P_3	<0.7, 0.1>	<0.9, 0.1>	<0.3, 0.6>	<0.1, 0.8>
P_4	<0.6, 0.1>	<0.5, 0.2>	<0.2, 0.7>	<0.2, 0.5>

Table 7 Mediative fuzzy medical information between symptoms and disease

Symptoms\Diseases	D_1	D_2	D_3	D_4
S_1	<0.8, 0.1>	<0.5, 0.3>	<0.5, 0.4>	<0, 0.7>
S_2	<0.7, 0.2>	<0.4, 0.3>	<0.2, 0.5>	<0.1, 0.8>
S_3	<0.6, 0.3>	<0.9, 0.1>	<0.3, 0.5>	<0.1, 0.7>
S_4	<0.8, 0.2>	<0.5, 0.2>	<0.6, 0.2>	<0.3, 0.6>

Table 8 Mediative fuzzy medical information between patient and disease

Patients\Diseases	D_1	D_2	D_3	D_4
P_1	**0.9383**	0.7294	0.9417	0.6326
P_2	**0.9930**	0.9889	0.5514	0.6913
P_3	0.8297	0.6319	**0.8443**	0.6555
P_4	**0.9865**	0.8605	0.9491	0.7432

Acknowledgements The authors are thankful to the 'University Grants Commission (U.G.C.)' and to the Higher Education Department of Uttar Pradesh Government for financial support under the research and development scheme, file number: 47/2021/606/sattar-4-2021-4(56)/2020. for the economic support.

References

Atanassov K (1986) Intuitionistic fuzzy sets. Fuzzy Sets Syst 20:87–96

Atanassov KT (1999) Intuitionistic fuzzy sets: theory and applications. Physica-Verlag, Heidelberg

Boran FE, Akay D (2014) A biparametric similarity measure on intuitionistic fuzzy sets with applications to pattern recognition. Inf Sci 255(10):45–57

Chen SM, Chang CH (2016) Fuzzy multi-attribute decision making based on transformation techniques of intuitionistic fuzzy values and intuitionistic fuzzy geometric averaging operators. Information Science 352:133–149

Chen SM, Cheng SH, Lan TC (2016a) Multi-criteria decision making based on the TOPSIS method and similarity measures between intuitionistic fuzzy values.Information Sciences: an International Journal 367

Chiang DA, Lin NP (1999) Correlation of fuzzy sets. Fuzzy Sets and Syst 102(2):221–226

Dhiman N, Sharma MK (2019) Mediative sugeno's-TSK fuzzy logic-based screening analysis to diagnosis of heart disease. Appl Math 10:448–467

Dhiman N, Sharma MK (2019) Mediative fuzzy logic approach based on sugeno's-TSK model for the diagnosis of diabetes. 24thInternational conference of CONIAPS in India, 9–11, Aug.

Dhiman N, Sharma MK (2019) Mediative Multi-Criteria Decision Support System for Various Alternatives Based on Fuzzy Logic. IJRTE 8:7940–7946

Ejegwa PA, Onyeke IC, Adah V (2020) A pythagorean fuzzy algorithm embedded with a new correlation measure and its application in diagnostic processes. Granul Comput

Ejegwa PA (2020) Modified and generalized correlation coefficient between intuitionistic fuzzy sets with applications. Notes Intuitionistic Fuzzy Sets 26: 8–22

Gerstenkorn T, Manko J (1991) Correlation of intuitionistic fuzzy sets. Fuzzy Set Syst 44(1):39–43

Garg H (2016) A novel correlation coefficients between Pythagorean fuzzy sets and its applications to decision making processes. Int J Intell Syst 31(12):1234–1252

Hong DH, Hwang SY (1995) Correlation of intuitionistic fuzzy sets in probability spaces. Fuzzy Sets Syst 75(1):77–81

Montiel O, Castillo O. Melin P, Rodríguez DA and Sepúlveda R (2005a) Reducing the cycling problem in evolutionary algorithms. In: Proceedings of ICAI-2005, pp 426–432

Montiel O, Castillo O, Melin P, Rodríguez DA, Sepúlveda R (2008) Meditative fuzzy logic: a new approach forcontradictory knowledge management. Soft Comput 12:251–256

Murthy CA, Pal SK, Majumder DD (1985) Correlation between two fuzzy membership functions. Fuzzy Sets Syst 17(1):23–38

Mitchell HB (2004) A correlation coefficient for intuitionistic fuzzy sets. Int J Intell Syst 19(5):483–490
Sanchez E (1979) Medical diagnosis and composite fuzzy relations. In Gupta MM,Ragade RK, Yager RR (eds) Advances in fuzzy set theory and applications. Amsterdam, North-Holland, pp 437–444
Szmidt E, Kacprzyk J (2001) Intuitionistic fuzzy sets in some medical applications. Note IFS 7(4):58–64
Szmidt E, Kacprzyk J (2004) Medical diagnostic reasoning using a similarity measure for intuitionistic fuzzy sets. Note IFS 10(4):61–69
Szmidt E, Kacprzyk J (2010) Correlation of intuitionistic fuzzy sets. In: Hullermeier E, Kruse R, Hoffmann (eds) IPMU, LNAI 6178. Springer, Berlin, Heidelberg, pp 169–177
Thao NX (2018) A new correlation coefficient of the intuitionistic fuzzy sets and its application. J Intell Fuzzy Syst 35(2):1959–1968
Thao NX (2019) A new correlation coefficient of the Pythagorean fuzzy sets and its applications. Soft Comput
Xu S, Chen J, Wu JJ (2008) Cluster algorithm for intuitionistic fuzzy sets. Inf Sci 178:3775–3790
Yager RR (2013) Pythagorean membership grades in multicriteria decision making. Technical Report MII-3301 Machine Intelligence Institute, Iona College, New Rochelle, NY
Yager RR (2014) Pythagorean membership grades in multicriteria decision making. IEEE Trans Fuzzy Syst 22(4):958–965
Yager RR (2016) Properties and applications of Pythagorean fuzzy sets. Springer, Berlin
Yager RR, Abbasov AM (2013) Pythagorean membership grades, complex numbers and decision making. J Intell Fuzzy Syst 28(5):436–452
Zadeh LA (1965) Fuzzy sets. Inf Control 8:338–356
Zhang Z, Chen SM, Wang C (2020) Group decision making with incomplete intuitionistic multiplicative preference relations. Inf Sci 516:560–571
Zeng S, Chen SM, Kuo LW (2019) Multiattribute decision making based on novel score function of intuitionistic fuzzy values and modified VIKOR method. Inf Sci 488:76–92
Zeng W, Li H (2007) Correlation coefficient of intuitionistic fuzzy sets. J Indust Eng Int 3(5):33–40

Chapter 15
On the Parametric Representation of Type-2 Interval and Ranking: Its Application in the Unconstrained Non-linear Programming Problem Under Type-2 Interval Uncertainty

Subhajit Das, Md Sadikur Rahman, Sanat Kumar Mahato, Ali Akbar Shaikh, and Asoke Kumar Bhunia

Abstract This chapter aims to introduce the parametric representation of the Type-2 interval and its ranking. This chapter also wishes to derive the optimality criteria of the imprecise unconstrained optimization problem in a Type-2 interval environment with these concepts. For this purpose, at first, by recapitulating the idea of Type-2 interval introduced by Rahman et al. (Rahman MS, Shaikh AA, Bhunia AK (2020d)), the parametric representation of Type-2 interval is proposed. Then an order relation on the set of Type-2 intervals is introduced. Using this order relation, the maximizer and minimizer of an unconstrained Type-2 interval-valued optimization problem are defined. Then the optimality conditions (both necessary and sufficient) for the said optimization problem are derived. Finally, the obtained optimality results are illustrated by some numerical examples.

Keywords Type-2 interval parametric form · Type-2 interval Order relation · Optimizer · Type-2 interval optimization

---

S. Das · M. S. Rahman (✉) · A. A. Shaikh · A. K. Bhunia
Department of Mathematics, The University of Burdwan, Burdwan 713104, India

A. K. Bhunia
e-mail: akbhunia@math.buruniv.ac.in

S. K. Mahato
Department of Mathematics, Sidho-Kanho-Birsha University, Purilia 723104, India

I. Ali et al. (eds.), *Computational Modelling in Industry 4.0*,
https://doi.org/10.1007/978-981-16-7723-6_15

1 Introduction

Due to the uncertainty and inexactness, the parameters of most real-life problems, especially optimization problems, are not precise. So, the study of optimality conditions of an imprecise optimization problem is an important research area. To tackle the imprecise optimization problems, several researchers used various approaches, viz. fuzzy, stochastic, fuzzy-stochastic and interval approaches, etc.

In the stochastic approach, the flexible parameters of an imprecise optimization problem are presented in the shape of random variables with proper probability distribution function. Whereas in the fuzzy approach, the flexible parameters are presented in fuzzy sets or fuzzy numbers with appropriate membership function. In the fuzzy-stochastic approach, some of the flexible parameters are considered in the form of fuzzy sets or fuzzy numbers, and others considered random. On the other hand, in the interval approach, the optimization problem's imprecise parameters are presented in the form of intervals. In the existing literature, a lot of research works on these approaches are available. Among those, some interesting works are reported here:

With some theoretical developments, Catoni (2004), Schneider and Kirkpatrick (2007), and Powell (2019) accomplished their works on stochastic optimization. On the other side, Heyman and Sobel (2004), Ziemba and Vickson (2014), and Tang et al. (2020) used the theory of stochastic optimization to analyse the mathematical models in operations research. In the area of fuzzy optimization, Tang et al. (2004), Wu (2004), Lodwick and Kacprzyk (2010), Heidari et al. (2016), and Anter and Ali (2020) established some applicable theories to enrich this area. Wang and Watada (2012) and Farrokh et al. (2018) used fuzzy and stochastic optimization theories to analyse some supply chain/inventory models. In the area of interval optimization, Chen and Wu (2004), Bhurjee and Panda (2016), Ghosh et al. (2019), and Rahman et al. (2020c) derived several helpful techniques to solve interval optimization problems. Rahman et al. (2020a, 2020b) solved some inventory problems in an interval environment with these concepts.

Recently, generalizing the interval approach by taking the flexibility rather than fixing the interval's bounds, Rahman et al. (2020a) introduced a new representation of intervals named Type-2 interval. In this representation, both the bounds belong to two different intervals. In this approach, a flexible parameter can be presented in the form $A = [a_L, a_U]$, where $a_L \in \left[\underline{a}_L, \overline{a}_L\right]$ and $a_U \in \left[\underline{a}_U, \overline{a}_U\right]$. Thus, in this approach, the imprecise parameter can be expressed as $A = \left[\left(\underline{a}_L, \overline{a}_L\right), \left(\underline{a}_U, \overline{a}_U\right)\right]$. In this area, till now, no theoretical development had been done.

This chapter's main objective is to introduce the concepts of parametrization of Type-2 interval and its order relation. Then using these concepts, the definition of the optimizer and Type-2 interval-valued support function is proposed. After that, the optimality conditions of an unconstrained Type-2 interval-valued optimization problem are derived. Finally, these theoretical results are illustrated with some numerical examples.

2 The Concept of Type-2 Interval

2.1 Type-2 Interval in Parametric Form:

The concept of Type-2 interval introduced by Rahman et al. (2020) is denoted by $\left[\left(\underline{a}_L, \overline{a}_L\right), \left(\underline{a}_U, \overline{a}_U\right)\right]$ and is defined by employing Type-1 intervals as follows:

$$\left[\left(\underline{a}_L, \overline{a}_L\right), \left(\underline{a}_U, \overline{a}_U\right)\right] = \left\{[a_L, a_U] : a_L \in \left[\underline{a}_L, \overline{a}_L\right] \text{and } a_U \in \left[\underline{a}_U, \overline{a}_U\right]\right\}.$$

Now, we have defined the parametric representation of the Type-2 interval in the following definitions. This representation is defined in two steps: first step parametrization and second step parametrization of Type-2 interval.

Definition 1: Let $A_2 = \left[\left(\underline{a}_L, \overline{a}_L\right), \left(\underline{a}_U, \overline{a}_U\right)\right]$ be a Type-2 interval. Then

(i) the first step parametrization of A_2 is defined by the set of Type-1 intervals as follows:

$$A_2^1 = \left\{\begin{array}{l}[a_L(r_1), a_U(r_2)] : a_L(r_1) = \underline{a}_L + r_1\left(\bar{a}_L - \underline{a}_L\right),\\ a_U(r_2) = \underline{a}_U + r_2\left(\bar{a}_U - \underline{a}_U\right) \text{ and } r_1, r_2 \in [0, 1]\end{array}\right\}$$

(ii) the second step, parametrization of A_2 is defined as

$$A_2^2 = \left\{\begin{array}{l}a(r_1, r_2, r_3) : a(r_1, r_2, r_3) = a_L(r_1) + r_2(a_U(r_3) - a_L(r_1))\\ \text{and } r_1, r_2, r_3 \in [0, 1]\end{array}\right\}$$

Definition 2: Let $A_2 = \left[\left(\underline{a}_L, \overline{a}_L\right), \left(\underline{a}_U, \overline{a}_U\right)\right]$ and $B_2 = \left[\left(\underline{b}_L, \overline{b}_L\right), \left(\underline{b}_U, \overline{b}_U\right)\right]$ be two Type-2 intervals with their second step parametrization $A_2^2 = \left\{a\left(r_1, r_2, r_3\right) : r_1, r_2, r_3 \in \left[0, 1\right]\right\}$ *and* $B_2^2 = \left\{b\left(r_1, r_2, r_3\right) : r_1, r_2, r_3 \in \left[0, 1\right]\right\}$, respectively. Then $A_2 = B_2$ iff $a(r_1, r_2, r_3) = b(r_1, r_2, r_3), \ \forall r_1, r_2, r_3 \in [0, 1]$.

Example 1: Let us consider a Type-2 interval $A_2 = [(2, 4), (5, 7)]$.

Then its first step, parametrization, is $A_2^1 = \{[2 + 2r_1, 5 + 2r_2] : r_1, r_2 \in [0, 1]\}$.

Therefore, its second step, parametrization, is $A_2^2 = \{2 + 2r_1 + r_2\left(3 + 2r_3 - 2r_1\right) : r_1, r_2, r_3 \in \left[0, 1\right]\}$.

2.2 Order Relation of Type-2 Intervals:

Here, a new type of order relation has been introduced on the set of all Type-2 intervals by using the four different centres of a Type-2 interval which is defined in the **Definition 3**.

Definition 3: Let $A_2 = [(\underline{a}_L, \overline{a}_L), (\underline{a}_U, \overline{a}_U)]$ be a Type-2 interval with second step parametrization $A_2^2 = \{a(r_1, r_2, r_3) : a(r_1, r_2, r_3) = a_L(r_1) + r_2(a_U(r_3) - a_L(r_1))$ and $r_1, r_2, r_3 \in [0, 1]\}$. Then a set of support of A_2 be defined by the set of four elements $\{A_{S1}, A_{S2}, A_{S3}, A_{S4}\}$,

where

$$A_{S1} = \frac{a(1,1,1) + a(1,1,0) + a(1,0,0) + a(0,0,0)}{4}$$
$$A_{S2} = \frac{a(1,1,1) + a(1,1,0) + a(1,0,0)}{3}$$
$$A_{S3} = \frac{a(1,1,1) + a(1,1,0)}{2}$$
$$A_{S4} = a(1,1,1)$$

Definition 4: Let $A = [(\underline{a}_L, \overline{a}_L), (\underline{a}_U, \overline{a}_U)]$ and $B = [(\underline{b}_L, \overline{b}_L), (\underline{b}_U, \overline{b}_U)]$ be two Type-2 intervals with second step parametrization $\{a(r_1, r_2, r_3) : r_1, r_2, r_3 \in [0, 1]\}$ and $\{b(r_1, r_2, r_3) : r_1, r_2, r_3 \in [0, 1]\}$. Then A is said to be less or equal to B denoted by $A \leq_2 B$ if the following conditions hold:

$$A \leq_2 B \Leftrightarrow \begin{cases} A_{S1} < B_{S1}, \text{ when } A_{S1} \neq B_{S1} \\ A_{S2} < B_{S2}, \text{ when } A_{S1} = B_{S1} \text{ and } A_{S2} \neq B_{S2} \\ A_{S3} < B_{S3}, \text{ when } A_{S2} = B_{S2} \text{ and } A_{S3} \neq B_{S3} \\ A_{S4} \leq B_{S4}, \text{ when } A_{S3} = B_{S3} \end{cases}$$

where $A_{S1}, A_{S2}, A_{S3}, A_{S4}$ are defined in **Definition 3**

Definition 5: Let us consider two Type-2 intervals $A = [(\underline{a}_L, \overline{a}_L), (\underline{a}_U, \overline{a}_U)]$ and $B = [(\underline{b}_L, \overline{b}_L), (\underline{b}_U, \overline{b}_U)]$. Then $A \geq_2 B$ iff $B \leq_2 A$.

Remark-1: The order relations $\leq_2$ and $\geq_2$ on the set of Type-2 intervals satisfy the reflexive, anti-symmetric, and transitive properties. Thus $\leq_2$ and $\geq_2$ are the partial order relations.

Example 2: Compare the following pair of Type-2 intervals using the above definitions.

$$\text{(i) } A = [(-3, -1), (2, 5)],\ B = [(-1, 3), (6, 7)]$$
$$\text{(ii) } A = [(-1, 3), (5, 9)],\ B = [(1, 2), (6, 7)]$$

Solution:

(i) The second step parametrization of A and B are

$$\{a(r_1, r_2, r_3) = -3 + 2r_1 + r_2(5 - 2r_1 + 3r_3) : r_1, r_2, r_3 \in [0, 1]\},$$
$$\{b(r_1, r_2, r_3) = -1 + 4r_1 + r_2(2 - 4r_1 + r_3) : r_1, r_2, r_3 \in [0, 1]\}$$

Here, $A_{S1} = \frac{a(1,1,1)+a(1,1,0)+a(1,0,0)+a(0,0,0)}{4} = \frac{3}{4}$

$$B_{S1} = \frac{b(1, 1, 1) + b(1, 1, 0) + b(1, 0, 0) + b(0, 0, 0)}{4} = \frac{15}{4}$$

Since $A_{S1} < B_{S1}$, thus, $A \preceq_2 B$.

(ii) The second step parametrization of A and B are

$$\{a(r_1, r_2, r_3) = -1 + 4r_1 + r_2(6 - 4r_1 + 4r_3) : r_1, r_2, r_3 \in [0, 1]\},$$
$$\{b(r_1, r_2, r_3) = 1 + r_1 + r_2(5 - r_1 + r_3) : r_1, r_2, r_3 \in [0, 1]\}$$

Here,

$$A_{S1} = \frac{a(1, 1, 1) + a(1, 1, 0) + a(1, 0, 0) + a(0, 0, 0)}{4} = 4$$
$$= \frac{b(1, 1, 1) + b(1, 1, 0) + b(1, 0, 0) + b(0, 0, 0)}{4} = B_{S1}$$

and

$$A_{S2} = \frac{a(1, 1, 1) + a(1, 1, 0) + a(1, 0, 0)}{3} = \frac{17}{3} > 5$$
$$= \frac{b(1, 1, 1) + b(1, 1, 0) + b(1, 0, 0)}{3} = B_{S2}.$$

Hence, $A \geq_2 B$.

3 Optimality of Unconstrained Type-2 Interval-Valued Optimization Problem

3.1 *Type-2 Interval-Valued Function and Its Parametrized Form*

Let the set of all Type-2 intervals be denoted by $I_2(\mathbb{R})$, i.e., $I_2(\mathbb{R}) = \{[(\underline{a}_L, \overline{a}_L), (\underline{a}_U, \overline{a}_U)] : \underline{a}_L, \overline{a}_L, \underline{a}_U, \overline{a}_U \in \mathbb{R}\}$.

Now a Type-2 interval-valued function of several variables (say n variables) is a function $H_2 : S \subseteq \mathbb{R}^n \to I_2(\mathbb{R})$ given by $H_2(x) = [(\underline{h}_L(x), \overline{h}_L(x))S]$, $x \in (\underline{h}_U(x), \overline{h}_U(x))$.

Definition 6: The first parametrized representation of H_2 is defined as $H_2(x) = \{[h_L(x, r_1), h_U(x, r_2)] : r_1, r_2 \in [0, 1]\}$ where $h_L(x, r_1) = \underline{h}_L(x) + r_1(\overline{h}_L(x) - \underline{h}_L(x))$ and $h_U(x, r_1) = \underline{h}_U(x) + r_2(\overline{h}_U(x) - \underline{h}_U(x))$.

Definition 7: The second parameterized representation of the Type-2 interval-valued function H_2 is defined as $H_2(x) = \left\{ \begin{array}{l} h(x, r_1, r_2, r_3) : h(x, r_1, r_2, r_3) = h_L(x, r_1) \\ +r_2(h_U(x, r_3) - h_L(x, r_1)) \text{ and } r_1, r_2, r_3 \in [0, 1] \end{array} \right\}$.

where $h_L(x, r_1) = \underline{h}_L(x) + r_1(\overline{h}_L(x) - \underline{h}_L(x))$ and $h_U(x, r_1) = \underline{h}_U(x) + r_2(\overline{h}_U(x) - \underline{h}_U(x))$.

A special type of Type-2 interval-valued function is defined as follows:

$$H_2 : S \subseteq \mathbb{R}^n \to I_2(\mathbb{R}) \text{ given by } H_2(x) = \sum_{i=1}^{k} \left[(\underline{a}_{iL}, \overline{a}_{iL}), (\underline{a}_{iU}, \overline{a}_{iU}) \right] h_i(x), \quad x \in S. \tag{1}$$

where $h_i : S \to \mathbb{R}$, $i = 1, 2, ..., k$

Definition 8: The first parametrized representation of the function H_2 given in (1) is defined by $H_2(x) = \sum_{i=1}^{k} \{[a_{iL}(r_{1i}), a_{iU}(r_{2i})] : r_{1i}, r_{2i} \in [0, 1]\} h_i(x)$, $a_{iL}(r_1) = \underline{a}_{iL} + r_1(\overline{a}_{iL} - \underline{a}_{iL})$ and $a_{iU}(r_2) = \underline{a}_{iU} + r_2(\overline{a}_{iU} - \underline{a}_{iU})$

Definition 9: The second parameterized representation of the function H_2 given in (1) is defined by $H_2(x) = \sum_{i=1}^{k} \{a_i(r_{1i}, r_{2i}, r_{3i}) : r_{1i}, r_{2i}, r_{3i} \in [0, 1]\} h_i(x)$, $a_i(r_{1i}, r_{2i}, r_{3i}) = a_{iL}(r_{1i}) + r_{2i}(a_{iU}(r_{3i}) - a_{iL}(r_{1i}))$

3.2 Standard Form Type-2 Interval-Valued Optimization Problem

The standard form of unconstrained Type-2 interval maximization problem is as follows:

$$Maximize/Minimize \ \ H_2(x) \tag{2}$$

subject to $x \in S \subseteq \mathbb{R}^n$

where $H_2(x)$ may be represented by either

$H_2(x) = \left[\left(\underline{h}_L(x), \bar{h}_L(x)\right), \left(\underline{h}_U(x), \bar{h}_U(x)\right)\right]$ or

$$H_2(x) = \sum_{i=1}^{n} \left[\left(\underline{a}_{iL}, \bar{a}_{iL}\right), \left(\underline{a}_{iU}, \bar{a}_{iU}\right)\right] g_i(x)$$

and $\underline{h}_L, \bar{h}_L, \underline{h}_U, \bar{h}_U, g_i : S \to \mathbb{R}, i = 1, \ldots, n.$

The corresponding second step parametric form of (2) is as follows:

$$Maximize/Minimize\ h(r_1, r_2, r_3, x) \tag{3}$$

subject to $x \in S \subseteq \mathbb{R}^n, \quad r_i \in [0, 1]$

where

either, $h(r_1, r_2, r_3, x) \ = h_L(x) + p(h_U(x) - h_L(x))$

$$or\ h(r_1, r_2, r_3, x) \ = \sum_{i=1}^{n} a_i(r_1, r_2, r_3) g_i(x) \text{ and}$$

and $a_i(r_1, r_2, r_3) = a_{iL}(r_1) + r_2(a_{iU}(r_3) - a_{iL}(r_1))$,

$a_{iL}(r_1) = \underline{a}_{iL} + r_1\left(\bar{a}_{iL} - \underline{a}_{iL}\right), \ a_{iU}(r_2) = \underline{a}_{iU} + r_2\left(\bar{a}_{iU} - \underline{a}_{iU}\right)$

Definition 10: The support of the optimization problem (3) is the set of four real-valued functions $\{H_{S1}, H_{S2}, H_{S3}, H_{S4}\}$ such that.

$$H_{S1}(x) = \frac{h(0,0,0,x) + h(0,0,1,x) + h(0,1,1,x) + h(1,1,1,x)}{4}$$

$$H_{S2}(x) = \frac{h(0,0,1,x) + h(0,1,1,x) + h(1,1,1,x)}{3}$$

$$H_{S3}(x) = \frac{h(0,1,1,x) + h(1,1,1,x)}{2}$$

$$H_{S4}(x) = h(1,1,1,x)$$

Definition 11: The point $x^* \in S$ will be a local maximizer of maximization problem (2) if $\exists$ a $\delta > 0$ such that $H_2(x^*) \geq_2 H_2(x), \ \forall x \in N(x^*, d) \cap S$,

where $N(x^*, d)$ is a neighbourhood with a centre at x^* and radius d.

Definition 12: The point $x^* \in S$ will be a global maximizer of the maximization problem (3) if $H_2(x^*) \geq_2 H_2(x), \ \forall x \in S$.

Note 1: Similarly, the local and global minimizer of the problem (3) can be defined.

Note 2: The inequality $\geq_2$ used in **Definition 11 and 12** can be written explicitly as follows:

$$H_2(x^*) \geq_2 H_2(x) \Leftrightarrow \begin{cases} H_{S1}(x^*) > H_{S1}(x) \text{ when } H_{S1}(x^*) \neq H_{S1}(x) \\ H_{S2}(x^*) > H_{S2}(x) \text{ when } H_{S1}(x^*) = H_{S1}(x) \\ \text{and when } H_{S2}(x^*) \neq H_{S2}(x) \\ H_{S3}(x^*) > H_{S3}(x) \text{ when } H_{S2}(x^*) = H_{S2}(x) \\ \text{and when } H_{S3}(x^*) \neq H_{S3}(x) \\ H_{S4}(x^*) \geq H_{S4}(x) \text{ when } H_{S3}(x^*) = H_{S3}(x) \end{cases}$$

4 Optimality Conditions of Unconstrained Type-2 Interval-Valued Optimization Problem

4.1 Necessary Condition

Theorem 1: If $x^* \in S$ be a local optimizer (maximizer or minimizer) of the unconstrained Type-2 interval-valued optimization problem (2) or (3), then the following conditions are satisfied:

$$\begin{cases} \nabla H_{S1}(x^*) = 0 \text{ when } H_{S1}(x) \text{ is nonconstant} \\ \nabla H_{S2}(x^*) = 0 \text{ when } H_{S1}(x) \text{ is constant and } H_{S2}(x) \text{is nonconstant} \\ \nabla H_{S3}(x^*) = 0 \text{ when } H_{S2}(x) \text{ is constant and } H_{S3}(x) \text{ is nonconstant} \\ \nabla H_{S4}(x^*) = 0 \text{ when } H_{S3}(x) \text{ is constant} \end{cases}$$

Proof: Here, this theorem has been proved for the minimization case only. The proof is also similar to the maximization case.

Let $x^* \in T$ be a local minimizer of $H_2(x)$.
From the definition of the local minimizer, $H_2(x^*) \leq_2 H_2(x), \forall x \in S \cap N(x^*, d)$.
Then by **Definition 4**, it follows that $\forall x \in S \cap N(x^*, d)$,

$$\begin{cases} H_{S1}(x^*) < H_{S1}(x), \text{ when } H_{S1}(x^*) \neq H_{S1}(x) \\ H_{S2}(x^*) < H_{S2}(x), \text{ when } H_{S1}(x^*) = H_{S1}(x) \\ H_{S3}(x^*) < H_{S3}(x), \text{ when } H_{S1}(x^*) = H_{S1}(x) \text{ and } H_{S2}(x^*) = H_{S2}(x) \\ H_{S4}(x^*) < H_{S4}(x), \text{ when } H_{S2}(x^*) = H_{S2}(x) \text{ and } H_{S3}(x^*) = H_{S3}(x) \end{cases}$$

It follows that, $\forall x \in S \cap N(x^*, d)$

$$\begin{cases} H_{S1}(x^*) \le H_{S1}(x), & \text{when } H_{S1}(x) \text{ is non - constant} \\ H_{S2}(x^*) \le H_{S2}(x), & \text{when } H_{S1}(x) \text{ is constant and } H_{S2}(x) \text{ is non - constant} \\ H_{S3}(x^*) \le H_{S3}(x), & \text{when } H_{S2}(x) \text{ is constant and } H_{S3}(x) \text{ is constant} \\ H_{S4}(x^*) \le H_{S4}(x), & \text{when } H_{S3}(x) \text{ is constant} \end{cases}$$

Then by the necessary conditions of optimality $H_{S1}(x), H_{S2}(x), H_{S3}(x)$, and $H_{S4}(x)$, we have

$$\begin{cases} \nabla H_{S1}(x^*) = 0 \text{ when } H_{S1}(x) \text{ is nonconstant} \\ \nabla H_{S2}(x^*) = 0 \text{ when } H_{S1}(x) \text{is constant and } H_{S2}(x) \text{is nonconstant} \\ \nabla H_{S3}(x^*) = 0 \text{ when } H_{S2}(x) \text{is constant and } H_{S3}(x) \text{ is nonconstant} \\ \nabla H_{S4}(x^*) = 0 \text{ when } H_{S3}(x) \text{ is constant.} \end{cases}$$

Definition 13: Let us consider a twice differentiable real-valued function $f : S \to \mathbb{R}$ on a nonempty open set $S \subseteq \mathbb{R}^n$. Then the Hessian matrix of f is denoted by $\nabla^2 f(x)$ and is defined by the $n \times n$ matrix of second-order partial derivatives of f, i.e.$\nabla^2 f(x) = \left(\frac{\partial^2 f}{\partial x_i \partial x_j}\right)_{n \times n}$.

4.2 Sufficient Conditions

Theorem 2: Suppose each member of the support of the unconstrained Type-2 interval-valued optimization problem (2) is continuously differentiable up to second-order, i.e. $\nabla^2 H_{S1}(x)$, $\nabla^2 H_{S2}(x)$, $\nabla^2 H_{S3}(x)$ and $\nabla^2 H_{S4}(x)$ exist. Further, assume that $x^* \in S$ satisfies the conditions of **Theorem 1**.

Then (*i*) $x^* \in S$ be a local minimizer of the optimization problem (2), if

$$\begin{cases} \nabla^2 H_{S1}(x^*) \text{ is positive definite when } H_{S1}(x) \text{ is nonconstant} \\ \nabla^2 H_{S2}(x^*) \text{ is positive definite when } H_{S1}(x) \text{ is constant} \\ \text{and } H_{S2}(x) \text{ is nonconstant} \\ \nabla^2 H_{S3}(x^*) \text{ is positive definite when } H_{S2}(x) \text{ is constant} \\ \text{and } H_{S3}(x) \text{ is nonconstant} \\ \nabla^2 H_{S4}(x^*) \text{ is positive definite when } H_{S3}(x) \text{ is constant} \end{cases}$$

(*ii*) $x^* \in S$ be a local maximizer of the optimization problem (2), if

$$\begin{cases} \nabla^2 H_{S1}(x^*) \text{ is negetive definite when } H_{S1}(x) \text{ is nonconstant} \\ \nabla^2 H_{S2}(x^*) \text{ is negetive definite when } H_{S1}(x) \text{ is constant} \\ \text{and } H_{S2}(x) \text{ is nonconstant} \\ \nabla^2 H_{S3}(x^*) \text{ is negetive definite when } H_{S2}(x) \text{ is constant} \\ \text{and } H_{S3}(x) \text{ is nonconstant} \\ \nabla^2 H_{S4}(x^*) \text{ is negetive definite when } H_{S3}(x) \text{ is constant} \end{cases}$$

Proof: (*i*) When $H_{S1}(x) \neq$ constant,.

$\nabla H_{S1}(x^*) = 0$ and $\nabla^2 H_{S1}(x^*)$ is positive definite, then $x^* \in S$ will be a local minimizer of $H_{S1}(x)$.

Then $\exists\, d_1 > 0$, such that $H_{S1}(x^*) \leq H_{S1}(x)$, $\forall x \in S \cap N(x^*, d)$.

where $N(x^*, d_1)$ is an open ball with centre at x^* and radius d_1.

when $H_{S1}(x) =$ constant and $H_{S2}(x) \neq$ constant.

$\nabla H_{S2}(x^*) = 0$ and $\nabla^2 H_{S2}(x^*)$ is positive definite, then $x^* \in S$ will be a local minimizer of $H_{S2}(x)$.

$$\text{Then } \exists\, d_2 > 0, \text{ such that } H_{S2}(x^*) \leq H_{S2}(x),\ \forall x \in S \cap N(x^*, d_2)$$

where $N(x^*, d_2)$ is an open ball with centre at x^* and radius d_2.

when $H_{S1}(x) =$ constant, $H_{S2}(x) =$ constant and $H_{S3}(x) \neq$ constant,.

$\nabla H_{S3}(x^*) = 0$ and $\nabla^2 H_{S3}(x^*)$ is positive definite, then $x^* \in S$ will be a local minimizer of $H_{S3}(x)$.

$$\text{Then } \exists\, d_3 > 0, \text{ such that } H_{S3}(x^*) \leq H_{S3}(x),\ \forall x \in S \cap N(x^*, d_3)$$

where $B(x^*, d_3)$ is an open ball with centre at x^* and radius d_3.

When $H_{S1}(x) =$ constant, $H_{S2}(x) =$ constant and $H_{S3}(x) =$ constant,.

$\nabla H_{S4}(x^*) = 0$ and $\nabla^2 H_{S4}(x^*)$ is positive definite, then $x^* \in S$ will be a local minimizer of $H_{S4}(x)$.

$$\text{Then } \exists\, d_4 > 0, \text{ such that } H_{S4}(x^*) \leq H_{S4}(x),\ \forall x \in S \cap N(x^*, d_4)$$

where $N(x^*, d_4)$ is an open ball with centre at x^* and radius d_4.

$$\text{Let us take } d = \min\{d_1, d_2, d_3, d_4\}$$

Then combining the above conditions, we obtain

$$\forall x \in S \cap N(x^*, d)$$

$$\begin{cases} H_{S1}(x^*) \leq H_{S1}(x), & \text{when } H_{S1}(x) \text{ is nonconstant} \\ H_{S2}(x^*) \leq H_{S2}(x), & \text{when } H_{S1}(x) \text{ is constant } H_{S2}(x) \text{ is nonconstant} \\ H_{S3}(x^*) \leq H_{S3}(x), & \text{when } H_{S2}(x) \text{ is constant and } H_{S3}(x) \text{ is nonconstant} \\ H_{S4}(x^*) \leq H_{S4}(x), & \text{when } H_{S3}(x) \text{ is constant.} \end{cases}$$

So, by **Definition 4**, we get

$$H_2(x^*) \leq_2 H_2(x),\ \forall x \in S \cap N(x^*, d).$$

Therefore $x^* \in S$ is the local minimizer of the Type-2 interval-valued function H_2.

(*ii*) The proof of maximization case can be derived similarly.

Example 3: Let us consider the following function for optimization.

$$H_2(x_1, x_2) = \left[\left(-6\left(x_1^2 + x_2^2\right) - 2,\ \left(x_1^2 + x_2^2\right)\right),\ \left(\left(x_1^2 + x_2^2\right) + 3,\ 6\left(x_1^2 + x_2^2\right) + 6\right)\right] \tag{4}$$

Solution: Here $H_{S1}(x_1, x_2) = \frac{x_1^2+x_2^2}{2} + \frac{7}{4} \neq \text{constant}$.

So, if (x_1^*, x_2^*) be an optimizer of H_2, then from the necessary condition 4.1, it follows that $\nabla H_{S1}(x_1^*, x_2^*) = 0$ which gives $(x_1^*, x_2^*) \equiv (0, 0)$.

$$\text{Now, } \nabla^2 H_{S1}(0, 0) = \begin{pmatrix} 1 & 0 \\ 0 & 1 \end{pmatrix} \text{ which is a positive definite matrix.}$$

Therefore, from sufficient condition 4.2, it follows that $(x_1^*, x_2^*) \equiv (0, 0)$ is a minimizer of (4).

Example 4: Let us take the following interval optimization problem.

$$\begin{aligned} \text{Minimize } H_2(x_1, x_2) =& [(1, 2), (4, 5)]x_1^2 - [(1, 3), (5, 6)]x_1x_2 \\ & + [(2, 3), (6, 7)]x_2^2 + [(-1, 1), (3, 4)] \\ \text{subject to } (x_1, x_2) \in \mathbb{R}^2 & \end{aligned} \tag{5}$$

Solution: The corresponding second step parametrization of the objective function is given by

$h(p_1, p_2, p_3, x_1, x_2) = A(p_1, p_2, p_3)x_1^2 - B(p_1, p_2, p_3)x_1x_2 + C(p_1, p_2, p_3)x_2^2 + D(p_1, p_2, p_3)$

where

$A(p_1, p_2, p_3) = (1 + p_1 + 3p_2 + p_2p_3 - p_1p_2)$

$B(p_1, p_2, p_3) = (1 + 2p_1 + 4p_2 + p_2p_3 - 2p_1p_2)$

$C(p_1, p_2, p_3) = (2 + p_1 + 4p_2 + p_2p_3 - p_1p_2)$

$D(p_1, p_2, p_3) = (-1 + 2p_1 + 4p_2 + p_2p_3 - 2p_1p_2)$

$p_1, p_2, p_3 \in [0, 1]$

Now,

$$H_{S1}(x_1, x_2) = \frac{h(1, 1, 1, x_1, x_2) + h(1, 1, 0, x_1, x_2) + h(1, 0, 0, x_1, x_2) + h(0, 00, x_1, x_2)}{4}$$

$$= \frac{(13x_1^2 - 15x_1x_2 + 15x_2^2 + 7)}{4} \neq \text{constant},$$

Thus, from the optimality conditions of $H_{S1}(x_1, x_2)$ we get

$$\nabla H_{S1}(x_1, x_2) = (0, 0)$$

which implies $(x_1, x_2) = (0, 0)$.

And $\nabla^2 H_{S1}(0, 0) = \begin{pmatrix} 26 & -15 \\ -15 & 30 \end{pmatrix}$ which is positive definite.

Hence, the optimization problem (5) has a minimum value at $(x_1, x_2) = (0, 0)$, and the minimum value is $H_2(0, 0) = [(-1, 1), (3, 4)]$.

5 Application

In this section, the optimal policy of the classical EOQ model with Type-2 interval-valued inventory costs are derived as an application of the optimality theory of Type-2 interval-valued function that are derived in this chapter. To formulate the model, some essential notations and assumptions are given below:

5.1 *Notations*

Notation	Description
$q(t)$	Inventory level at time t
Q	Initial inventory level
D	Demand rate

(continued)

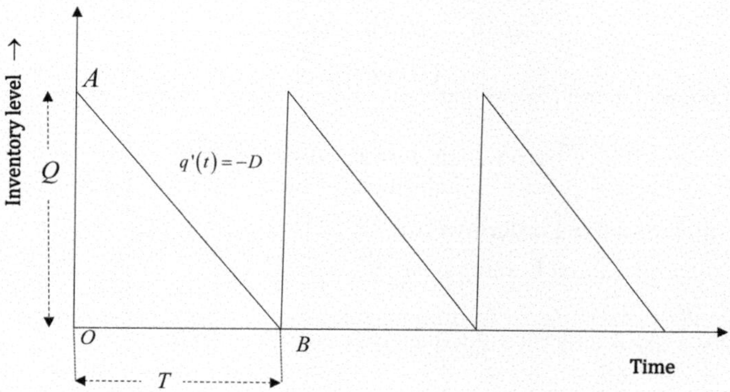

Fig. 1 Inventory level

(continued)

Notation	Description
T	Cycle length
$O_2 = [(\underline{O}_L, \overline{O}_L), (\underline{O}_U, \overline{O}_U)]$	Type-2 interval-valued ordering cost
$H_2 = [(\underline{h}_L, \overline{h}_L), (\underline{h}_U, \overline{h}_U)]$	Type-2 interval-valued holding cost

5.2 Assumptions

(i). Single item is being delivered during per order.
(ii). A known constant demand rate of D units per unit time.
(iii). The order quantity (Q) to replenish inventory arrives all at once just when desired, namely, when the inventory level drops to 0.
(iv). Taking uncertainty under consideration, all the cost components (namely, ordering cost, holding cost) are taken Type-2 interval-valued.
(v). System deals with a constant lead time with planned shortages are not allowed (Fig. 1).

5.3 Model Formulation

The Rate of Change of Inventory Level is Governed by the Differential Equation

$$\frac{dq(t)}{dt} = -D \tag{6}$$

With the conditions

$$q(0) = Q \text{ and } q(T) = 0. \tag{7}$$

Solving (6) and using (7), we get

$$q(t) = D(T - t),\ 0 \le t \le T \tag{8}$$

Type-2 interval-valued ordering cost: $O_2 = \left[\left(\underline{O}_L, \overline{O}_L\right), \left(\underline{O}_U, \overline{O}_U\right)\right]$.

Tpye-2 interval-valued holding cost: $HC_2 = \int_0^T \left[\left(\underline{h}_L, \overline{h}_L\right), \left(\underline{h}_U, \overline{h}_U\right)\right] q(t)\,dt = \frac{1}{2}\left[\left(\underline{h}_L, \overline{h}_L\right), \left(\underline{h}_U, \overline{h}_U\right)\right] DT^2$.

Therefore Type-2 interval-valued average cost is given by

$$\begin{aligned}
AC_2(T) &= \frac{1}{T}\left[\left(T\underline{C}(T), T\overline{C}(T)\right), \left(T\underline{C}(T), T\overline{C}(T)\right)\right] \\
&= \frac{1}{2T}\left[\left(\underline{h}_L, \overline{h}_L\right), \left(\underline{h}_U, \overline{h}_U\right)\right] DT^2 + \left[\left(\underline{O}_L, \overline{O}_L\right), \left(\underline{O}_U, \overline{O}_U\right)\right] \\
&= \left[\left(\frac{\underline{h}_L DT}{2} + \frac{\underline{O}_L}{T}, \frac{\overline{h}_L DT}{2} + \frac{\overline{O}_L}{T}\right), \left(\frac{\underline{h}_U DT}{2} + \frac{\underline{O}_U}{T}, \frac{\overline{h}_U DT}{2} + \frac{\overline{O}_U}{T}\right)\right]
\end{aligned}$$

Hence, we obtain a Type-2 interval-valued unconstrained optimization problem

$$Minimize\ AC_2(T)$$

Now, the second step parametric form of $AC_2(T)$ is given by $AC_2(T, r_1, r_2, r_3)$

$$\begin{aligned}
&= \left(\frac{\underline{h}_L DT}{2} + \frac{\underline{O}_L}{T} + r_1\left(\frac{\bar{h}_L DT}{2} + \frac{\bar{O}_L}{T} - \frac{\underline{h}_L DT}{2} - \frac{\underline{O}_L}{T}\right)\right) \\
&+ r_2 \left(\begin{aligned} &\frac{\underline{h}_U DT}{2} + \frac{\underline{O}_U}{T} + r_3\left(\frac{\bar{h}_U DT}{2} + \frac{\bar{O}_U}{T} - \frac{\underline{h}_U DT}{2} - \frac{\underline{O}_U}{T}\right) \\ &-\frac{\underline{h}_L DT}{2} - \frac{\underline{O}_L}{T} - r_1\left(\frac{\bar{h}_L DT}{2} + \frac{\bar{O}_L}{T} - \frac{\underline{h}_L DT}{2} - \frac{\underline{O}_L}{T}\right)\end{aligned}\right)
\end{aligned}$$

$$\begin{aligned}
(AC_2(T))_{s1} &= \frac{AC_2(T, 0, 0, 0) + AC_2(T, 1, 0, 0) + AC_2(T, 1, 1, 0) + AC_2(T, 1, 1, 1)}{4} \\
&= \frac{\left(\underline{h}_L + \overline{h}_L + \underline{h}_U + \overline{h}_U\right) DT^2 + 2\left(\underline{O}_L + \overline{O}_L + \underline{O}_U + \overline{O}_U\right)}{8T}
\end{aligned}$$

Now, using **Theorem 1**, we obtain

$$\begin{aligned}
&\nabla (AC_2(T))_{s1} = 0 \\
&\Rightarrow T^* = \sqrt{\frac{2\left(\underline{O}_L + \overline{O}_L + \underline{O}_U + \overline{O}_U\right)}{D\left(\underline{h}_L + \overline{h}_L + \underline{h}_U + \overline{h}_U\right)}}
\end{aligned} \tag{9}$$

Now, $\nabla^2(AC_2(T))_{s1} = \frac{(\underline{O}_L+\overline{O}_L+\overline{O}_U+\underline{O}_U)}{2T^3} > 0$.
Therefore, using **Theorem 2**, we get the minimum value of average cost,

$$AC_2(T^*)_{\min} = \frac{1}{T}[(T\underline{C}(T^*), T\overline{C}(T^*)), (T\underline{C}(T^*), T\overline{C}(T^*))]$$
$$= \frac{1}{2T}[(\underline{h}_L, \overline{h}_L), (\underline{h}_U, \overline{h}_U)]DT^2 + [(\underline{O}_L, \overline{O}_L), (\underline{O}_U, \overline{O}_U)]$$
$$= \left[\left(\frac{\underline{h}_L DT^*}{2} + \frac{\underline{O}_L}{T^*}, \frac{\overline{h}_L DT^*}{2} + \frac{\overline{O}_L}{T^*}\right), \left(\frac{\underline{h}_U DT^*}{2} + \frac{\underline{O}_U}{T^*}, \frac{\overline{h}_U DT^*}{2} + \frac{\overline{O}_U}{T^*}\right)\right] \tag{10}$$

And the optimal initial inventory level is given by

$$Q^* = DT^* = \sqrt{\frac{2D(\underline{O}_L + \overline{O}_L + \underline{O}_U + \overline{O}_U)}{(\underline{h}_L + \overline{h}_L + \underline{h}_U + \overline{h}_U)}} \tag{11}$$

5.4 *Numerical Illustration*

To illustrate the optimal policy of the proposed model, a numerical example is considered as follows:

Example 5: The values of parameters for these examples are given below:

$$D = 350, \left[\left(\underline{O}_L, \bar{O}_L\right), \left(\underline{O}_U, \bar{O}_U\right)\right] = [(250, 260), (290, 300)],$$
$$\left[\left(\underline{h}_L, \bar{h}_L\right), \left(\underline{h}_U, \bar{h}_U\right)\right] = [(3, 4), (7, 8)].$$

Solution:

The optimal cycle, lot-size, and average cost are obtained by using Eqs. (9)-(11). The optimal values of these inventory parameters are

$$T^* = 0.142857, \quad Q^* = 17500.$$
$$\left[\left(A\underline{C}_L^*, A\overline{C}_L^*\right), \left(A\underline{C}_U^*, A\overline{C}_U^*\right)\right] = [(1825, 1920), (2205, 2300)]$$

6 Conclusion

In this chapter, the idea of the parametric representation of the Type-2 interval has been introduced. Then, using this concept, a new definition of Type-2 interval order relation has been introduced. Then, optimality conditions (necessary and sufficient) of an unconstrained Type-2 interval-valued optimization have been derived as an application of Type-2 interval ranking.

Future research may derive the optimality conditions of non-linear constrained optimization problems of Type-2 interval-valued functions. Besides one may apply the concept of this work in the economical modelling, bio-economical modelling under imprecise circumstances.

References

Anter AM, Ali M (2020) Feature selection strategy based on hybrid crow search optimization algorithm integrated with chaos theory and fuzzy c-means algorithm for medical diagnosis problems. Soft Comput 24(3):1565–1584

Bhurjee AK, Panda G (2016) Sufficient optimality conditions and duality theory for interval optimization problem. Ann Oper Res 243(1–2):335–348

Catoni O (2004) Statistical learning theory and stochastic optimization: Ecole d'Eté de Probabilités de Saint-Flour XXXI-2001. Springer

Chen SH, Wu J (2004) Interval optimization of dynamic response for structures with interval parameters. Comput Struc 82(1):1–11

Farrokh M, Azar A, Jandaghi G, Ahmadi E (2018) A novel robust fuzzy stochastic programming for closed loop supply chain network design under hybrid uncertainty. Fuzzy Sets Syst 341:69–91

Ghosh D, Singh A, Shukla KK, Manchanda K (2019) Extended Karush-Kuhn-Tucker condition for constrained interval optimization problems and its application in support vector machines. Inf Sci 504:276–292

Heidari M, Zadeh MR, Fard OS, Borzabadi AH (2016) On unconstrained fuzzy-valued optimization problems. Int J Fuzzy Syst 18(2):270–283

Heyman DP, Sobel MJ (2004) Stochastic models in operations research: stochastic optimization (eds). Courier Corp

Lodwick WA, Kacprzyk J (2010) Fuzzy optimization: recent advances and applications (eds). Springer

Powell WB (2019) A unified framework for stochastic optimization. Eur J Oper Res 275(3):795–821

Rahman MS, Manna AK, Shaikh AA, Bhunia AK (2020a) An application of interval differential equation on a production inventory model with interval-valued demand via centre-radius optimization technique and particle swarm optimization. Int J Intel Syst 35(8):1280–1326

Rahman MS, Duary A, Shaikh AA, Bhunia AK (2020b) An application of parametric approach for interval differential equation in inventory model for deteriorating items with selling-price-dependent demand. Neural Comput Appl 1–17

Rahman MS, Shaikh AA, Bhunia AK (2020c) Necessary and sufficient optimality conditions for nonlinear unconstrained and constrained optimization problem with interval valued objective function. Comput Ind Eng 147:106634

Rahman MS, Shaikh AA, Bhunia AK (2020d) On Type-2 interval with interval mathematics and order relations: its applications in inventory control. Int J Syst Sci 1–13

Schneider J, Kirkpatrick S (2007) Stochastic optimization. Springer Sci Bus Media

Tang J, Wang DW, Fung RY, Yung KL (2004) Understanding of fuzzy optimization: theories and methods. J Syst Sci Comp 17(1):117–136
Tang J, Egiazarian K, Golbabaee M, Davies M (2020) The practicality of stochastic optimization in imaging inverse problems. IEEE Tran Comput Ima 6:1471–1485
Trujillo-Cabezas R, Verdegay JL (2020) Integrating soft computing into strategic prospective methods: towards an adaptive learning environment supported by futures studies. Springer Nature
Wang S, Watada J (2012) Fuzzy stochastic optimization: theory, models and applications. Springer Sci Bus Media
Wu HC (2004) Duality theory in fuzzy optimization problems. Fuzzy Optim Dec Mak 3(4):345–365
Ziemba WT, Vickson RG (2014) Stochastic optimization models in finance (eds). Acad Press

Chapter 16
State of Energy Estimation of Li-Ion Batteries Using Deep Neural Network and Support Vector Regression

Pradeep Kumar, Yasser Rafat, Paolo Cicconi, and Mohammad Saad Alam

Abstract Efficient management of the power and energy output of a high voltage battery pack requires a precise estimation of the State of Energy (SOE). For the accurate estimation of SOE, this work presents two data-driven methods as Deep Neural Network (DNN) and a regression model, i.e. Support Vector Regression (SVR). The effectiveness of the SOE estimation was compared, analysed, and studied through these models under similar conditions. For performance enhancement of estimation, a modified algorithm based on the grid search of optimized hyperparameters was proposed and evaluated in both the models. For training of the model at subsequent thermal ranges, two case studies were performed using US06, UDDS, LA92, and HWFET drive cycles and at four different temperature levels (−10, 0, 10, and 25 °C), for each cycle. The results indicate that the DNN method has provided enhanced performance for State of Energy Estimation as compared to the regression models of ML, i.e. SVR. This work highlights the prevailing challenges in the industry and proposes the potential recommendation for Battery Management System (BMS) development and SOE estimation in next-generation EV applications.

Keywords The State of Energy · Deep neural network · Support vector regression · Hyperparameters optimization · Regression

---

P. Kumar (✉)
Department of Electrical and Computer Engineering, University of Windsor, Windsor, ON N9B 3P4, Canada
e-mail: kumar84@uwindsor.ca

Y. Rafat
Sustainable Energy and Acoustics Research Laboratory, SEARL, Mechanical Engineering Department, Aligarh Muslim University, Aligarh 202002, India

P. Cicconi
Università Degli Studi Roma Tre, Rome, Italy

Y. Rafat · M. S. Alam
Centre of Advanced Research in Electrified Transportation, CARET, Aligarh Muslim University, Aligarh, India

I. Ali et al. (eds.), *Computational Modelling in Industry 4.0*,
https://doi.org/10.1007/978-981-16-7723-6_16

1 Introduction

With the growing problems such as the depletion of energy resources and the problem of global warming caused by use of internal conventional engines-based vehicles, electrical vehicles (EVs) have attracted the very high attention of people (Tie and Tan 2013; Xia et al. 2017). In recent years, a lot of development has taken place in the electrochemical energy storage system. Different energy storage systems have been proposed for the use in EVs such as Nickel/Metal Hydride (NiMH) battery, Lithium ion (Li-ion) battery, Fuel cells, ultra-capacitors, etc. (Iclodean et al. 2017). With the advancement and improvement in the properties of Li-ion batteries such as energy density, low self-discharge rate, long cycle life, and safety performance, they become widely applicable in EVs, electronics, mobile devices, etc. (Lu et al. 2013). A Li-ion battery module consists of an array of Li-ion cells. Temperature management is one of the issues of the Li-ion batteries which affect performance and safety (Gandoman et al. 2019; Kumar et al. 2020). The literature shows several solutions to improve the cooling effects by optimized battery layout (Qian et al. 2019), Heat Pipe (Lu et al. 2020), Phase Change Materials (Jaguemont et al. 2018), modular and mixed solutions (Cicconi et al. 2020). This paper aims to estimate the level of State of Energy (SOE) considering the achieved cell temperature with the operation parameters, i.e. voltage and current. The estimation of the cell state parameters such as State of Charge (SOC), Open Circuit Voltage (OCV), State of Energy (SOE), etc., is very important for the accurate functioning of the system and for the long life of batteries. These state parameters are also important in the estimation of driving range of electric vehicles which is also one of the complex issues as studied by Ronan German et al. (2020). However, the correct estimation of these state parameters is quite complex and difficult due to the non-linear behaviour of the electrochemical processes in Li-ion cells (Hafsaoui and Sellier 2010).

Traditionally, SOC estimation is performed for the identification of residual energy and the protection of batteries from being overcharged and discharged. From the recent literature review, it is well shown that different researchers had used a wide variety of techniques for the accurate estimation of SOC such as current integral method (Ng et al. 2009), proportional-integral (PI) method (Xu et al. 2014), electrical model-based method (Plett 2004b, c; He et al. 2013; Zhong et al. 2014), Sliding-mode observer based (Kim 2006, 2010), Kalman filter-based algorithms (Plett 2004a, b, c; Xu et al. 2012; Xiong et al. 2014), and neural network model (Kang et al. 2014) methods. With the increasing demand of Li-ion batteries for different applications, there is increased demand of use of Battery Management System (BMS) as well. Therefore, it is more and more important to accurately analyse the battery states for the proper functioning of batteries. But Liu et al. (2014) reviewed several disadvantages in the use of SOC for the estimation of residual energy. Because SOC is defined as the ratio of available capacity to the maximum stored charge in the battery, i.e. nominal capacity, from which it can be seen that there is no representation of the State of Energy. SOC clearly gives the information only about the residual capacity, not about the energy. That's why there is the need to estimate SOE independently. SOE is defined as the available energy to the maximum stored energy in Li-ion battery. Some researchers (Shen 2007; Hausmann and Depcik 2013; Waag and Sauer 2013; Zheng et al. 2013) used residual available capacity instead of SOC for the estimation of SOE. Waag and Sauer (2013) used the estimation of battery electromotive force for the identification of the battery capacity and SOC. While Shen et al. (2007) defined state of available capacity of battery, instead of SOC for the estimation of battery residual capacity. These are some of the few works which used the capacity instead of SOC in the study. It is also important to study the effect of discharge current and temperature, since at the same SOC; SOE can be different because discharge efficiency is dependent on discharge current and temperature (Liu et al. 2014). Few researchers worked in this direction to understand the working of Li-ion batteries, as Wang et al. (2014a, b) show the study on the estimation of electronic conductivity of $LiFePO_4$ cells at different temperatures to understand the low-temperature electrochemical performance at carbon-coated and uncoated cathodes. Yi et al. (2013) developed a model for the study of behaviour of Li-ion battery of temperature dependence on discharge in low ambient temperature. These studies prove that for the accurate estimation of SOE, it is necessary to consider the effect of discharge current and temperature.

In recent years, different researchers developed different techniques for the estimation of SOC and also proved well with the results. But SOC is different from SOE because SOE is the product of the residual battery capacity per OCV. The trend of variation of SOE is different from SOC. Similar to the estimation of SOC, few researchers developed systematic methods for the SOE measurement (Mamadou et al. 2012, 2019; Liu et al. 2014). In these studies, direct evaluation techniques were shown for the residual energy of battery with the consideration of battery discharge states. Some other researchers such as Stockar et al. (2011) and Kermani et al. (2011) presented the method of power integral for the estimation of SOE. But these studies were not found very accurate for the SOE measurement due to the measurement noises in the current and voltages of battery. Wang et al. (2014a, b) proposed the joint estimation technique of SOC and SOE to minimize the negative effect of power integral method. But in this method SOE estimation accuracy depends on the SOC estimation accuracy. To overcome this issue, Zhang et al. (2015) proposed the method of model-based joint estimation of SOC and SOE.

From the literature reviews, it can be seen that there is still a lot of possibility of research for the estimation of SOE. Since in this area, research has not been done as far as compared to the SOC estimation. Also, it was found out that SOE estimation is somewhat hard to estimate using mathematical relations and equations. So, there is a need to develop a method which is less complex and computationally efficient to be employed in real BMS condition. Therefore, this paper presents the two data-driven approaches for the estimation of SOE such as a regression model, i.e. SVR and DNN. A modified DNN-based method is shown for the accurate estimation of SOE. The modified method consists of the two simultaneous processes, i.e. first the optimization of hyperparameters and then the DNN model development which will be used for the prediction. In the neural networks, the critical task is the optimization of hyperparameters which is accurately done using a grid-based approach. In the proposed grid-based approach, grids were developed based on the Root Mean Square Error (RMSE), training time, and hyperparameters such as the number of neurons and number of layers. Optimized values of the number of neurons and number of hidden layers were chosen when minimum RMSE with suitable training time was recorded. The same process was used in the development of regression model using SVR for the estimation of SOE. Grid-based approach was used in the modified algorithm of SVR model for the searching of optimized hyperparameters such as regularization parameter (C) and gamma (γ).

The main aim of this study is to develop an efficient model for the estimation of SOE and to overcome the shortcomings of regression model. For the implementation of such task, in this study two data-driven methods were used for the development of SOE estimation model, DNN and SVR. The results obtained from the DNN model supports that it is more efficient in the estimation as compared to the SVR. The results shown in Sect. 4 clearly state that the DNN model is fully efficient in overcoming the shortcoming of regression model. DNN model shows significant results in the prediction of SOE at different drive cycles and at different temperatures as well.

2 Proposed Methodology for SOE Estimation

The SOE of battery can be defined by Eq. (1) as the ratio of residual energy to the maximum available energy, where residual energy is denoted by $E_{res,k}$, maximum energy by E_M, battery energy efficiency as η_e, battery terminal voltage and current at kth time instant as $V_{t,k} and I_k$, T_s is the time sampling period. The methodological approach for the estimation of SOE is described in Fig. 1. The approach is focused on the accurate estimation of SOE followed by the selection of optimized hyperparameters for the respective estimation learning models. As shown in Fig. 1, raw data was collected from the publicly available dataset (Kollmeyer et al. 2020). The data for SOE was processed from this dataset using the Eq. (1). Data cleansing was performed in the next step such as resampling of the dataset to 1 Hz frequency and the removal of errors in data (such as Voltage, Current spikes). After the data cleansing, the further dataset was developed according to the respective case studies of this research. In this study, two estimation learning model was selected, i.e. DNN and SVR. Optimization of hyperparameters was the main step after the choosing of the training dataset, validation dataset, and the selection of model. Hyperparameters are responsible for estimation accuracy, computation time, and computational speed. The processes of estimation in both models are shown in detail in Sect. 2.1 and 2.2. The performance evaluation was done on the basis of RMSE and MAE. The evaluation accuracy is discussed in the result section, i.e. Sect. 4.

$$SOE_{k+1} = \begin{cases} SOE_k - \frac{E_{res,k}}{E_M} \\ SOE_k - \frac{\eta_e V_{t,k} I_k T_s}{E_M} \end{cases} \tag{1}$$

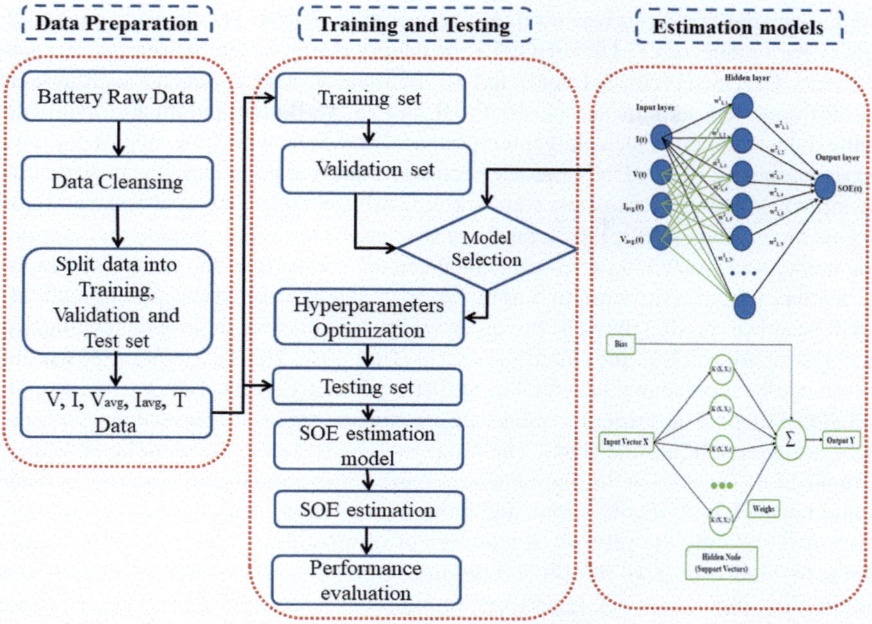

Fig. 1 The proposed methodological approach for SOE estimation

2.1 *DNN Architecture*

DNN is part of the family of machine learning methods based on artificial neural networks (ANN). In the present time, DNN has wide applications, there are many examples available where DNN architecture shows good response over the conventional algorithms such as DNN showed the level of accuracy more the human level such as in 2015 Microsoft Research's deep neural algorithm won the ImageNet challenge with the error of about 3.57% (He et al. 2016).

DNN is considered fast and efficient in comparison to the conventional methods for the estimation of SOE, since in conventional method expertise is required in battery chemistry and to mathematical model the battery behaviours. On the other hand, DNN doesn't involve such complexity for the SOE estimation. DNN consists of three layers, i.e. input layer, hidden layer, and output layer. The input and output layer remains fixed while the hidden layer can be varied to find out the optimum value where minimum error can obtain. Along with the layers, nodes or it can be said as neurons are also present in the respective layers to capture the non-linearity between the system input and output. The optimum value of nodes is also required to be found with the hidden layers for the development of the efficient and computational fast model.

DNN learns the battery functioning by mapping the input parameters with the output parameters. In this study for the creation and validation of DNN model a publicly available dataset was used (Kollmeyer et al. 2020). The dataset consists of battery parameters of LG 18650HG2 Li-ion battery at four different drive cycles such as US06, UDDS, HWFET, LA92, and all the data of drive cycles are available at four different temperature set, i.e. 10, 0, 10, and 25 °C. Battery parameters available in the dataset are current, voltage, temperature, and SOE with time step of 1 s. For the development of the DNN model, vector of battery parameters was provided to the input layer where parameters were current, voltage, temperature, average current, and average voltage, while the output layer was having the single parameter, i.e. SOE. The model was created in such a way that it can effectively map the variation in input parameters with the variation in output, i.e. SOE and by using this learning method, it will be able to predict the SOE for any new input parameters value. Mathematically the input vector can be represented as—$X(t) = [I(t), V(t), T(t), I_{avg}(t), V_{avg}(t)]$ and the output can be represented as—$Y(t) = SOE(t)$, where $I(t)$, $V(t)$, $T(t)$, $I_{avg}(t)$, $V_{avg}(t)$, and SOE(t) denote the current, voltage, temperature, average current, average voltage, and estimated SOE at time step t. The input vector X(t) fed to the input layer to map with the SOE(t) values at the output layer. Figure 2 illustrates the architecture of deep neural network with input, output, and hidden layers along with the neurons in each layer. It is showing the process of mapping of observables with the desired output. The computational speed of offline training of DNN is fast because to capitulate the output Y(t), input vector X(t) perform certain matrix multiplication whereas other strategies are not so computational fast since it involves partial differential equations. The hidden layer activation is represented by Eq. (2). Since DNN is matrix-based,

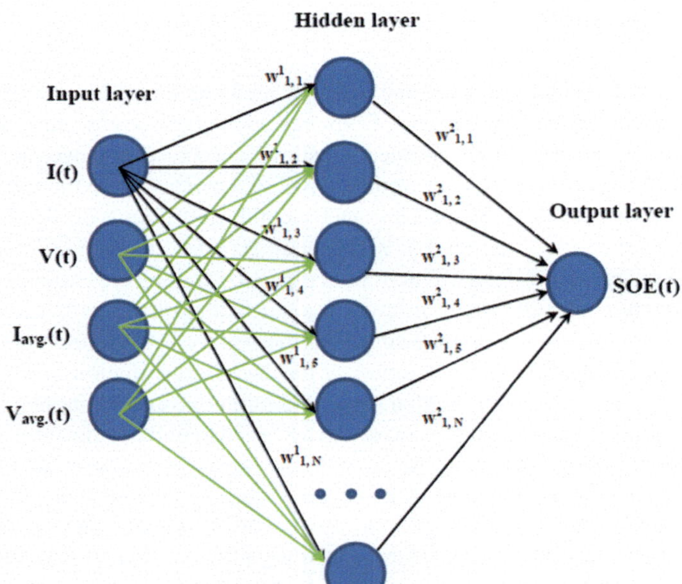

Fig. 2 Architecture of 2-layer DNN with representation of input, hidden and output layer

there are certain variables which need to be defined such as $w^{L}_{j,k}$ (which denotes the weights connection between neuron j in layer L-1 and neuron k in layer l), b^{L}_{k} (which denotes the bias), and h^{L}_{k} (which denotes the activation, respectively, of neuron k in layer L).

$$h_k^L = \sigma\left(\sum_k \left(w_{j,k}^L h_k^{L-1}(t) + b_k^L\right)\right) \tag{2}$$

Reading Eq. (2), σ is the activation function. Due to simplicity in training and testing, the non-linearity or activation function used for hidden layer is Rectified Linear Unit (ReLU) which can be given by Eq. (3).

$$\sigma(x) = \begin{cases} 0 for x < 0 \\ x for x \geq 0 \end{cases} \tag{3}$$

Similarly, at the output layer, SOE can be calculated using Eq. (4).

$$SOE(t) = \eta\left(\sum_k \left(w_{j,k}^L h_k^{L-1}(t) + b_k^L\right)\right) \tag{4}$$

where L is the last hidden layer of network and η is the activation function for output layer which was chosen as Sigmoid function and can be represented by Eq. (5).

$$\eta(x) = \frac{1}{(1 + \exp(-x))} \tag{5}$$

2.1.1 DNN Hyperparameters and Its Optimization

Effective modelling of DNN requires selection of optimum values of hyperparameters because the computational speed and performance of model depends upon these hyperparameters. There are number of hyperparameters available such as hidden layers, neurons, learning rate, optimization algorithm, activation functions, etc. But the optimization of all the hyperparameters at the same time will not be a time-effective approach; therefore the most important hyperparameters were optimized in this study which was number of hidden layers and number of neurons in hidden layer. Rest of the hyperparameters were selected based on the literature survey such as learning rate was kept 0.001, optimization algorithm was chosen as Adam due to its good performance, ReLU activation function was used for hidden layers and Sigmoid function for output layer. Optimization algorithm developed in this study for the hidden layers and neurons selection is shown in Sect. 2.1.2. Grid-based approach was used for developing the optimization algorithm and US06 drive cycle at 25

°C was used for the validation in this optimization technique. Range of values was chosen both for hidden layers and neurons, to search the combination where root mean square error (RMSE) minima will occur. The range of values checked for hidden layers was from 2 to 8, while the neurons set of values were (2, 4, 8, 16, and 32). Table 1 shows the values of RMSE and training time for the different combinations of hidden layer and neurons. From Table 1, it can be seen that the minimum RMSE occur at the combination of 3 hidden layers and 16 neurons with training time as 775.7467 s. Therefore, keeping in mind about the computational efficient concept, combination of 3 hidden layers and 16 neurons in each hidden layer was chosen for the further study. Table 2 shows the list of all the optimized hyperparameters used in this study.

Table 1 Grids showing the values of RMSE and training time for different combinations of hidden layers and neurons

Layers Neurons	2	3	4	5	6	7	8
Root mean Square Error (RMSE)							
2	0.173742	0.161347	0.286389	0.143908	0.145942	0.286164	0.287129
4	0.098428	0.110402	0.285939	0.027262	0.08073	0.170202	0.051682
8	0.089132	0.038056	0.030829	0.022372	0.017867	0.016731	0.015806
16	0.083765	**0.014997**	0.019165	0.017265	0.018367	0.016447	0.015747
32	0.040811	0.036811	0.031811	0.028821	0.021734	0.0199	0.017876
Training Time (seconds)							
2	640.9791	694.9241	719.4053	841.239	895.1712	939.1342	951.2096
4	568.7269	751.9329	582.862	770.4256	831.8109	864.1268	831.5526
8	582.5267	774.7212	592.3911	769.8025	893.7469	890.4075	913.9425
16	586.9675	**775.7467**	583.1441	657.4654	815.7815	871.2201	855.9458
32	590.2175	800.4176	608.9079	686.7024	976.5569	1007.908	852.5247

Table 2 Optimized values selected for the hyperparameters of DNN

Hyperparameters	Optimized values
Number of hidden layers	3
Number of neurons	16
Optimization algorithm	Adam
Activation function for hidden layers	ReLU
Activation function for output layer	Linear
Learning rate	0.001

2.1.2 DNN Training Strategy

This section gives the information regarding the strategy used for the training of DNN, with the insight about incorporation of optimization technique used for hidden layer and neurons along with the training. In this study, before starting the training of the model optimized value for number of hidden layers and neurons were found and the selection procedure is shown in Sect. 2.1.1. Appendix A.1 shows the modified algorithm used in this study for the development of effective DNN model. Algorithm starts with the importing of dataset used for the training. Dataset used in this study were at the different drive cycles such as US06, UDDS, HWFET and LA92. These drive cycles have the values for different parameters such as Current, Voltage, Temperature, Average Current, Average Voltage, and SOE at four different temperature set, i.e. −10, 0, 10, and 25 °C. Input parameters that will be fed into the input layers were put in a vector X and output parameter, i.e. SOE in Y. Since neural network have 3 layers, input, output, and hidden layers, input and output layers were ready as X and Y but number of hidden layers and number of neurons in each layers were still needed to be found and optimized. Therefore for finding the optimized values, range of values for hidden layer and neurons too were decided. RMSE was calculated for the different combinations of these values using the grid-based approach. Table 1 shows the RMSE and training time values for the different combinations. Finally, after the selection of optimized values of number of hidden layers and neurons the modelling of DNN begins. Optimization of these hyperparameters are utmost important because performance and efficiency of model depends on these parameters. In this study, the activation function of hidden layer was chosen to be ReLU, since from the literature study it was found to be most efficient among the other alternatives. The model used in this study updates the weights with the Adam optimization technique with learning rate 0.001. Mean Squared Error (MSE) loss function and Mean Absolute Error (MAE) metrics was chosen during the training of model. Performance analysis was done using the RMSE, as it can evaluate the percentage deviation of predicted values from original values. Training was done using the computer system with specifications of 4th generation core i5 processor and intel HD graphics. In all the cases of this study, model was trained for 1000 epochs to minimize error and to improve the model efficiency. Models were developed for 2 different cases, as in Case 1 model was trained with 3 drive cycles, i.e. UDDS, HWFET, and LA92 at all temperatures set and then performance evaluation was done using US06 drive cycle at different temperatures. In Case 2, the accuracy of model was analysed in mapping the temperature effect. All drive cycles were used for training but considering only three temperature set, i.e. −10, 0, and 25 °C. In this case evaluation was done using three drive cycles as UDDS, LA92, and HWFET at 10°C. The main purpose of considering these case studies were to evaluate the performance of model at some new temperature and new drive cycle which was not considered during the training. Results and comparison of these 2 cases are shown in Sect. 3. The performance of all the models was evaluated on the basis of MAE and RMSE error metrics.

2.2 *SVR Architecture*

In machine learning, Support Vector Machines (SVM) is the supervised learning models which analyse data for the classification and regression analysis. This study focuses on the SVR analysis technique. The functioning of SVR is in the way that it makes a complex relationship between input parameters and single output by optimizing the objective function, whose values get minimized when the predicted value from model approaches the target or original output value. For this case also the input parameters remain same as it was used for DNN, i.e. current, voltage, temperature, average current and average voltage. The major task of SVR is to correctly and precisely map the input parameters with the output value, i.e. SOE. For this study, it is not possible to have a linear relationship between the input and target values. Therefore, for this type of circumstance input values get linearly mapped with the output values in higher dimensional plane and the functions used for these transformations are known as Kernels (Müller et al. 2001). The available kernels functions in the SVR technique are sigmoid function, polynomial function, Radial Basis function (RBF), etc. Among the other available kernel function and based on the nature of the relationship between input and output parameters, RBF kernel was the most suitable for this study (Schölkopf et al. 1997). The architecture of the SVR is shown in Fig. 3, where an insight about the mapping of input and output parameters can be understood easily. From Fig. 3, it is shown that $K(x, x_n)$ is the output of the nth hidden node for the input X, also it is the mapping of input vector x and support vectors by precisely choosing kernel function (Chen and Yu 2007).

In the SVM technique, the two regression methods are available, i.e. ε—SVR and υ—SVR. Both methods differ from each other in a way that in ε – SVR methods

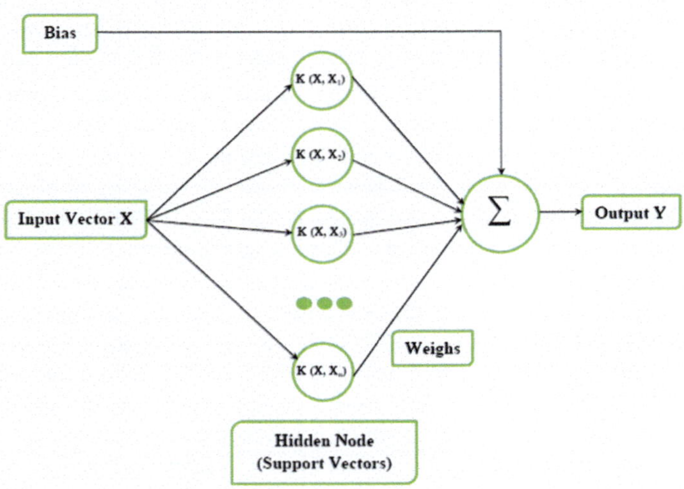

Fig. 3 υ—SVR architecture

there is no control on the number of data vectors from input dataset that become support vectors. While in υ—SVR method, there is the control on the limit of error tolerance where υ represents the upper bound on the fraction of error from training dataset and the lower bound on the fraction of support vectors. This control on limit of error tolerance can regulate by the use of regularization parameter C. For this study, authors find υ-SVR method more suitable since error tolerance in the prediction can be controlled.

2.2.1 SVR Hyperparameters and its Optimization

Similar to DNN modelling, the SVR modelling also involves hyperparameters handling and its optimization. In the SVR analysis the main hyperparameters which affect the performance and efficiency of model are gamma (γ), C, kernel function, and υ. The γ parameter shows the range of influence of input training data samples. Simply, it can be defined as the inverse of radius of influence of support vectors selected by model from the dataset. C is the regularization parameter which controls the trade-off between achieving the low training and low testing error. Kernel function is responsible for making the required relationship between input and output parameters. Due to the nonlinear relationship between the input and target parameters of this study, RBF was chosen as the kernel function. The use and functioning of υ is already explained in Sect. 2.2, it should be in the interval of (0, 1]. As the optimization of all the hyperparameters at the same time cannot be the efficient way to do the required study, therefore the default value of υ was chosen as 0.5, RBF kept as the kernel function, and then the optimum value was selected for the γ and C using the optimization technique similar to the DNN case. The grid-based approach similar to the one shown in Sect. 2.1.1 for the optimization of DNN hyperparameters was used for the optimization of C and γ. The range of values decided for the regularization parameter, C was –0.01, 0.1, 1, 10 and 100. While three values were chosen for gamma parameter for iterating with the combination in C, i.e. 0.1, 1, and 10. Table 3 shows the values of RMSE and training time for the different combinations of C and γ, validation drive cycle, i.e. US06 at 25°C was used for the testing. From these grids one combination of C and γ was chosen at which RMSE was found to be minimum and the training time was very optimum. The combination which satisfied these criteria was C as 0.1 and γ as 10. The RMSE obtained at this combination was 0.016497, which was minimum among rest of the grid combination and training time was 698.3137 s. All the parameters which were used in this modelling and their respective selected optimum values are shown in Table 4.

2.2.2 SVR Training Strategy

One of the most important parts of modelling in machine learning is the training strategy. This section will give the detail insight to the reader about the modified υ – SVR algorithm with the simultaneous optimization of hyperparameters. The

Table 3 Grids showing the values of RMSE and training time for different combinations of C and γ

C Gamma (γ)	0.01	0.1	1	10	100
Root Mean Squared Error (RMSE)					
0.1	0.099553	0.07286	0.065908	0.049061	0.040832
1	0.049851	0.03734	0.024061	0.017731	0.018968
10	0.031577	**0.016497**	0.017534	0.017223	0.007121
Training Time (seconds)					
0.1	667.7578	655.6799	671.5883	432.3391	1165.202
1	631.336	651.5541	407.5546	1106.363	7541.776
10	616.2116	**698.3137**	1557.31	11363.25	39771.38

Table 4 Optimized values selected for the hyperparameters of SVR

Hyperparameters	Optimized values
C	0.1
γ	10
υ	0.5
Kernel function	RBF

modified algorithm used for this method is shown in Appendix A.2. The details of the optimization of C and γ are already shown in Sect. 2.2.1. The datasets used for the training in this case are also similar with the DNN training. In this technique also 2 cases were studied such as in first case, three drive cycles, i.e. UDDS, LA92, and HWFET at temperature set of −10, 0, 10, and 25 °C was used for training. In this case the performance was evaluated by using the US06 drive cycle at all the same temperatures from −10 to 25 °C. In case study 2 as that of DNN training explained in Sect. 2.1.2, such that all the four cycles were fed for the training at three temperatures, i.e. −10, 0, and 25 °C. The validation was done using the three drive cycles as UDDS, LA92, and HWFET at 10°C. All these trainings were done using the same kernel function as RBF and nu (υ) as 0.5. Since in the SVR technique there is no involvement of epochs, therefore its computation time fully depends on the selection of these hyperparameters. Training was done using the same computer system with specifications of 4th generation core i5 processor and intel HD graphics. The performance of all the cases was studied on the basis of MAE and RMSE error metrics. The detailed comparison among the results of these cases is shown in Sect. 3.

Table 5 LG 18650HG2 cell parameters

Cell parameter	Specification
Chemistry	Li[NiMnCo]O_2 (H-NMC) / Graphite + SiO
Nominal Voltage	3.6 V
Charge	1.5A,4.2,50 mA End-Current (CC-CV) Normal
	4A, 4.2 V,100 mA End-Current (CC-CV) Fast
Discharge	2 V End Voltage, 20A Max Continuous Current
Nominal capacity	3.0 Ah
Energy density	240 Wh/Kg

3 Processing of Data for Training and Validation

3.1 Data Collection

In this study for the development of data-driven models and for the evaluation, the open source data was used available from Macmaster University (Kollmeyer et al. 2020). LG 18650HG2 battery was used in the testing for the gathering of this data as mentioned in the technical instruction of the open source data. The detailed information of battery used for the testing is shown in Table 5. The data are available for four different drive cycles, i.e. US06, UDDS, LA92, and HWFET tested at four different ambient temperature conditions as −10, 0, 10, and 25 °C. For this study all the required data for these drive cycles were gathered from this source (Kollmeyer et al. 2020) and then further data were prepared according to the cases of this particular study. The training and validation dataset prepared according to the cases as shown in Sect. 3.2.

3.2 Training and Validation Dataset

In this study, four drive cycles were used for the training in different combination as shown in Table 6. Simulation of battery cells by the use of these driving cycles is very time efficient as well as cost efficient. Use of these drive cycles in battery simulation makes it easier to monitor the behaviour of battery cells during real-world driving patterns without doing any extra experimentation on real physical driving vehicle. Driving cycles used in this study were Supplemental Federal Test Procedure

Table 6 Training and Validation dataset used in the study

Cases	Testing dataset	Validation dataset
Case 1	UDDS, LA92, and HWFET drive cycle at -10, 0, 10, and 25 °C	US06 drive cycle at -10, 0, 10, and 25 °C
Case 2	US06, UDDS, LA92, and HWFET drive cycle at -10, 0, and 25°C	UDDS, LA92, and HWFET drive cycle at 10 °C

(US06), Urban Dynamometer Driving Schedule (UDDS), Unified Driving Schedule (LA92), and the Highway Fuel Economy Driving Schedule (HWFET). These cycles are described in Appendix B. As an example, Fig. 4 shows the behaviour of the battery parameters, i.e. current, voltage, and temperature of US06 drive cycle at all ranges of temperature from −10 to 25 °C. Training and validation dataset used in

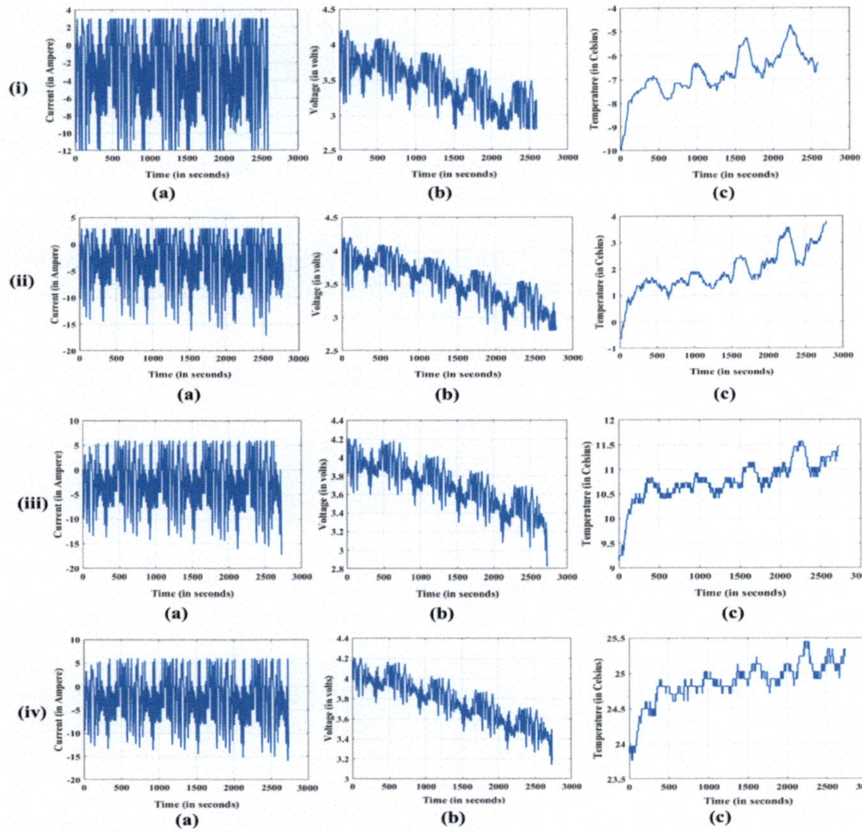

Fig. 4 The report of the US06 drive cycle in terms of: **a** Current, **b** Voltage, **c** Cell Temperature. This test was done at four conditions of room temperature: (i) −10 °C, (ii) 0 °C, (iii) 10 °C, (iv) 25 °C

different cases for the study in this research are shown in Table 6. Two cases were studied using DNN and then results of DNN were compared by another Machine Learning technique, i.e. SVR. These two cases are already explained in Sect. 2.1.2 and 2.2.2 according to the respective DNN and SVR training.

All the input vectors in each case were normalized before feeding for the training. Normalization technique was used to improve the convergence rate and to remove the negative influence. In this study min–max normalization was used since it retains the original data distribution pattern, only scaled the dataset in the range of [0, 1] as shown in Eq. (6) (Jain et al. 2005).

$$z_i^k = \frac{x_i^k - \min(x)}{\max(x) - \min(x)} i \in \{1, 2, \ldots\ldots, n\} \tag{6}$$

where x is the input vectors such as I(t), V(t), T(t), $I_{avg}(t)$, $V_{avg}(t)$, and n represents the total number of samples of data in the different drive cycles.

3.3 Evaluation Metrics

For the study of performance of trained models, some evaluation metrics were used such as MAE and RMSE. These evaluation or error metrics made the way of the comparison of different machine learning methods and different case study done in these methods in this research paper. These evaluation or error metrics can be represented by the Eqs. (7), (8), and (9).

$$MAE = \frac{1}{N}\sum_{k=1}^{N}\left(\left|SOE_k - SOE_k^*\right|\right) \tag{7}$$

$$MSE = \frac{1}{N}\sum_{k=1}^{N}\left(\left|SOE_k - SOE_k^*\right|\right)^2 \tag{8}$$

$$RMSE = \sqrt{\frac{1}{N}\sum_{k=1}^{N}\left(\left|SOE_k - SOE_k^*\right|\right)^2} \tag{9}$$

where SOE_k is the predicted value by model, $SOE_k{}^*$ is the actual value obtained from battery testing at time step k and N is the total number of training samples.

4 Results and Discussion

As mentioned in earlier section, the input vector fed in to the DNN and SVR can be represented as $X = [I(t), V(t), T(t), I_{avg}(t), V_{avg}(t)]$ and the output can be represented as $Y = SOE(t)$, where $I(t)$, $V(t)$, $T(t)$, $I_{avg}(t)$, $V_{avg}(t)$ and $SOE(t)$ denotes the current, voltage, average current, average voltage and estimated SOE at time step t. The drive cycles used for the training and validation are mentioned in Sect. 3.2 and were recorded at sampling frequency of 1 Hz. The following subSects. 4.1 and 4.2 will show the information about the results obtained after the training and validation according to different cases used in this study. Comparative analysis is shown in these sections for the DNN and SVR modelling for the prediction of SOE.

4.1 Case Study 1

In this case, as described earlier that three drive cycles namely UDDS, LA92, and HWFET were fed for training at all temperature sets as −10, 0, 10, and 25 °C. For the evaluation of the trained model US06 drive was used at all the temperature ranges. Comparative study is shown here between DNN and SVR technique of machine learning. Estimation performance of DNN and SVR was compared on the basis of error metrics, i.e. RMSE and MAE. Figure 5 depicts the prediction of SOE using both data-driven methods, i.e. DNN and SVR. It shows the comparative representation of SOE between actual SOE, predicted SOE using DNN, and predicted SOE using SVR. Figure 5a–d show the prediction of SOE using US06 drive cycle at four different temperature ranges, i.e. −10 °C, 0 °C, 10 °C, and 25 °C, respectively. Table 7 shows the values of RMSE and MAE for both the technique of machine learning, i.e. DNN and SVR.

4.2 Case Study 2

In case 1 the performance of model was analysed on a new drive cycle which was not fed for the training. In this case, model was evaluated using the drive cycles at temperatures set other than which fed for training. In this case all the drive cycles were used for the training but only at three temperature values, i.e. −10, 0, and 25 °C. Drive cycles at 10 °C were kept for the testing of developed model accuracy. Three drive cycles at 10 °C, i.e. UDDS, LA92, and HWFET was used for the validation

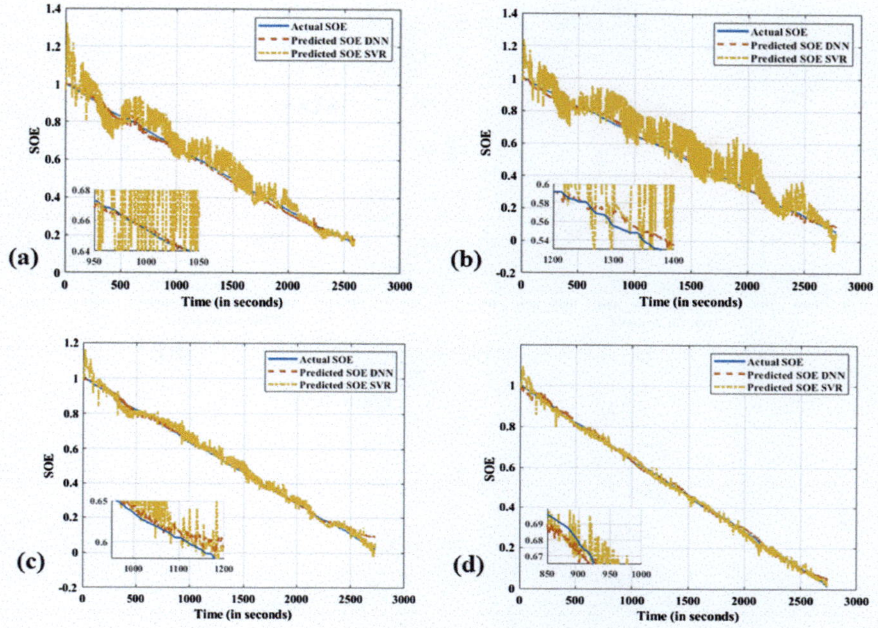

Fig. 5 Prediction of SOE using DNN and SVR technique on US06 drive cycle at following temperature range **a** at −10 °C temperature **b** at 0 °C temperature **c** at 10 °C temperature **d** at 25 °C temperature

Table 7 Evaluation results on the individual drive cycles

	DNN		SVR	
Testing Drive Cycle	RMSE	MAE	RMSE	MAE
US06 at −10°C	0.0148	0.0125	0.0602	0.0442
US06 at 0°C	0.0149	0.0118	0.0898	0.0672
US06 at 10°C	0.0178	0.0128	0.0287	0.0218
US06 at 25°C	0.012	0.010	0.0164	0.0122

Table 8 Evaluation results on the combination of drive cycles

	DNN		SVR	
Training Drive Cycle	RMSE	MAE	RMSE	MAE
UDDS at 10 °C	0.0154	0.0120	0.0207	0.0156
LA92 at 10 °C	0.0189	0.0162	0.0257	0.0177
HWFET at 10 °C	0.0118	0.0093	0.0339	0.0235

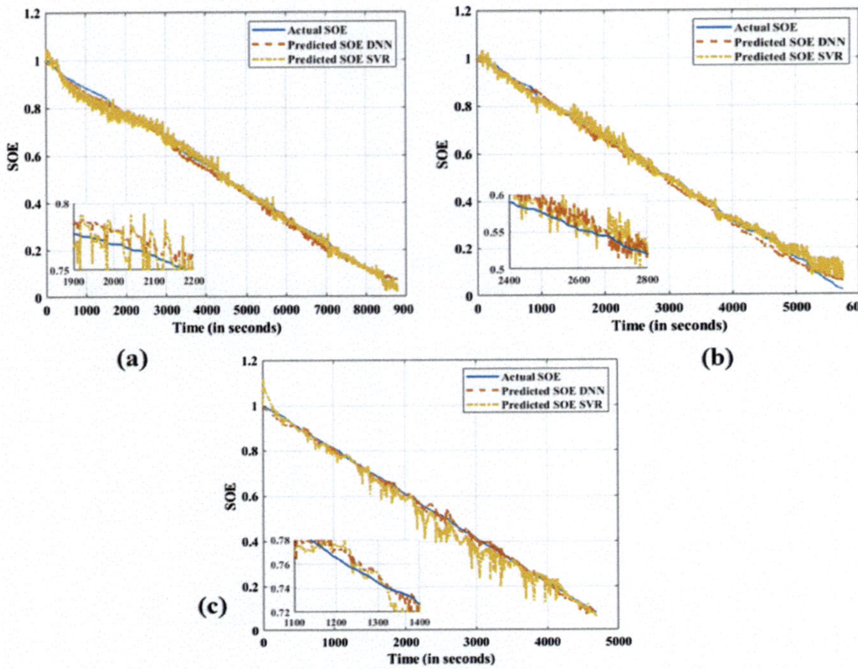

Fig. 6 Prediction of SOE using DNN and SVR technique on following drive cycle at 10 °C temperature **a** UDDS drive cycle **b** LA92 drive cycle **c** HWFET drive cycle

purpose. This same procedure was followed in both ML techniques. Figure 6 shows the predicted SOE behaviour for the different combinations of drive cycles in graphical way. In this case also, it was seen that the error in the prediction of SOE was high for SVR method in comparison to DNN. The RMSE and MAE error values are shown in Table 8 for both methods and it can be easily seen that DNN performance is significantly better as compared to SVR.

5 Conclusion

This work offers a unique contribution to the accurate estimation of SOE. In this study modified algorithm for DNN and SVR with optimization techniques are implemented. The effectiveness of DNN and SVR methods were compared for the prediction of SOE. Instead of using any random hyperparameters, this study proposed the approach to optimize the hyperparameters and then used the optimized values of hyperparameters in the subsequent prediction. During the optimization, it was noticed that the error in the prediction reduces as the hidden layers and number of neurons increases in the case of DNN but higher values of them make the model computationally less efficient. Similarly in the case of SVR technique, if the value of regularization parameter (C) and gamma (γ) increases then the training time increases significantly, this makes the model less time efficient. This work shows the unique method for the optimization of hyperparameters of both techniques, i.e. DNN and SVR. The prediction result of this work shows that the DNN model is more efficient in the prediction of SOE as compared to SVR. The results obtained in the different case studies show that the SOE prediction by DNN model at a drive cycle different from the one used in training is quite good. In the same way, DNN model is able to predict the SOE at different thermal ranges other than those used in training. The results of the case studies evidently prove that the prediction from DNN is far better than SVR. This work also suggests that the conventional regression model used for the estimation of SOE can be upgraded using the DNN model in the future. This work highlights the prevailing challenges in the industry and proposes the potential recommendation for BMS development and SOE estimation in next-generation EV applications.

Appendix A

A.1 Modified DNN Algorithm With the Optimization Technique

```
Import dataset   // Import the dataset required for the training
X = ['Current', 'Voltage', 'Average Current', 'Average Voltage'] // Input parameters
Y = ['SOE']  // Output parameters
Define range of values of hidden layers // range of values, example- 2, 3, 4……
Define range of values of neurons // range of neurons values, example- 1, 2, 4 ……
// Initializing optimization loop for hidden layers and neurons
For i in range of hidden layers values  // loop for the range of hidden layers values
        For j in range of neurons values // loop for the range of neurons values
        find RMSE  // rmse for the combination of hidden layer and neurons values
Hidden layers and neurons = (L, N) // Optimum values on the basis of minimum of rmse
Import model  // import the model of DNN
From layers import Dense  // import dense layer that will be add in model
model. add(Dense(units = N , input_dim = no. of input parameters,  activation = 'relu'))
// add the dense layer, where N is optimum value of neuron, input_dim is the no, of input
parameters and activation function is ReLU. More layers can be add in the similar manner
model.add(Dense(units = 1, activation = 'linear')) // Output layer with Linear activation
model.compile(optimizer = Adam(learning_rate = 0.01), loss = 'MSE', metrics = ['mae'])
// Compile the model with Adam optimizer, learning rate fixed to be 0.01, loss function be
MSE and metrics is MAE
model.fit(X, Y, epochs = 10000)  // fitting of model where epochs is the no. of iterations
Import testing dataset // import the dataset for the testing of performance of developed
model
Define X_test and Y_test  // define the X_test and Y_test in the similar manner as done for
training X and Y
Y_predict = model.predict(X_test)  // Calculate the predicted output
RMSE(Y_predict, Y_test)  // Finally calculate the rmse for predicted and actual output
```

A.2 Modified υ—SVR Algorithm With the Optimization Technique

```
Import dataset  // Import the dataset required for the training
X = ['Current', 'Voltage', 'Average Current', 'Average Voltage'] // Input parameters
Y = ['SOE'] // Output parameters
Define range of values of C // range of values selected were 0.01, 0.1, 1, 10, 100
Define range of values of γ // range of γ values selected 0.1, 1, 10
// Initializing optimization loop for hidden layers and neurons
For i in range of γ values  // loop for the range of γ values
        For j in range of C values // loop for the range of C values
        find RMSE  // rmse for the combination of C and γ values
C and γ = (c, gp) // Optimum values on the basis of minimum of rmse
Import model  // import the υ-SVR model of SVM
model = NuSVR(kernel = 'rbf', C = c, gamma = gp, nu = 0.5)  // fitting of model
Import testing dataset // import the dataset for the testing of performance of developed model
Define X_test and Y_test  // define the X_test and Y_test in the similar manner as done for training X and Y
Y_predict = model.predict(X_test)  // Calculate the predicted output
RMSE(Y_predict, Y_test)  // Finally calculate the rmse for predicted and actual output
```

Appendix B

B.1 US06 Drive Cycle Velocity Profile

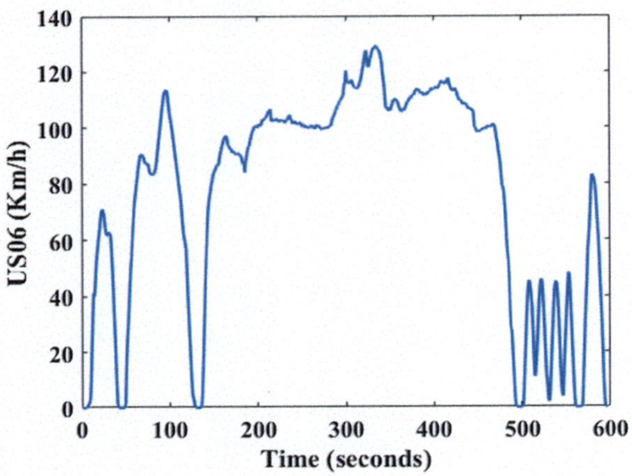

B.2 UDDS Drive Cycle Velocity Profile

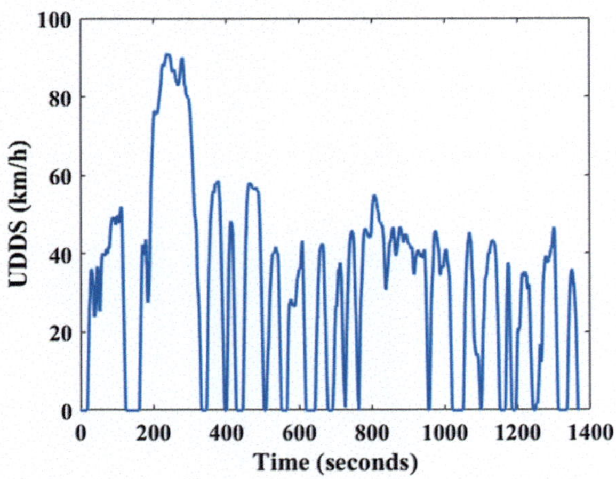

B.3 HWFET Drive Cycle Velocity Profile

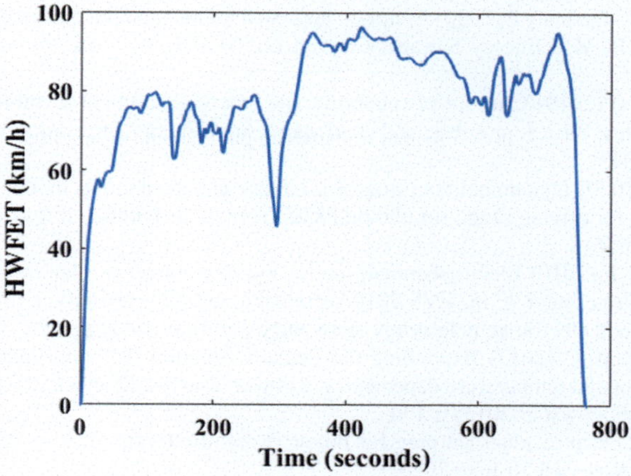

B.4 LA92 Drive Cycle Velocity Profile

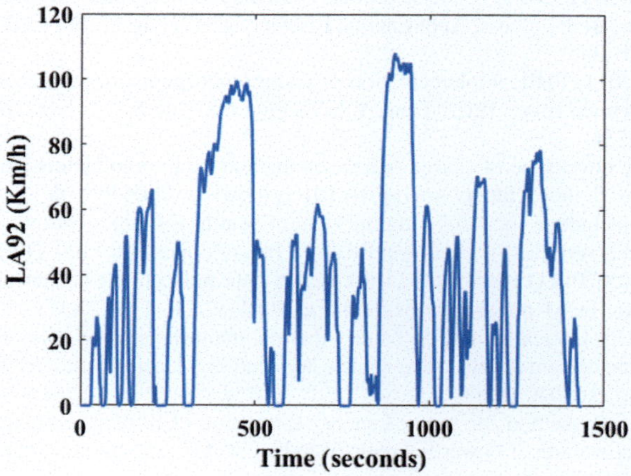

References

Chen ST, Yu PS (2007) Pruning of support vector networks on flood forecasting. J Hydrol, Elsevier 347(1–2):67–78. https://doi.org/10.1016/j.jhydrol.2007.08.029

Cicconi P, Kumar P, Varshney P (2020) A support approach for the modular design of li-ion batteries: a test case with PCM. J Energy Storage, Elsevier Ltd 31:101684. https://doi.org/10.1016/j.est.2020.101684

Gandoman FH et al (2019) Concept of reliability and safety assessment of lithium-ion batteries in electric vehicles: Basics, progress, and challenges. Appl Energy. https://doi.org/10.1016/j.apenergy.2019.113343

German R et al (2020) Dynamical coupling of a battery electro-thermal model and the traction model of an EV for driving range simulation. IEEE Trans Veh Technol. https://doi.org/10.1109/TVT.2019.2955856

Hafsaoui J, Sellier F (2010) Electrochemical model and its parameters identification tool for the follow up of batteries ageing. In: EVS 2010-sustainable mobility revolution: 25th world battery hybrid and fuel cell electric vehicle symposium and exhibition, vol 4, pp 386–395

Hausmann A, Depcik C (2013) Expanding the Peukert equation for battery capacity modeling through inclusion of a temperature dependency. J Power Sources, Elsevier 235:148–158. https://doi.org/10.1016/j.jpowsour.2013.01.174

He K et al (2016) Deep residual learning for image recognition. http://image-net.org/challenges/LSVRC/2015/. Accessed 30 June 2020

He Y et al (2013) A new model for state-of-charge (SOC) estimation for high-power Li-ion batteries. Appl Energy, Elsevier Ltd 101:808–814. https://doi.org/10.1016/j.apenergy.2012.08.031

Iclodean C et al (2017) Comparison of different battery types for electric vehicles. In: IOP conference series: materials science and engineering. https://doi.org/10.1088/1757-899X/252/1/012058

Jaguemont J et al (2018) Phase-change materials (PCM) for automotive applications: a review. Appl Therm Eng. https://doi.org/10.1016/j.applthermaleng.2017.12.097

Jain A, Nandakumar K, Ross A (2005) Score normalization in multimodal biometric systems. Pattern Recogn Pergamon 38(12):2270–2285. https://doi.org/10.1016/j.patcog.2005.01.012

Kang LW, Zhao X, Ma J (2014) A new neural network model for the state-of-charge estimation in the battery degradation process. Appl Energy, Elsevier Ltd 121:20–27. https://doi.org/10.1016/j.apenergy.2014.01.066

Kermani S et al (2011) PHIL implementation of energy management optimization for a parallel HEV on a predefined route. IEEE Trans Veh Technol 60(3):782–792. https://doi.org/10.1109/TVT.2011.2107534

Kim IS (2006) The novel state of charge estimation method for lithium battery using sliding mode observer. J Power Sources. https://doi.org/10.1016/j.jpowsour.2006.09.006

Kim IS (2010) A technique for estimating the state of health of lithium batteries through a dual-sliding-mode observer. IEEE Trans Power Electron. https://doi.org/10.1109/TPEL.2009.2034966

Kollmeyer P, et al (2020) LG 18650HG2 Li-ion battery data and example deep neural network xEV SOC estimator script. Mendeley 3.https://doi.org/10.17632/CP3473X7XV.3

Kumar P et al (2020) Critical review on battery thermal management and role of nanomaterial in heat transfer enhancement for electrical vehicle application. J Energy Storage 32:102003. https://doi.org/10.1016/j.est.2020.102003

Liu X et al (2014) 'A method for state of energy estimation of lithium-ion batteries at dynamic currents and temperatures. J Power Sources, Elsevier 270:151–157. https://doi.org/10.1016/j.jpowsour.2014.07.107

Lu L et al (2013) A review on the key issues for lithium-ion battery management in electric vehicles. J Power Sources. https://doi.org/10.1016/j.jpowsour.2012.10.060

Lu M et al (2020) Research progress on power battery cooling technology for electric vehicles. J Energy Storage. https://doi.org/10.1016/j.est.2019.101155

Mamadou K et al (2012) Definition of a state-of-energy indicator (SoE) for electrochemical storage devices: application for energetic availability forecasting. J Electrochem Soc, The Electrochemical Society 159(8):A1298–A1307. https://doi.org/10.1149/2.075208jes

Mamadou K et al (2019) The state-of-energy: a new criterion for the energetic performances evaluation of electrochemical storage devices. ECS Trans, the Electrochemical Society 25(35):105–112. https://doi.org/10.1149/1.3414008

Müller KR et al (2001) An introduction to kernel-based learning algorithms. IEEE Trans Neural Netw:181–201. https://doi.org/10.1109/72.914517

Ng KS et al (2009) Enhanced coulomb counting method for estimating state-of-charge and state-of-health of lithium-ion batteries. Appl Energy, Elsevier Ltd 86(9):1506–1511. https://doi.org/10.1016/j.apenergy.2008.11.021

Plett GL (2004a) Extended Kalman filtering for battery management systems of LiPB-based HEV battery packs—part 1 Background. J Power Sources. https://doi.org/10.1016/j.jpowsour.2004.02.031

Plett GL (2004b) Extended Kalman filtering for battery management systems of LiPB-based HEV battery packs—Part 2 modeling and Identification. J Power Sources. https://doi.org/10.1016/j.jpowsour.2004.02.032

Plett GL (2004c) Extended Kalman filtering for battery management systems of LiPB-based HEV battery packs—part 3 state and parameter estimation. J Power Sources. https://doi.org/10.1016/j.jpowsour.2004.02.033

Qian X et al (2019) Heat dissipation optimization of lithium-ion battery pack based on neural networks. Appl Therm Eng. https://doi.org/10.1016/j.applthermaleng.2019.114289

Schölkopf B et al (1997) Comparing support vector machines with gaussian kernels to radial basis function classifiers. IEEE Trans Signal Process 45(11):2758–2765. https://doi.org/10.1109/78.650102

Shen WX (2007) State of available capacity estimation for lead-acid batteries in electric vehicles using neural network. Energy Convers Manag Pergamon 48(2):433–442. https://doi.org/10.1016/j.enconman.2006.06.023

Stockar S et al (2011) Energy-optimal control of plug-in hybrid electric vehicles for real-world driving cycles. IEEE Trans Veh Technol 60(7):2949–2962. https://doi.org/10.1109/TVT.2011.2158565

Tie SF, Tan CW (2013) A review of energy sources and energy management system in electric vehicles. Renew Sustain Energy Rev. https://doi.org/10.1016/j.rser.2012.11.077

Waag W, Sauer DU (2013) Adaptive estimation of the electromotive force of the lithium-ion battery after current interruption for an accurate state-of-charge and capacity determination. Appl Energy, Elsevier Ltd 111:416–427. https://doi.org/10.1016/j.apenergy.2013.05.001

Wang F et al (2014a) 'Low-temperature electrochemical performances of LiFePO4 cathode materials for lithium ion batteries', J Taiwan Instit Chem Eng. Taiwan Institute of Chemical Engineers 45(4):1321–1330. https://doi.org/10.1016/j.jtice.2014.02.013

Wang Y, Zhang C, Chen Z (2014b) A method for joint estimation of state-of-charge and available energy of LiFePO4 batteries. Appl Energy. https://doi.org/10.1016/j.apenergy.2014.08.081

Xia G, Cao L, Bi G (2017) A review on battery thermal management in electric vehicle application. J Power Sources. https://doi.org/10.1016/j.jpowsour.2017.09.046

Xiong R et al (2014) A data-driven multi-scale extended Kalman filtering based parameter and state estimation approach of lithium-ion olymer battery in electric vehicles. Appl Energy. https://doi.org/10.1016/j.apenergy.2013.07.061

Xu J et al (2014) The state of charge estimation of lithium-ion batteries based on a proportional-integral observer. IEEE Trans Veh Technol. https://doi.org/10.1109/TVT.2013.2287375

Xu L, Wang J, Chen Q (2012) Kalman filtering state of charge estimation for battery management system based on a stochastic fuzzy neural network battery model. Energy Convers Manage Pergamon 53(1):33–39. https://doi.org/10.1016/j.enconman.2011.06.003

Yi J et al (2013) Modeling the temperature dependence of the discharge behavior of a lithium-ion battery in low environmental temperature. J Power Sources, Elsevier B.V, 244:143–148. https://doi.org/10.1016/j.jpowsour.2013.02.085

Zhang W, Shi W, Ma Z (2015) Adaptive unscented Kalman filter based state of energy and power capability estimation approach for lithium-ion battery. J Power Sources. https://doi.org/10.1016/j.jpowsour.2015.04.148

Zheng Y et al (2013) LiFePO4 battery pack capacity estimation for electric vehicles based on charging cell voltage curve transformation. J Power Sources, Elsevier 226:33–41. https://doi.org/10.1016/j.jpowsour.2012.10.057

Zhong L et al (2014) A method for the estimation of the battery pack state of charge based on in-pack cells uniformity analysis. Appl Energy, Elsevier Ltd 113:558–564. https://doi.org/10.1016/j.apenergy.2013.08.008

Chapter 17
Transportation and Parking Framework for Pollution-Free AMU Campus: A Case Study

Rafi Akhtar, Mohammad Arqam, Mohammad Saad Alam, Yasser Rafat, and Salman Hameed

Abstract Air pollution and climate change have become one of the major contemporary concerns for the entire world and if not addressed well the challenge is only going to surge in future that may be catastrophic for lives on the planet. Since the number of vehicles increasing rapidly transportation has become one of the major contributors to air pollution. Hence, there is a need to propose and develop a proper yet smart transportation and parking system to minimize the pollution and congestion problems. This chapter reveals the impacts of emission on campus life through providing an estimate of different pollutants emitted from the vehicles coming in and out of the campus. For a campus like Aligarh Muslim University (AMU), a different approach is required as the number of commutators does not vary largely and the major inflow is of two-wheelers. Appropriate locations for vehicle parking facility to be constituted have been identified to optimize the vehicle running time. Moreover, an IoT-based smart parking system framework for the pollution-free AMU campus has been conceptualized.

1 Introduction

From 'saving environment for future generations' to 'saving environment for the current generation itself,' environmental concern has become one of the most-talked issues in recent times. The outrage is righteous witnessing severe damage to lives on the planet caused by different forms of pollution; among which air pollution is the most prevalent in nature. Globally, nearly 90% people are currently breathing impure air and about 7 millions of them die every year because of related diseases (World Health Organisation 2021). Besides industries and construction works, transportation is the primary source of major pollutants emission namely CO, CO_2, oxides of

---

R. Akhtar · M. S. Alam · Y. Rafat · S. Hameed
Centre for Advanced Research in Electrified Transportation (CARET), Aligarh Muslim University, Aligarh, India

M. Arqam (✉)
School of Engineering and Built Environment, Griffith University, Brisbane, Australia
e-mail: mohammad.arqam@griffithuni.edu.au

I. Ali et al. (eds.), *Computational Modelling in Industry 4.0*,
https://doi.org/10.1007/978-981-16-7723-6_17

nitrogen and Sulphur, etc. As per the government sources, number of all types of motor vehicles in India increased from 55 million to 230 million during the period of 2001 to 2016 that include increase in number of cars from 5.3 million to 25.6 million and two-wheelers from 38.5 million to 169 million (Statistical Year book, India 2015).

Every year India witnesses a huge layer of smog, a phenomena directly related to air pollution, especially in winter. It causes adverse impact on health and economy, temporary effects include irritation, headache, and nausea and in long term lead to chronic respiratory and cardiovascular diseases like asthma (Yang et al. 2017, Zhang et al. 2017). Low visibility slows down traffic on railways and roads leading to significant economic losses. Moreover unwanted noise from vehicles is a well-known reason of annoyance and insomnia (Jariwala et al. 2017). Hence, to mitigate the side effects of transportation, it is essential to implement a smart transportation and parking system framework to reduce traffic congestion that eventually amplifies the pollutants emission.

Several studies have already been conducted concerning pollution caused by transportation and need to develop smart parking system to avoid congestion. Most of them used sensor-based technologies to let customer rider know where to park the vehicle based on the availability of the slots. Researchers in India recently used single board computer Raspberry Pi, GPS module, and ultrasonic sensors to send information to the customers about the availability of parking lots through a mobile application based on google map (Shinde et al. 2017). This approach is significantly reliable but separate sensors are needed for each parking space, hence costlier.

Similarly, another group of researchers used camera-based Raspberry sensors to collect data and utilized machine learning to distinguish between occupied and vacant parking space (Ling et al. 2017). This one-to-many sensor solution is comparatively cheaper and showed 91% accuracy in experiment that can be further improved as machine learning gets matured. The California State University researchers suggested a more versatile system to give more information like topological mapping using single node camera sensors (Telles and Meduri 2017). To ensure shortest path with minimum turns to be covered in the parking complex, Chinese researchers developed an improved ant colony algorithm (Wang et al. 2017). Pampa Sadhukhan also addressed the problem of improper parking and alarming authorities if someone does not park his vehicle properly (Sadukhan 2017). Lukmayang et al. (2018) proposed a system to let the vehicle owner know when and where it has to be parked to enhance security. (Khanna and Anand 2016) have also proposed an application-based smart parking system.

Most of the prior works are for public and paid parking with fluctuating number of customers every day and the availability was the major concern than location of the available space in the complex. For a campus like Aligarh Muslim University (AMU), a different approach is needed since the number of vehicles to be parked do not vary largely on the daily basis. In the varsity campus, availability is not the paramount apprehension rather the utmost anxiety is finding the appropriate space, especially for two-wheelers, from where one can also vacate easily. Therefore, this chapter reveals the impacts of emission on campus life through providing an estimate of different pollutants emitted from the vehicles coming in and out of the campus. Appropriate locations for vehicle parking facility to be constituted have been identified to optimize the vehicle running time. Moreover, an IoT-based smart parking system framework for the pollution-free AMU campus has been conceptualized.

2 Estimation of Pollutants Emission

The AMU Campus is a workplace of almost 28,000 students, 1342 teachers, and 5610 non-teaching staff (AMU website 2021). Most of them use their own motor vehicles for travelling inside the campus. In spite of having adequate space and road networks, unspecified routes and ill-managed parking system still need serious attention.

2.1 The Campus Map

The campus map shown in Fig. 1 is a screen-grab taken from google map and lines are drawn manually to show major routes to the specific faculty building which are, in turn, marked with alphabets. A, B, and C are the three major entrances. Different places are represented by alphabets according to the Table 1. Distance from entrances to any faculty are measured using google map's 'find route' feature.

Vehicle models taken into consideration to calculate the amount of emission are shown in Table 2. All data are for positive ignition, gasoline engines based on BS

Fig. 1 AMU campus map

Table 1 Different location and their representation

Location	Represented by
Chungi entrance	A
Medical entrance	B
Bab-e-Syed entrance	C
Faculty of engineering	D
Faculty of arts	E
Faculty of science	F
Faculty of business studies	G
Faculty of theology	H
Faculty of commerce	I
Faculty of social Science	J
Faculty of International Studies	K

Table 2 Vehicle models under consideration

Type	Model	Engine capacity (cc)
Motorcycle	Passion Pro	100
Motorcycle	Duke	300
Scooter	Alpha	150
Scooter	Activa	110
Motorcycle	Discover	135
Motorcycle	Apache	200
Car	Alto	800
Car	Wagon R	1000
Car	Swift Dzire	1400
Car	Duster	1600
Car	Safari	2500

IV Pollution Standard. To obtain the amount of a pollutant emitted from a particular vehicle, distance covered is multiplied by the emission factor (emission in grams per kilometre) and then doubled to account for both ways. For instance, 20 Activa Scooters are recorded to come to Engineering Faculty from **A** covering 0.5 km distance **AD**; upon multiplying the distance by emission factor of CO_2, i.e. 33.4 (India GHD Program 2015) and then doubling it, emission of CO_2 comes out to be 668 gram. Similarly, amount of emission are calculated for other paths **BD** and **CD** as 1336 gram and 2338 gram, respectively constituting total of 4322 gram of CO_2 emission by Activa Scooters to and from Engineering Faculty.

The emission factor of CO and NO_X are 1.403 and 0.39 (India light duty and motor cycle emissions 2018) and upon following similar calculation technique as before total emissions come out to be 182.39 gram CO and 50.7 gram NO_X for the exactly same case as that of CO_2. Fig. 2 shows emission from other vehicle models

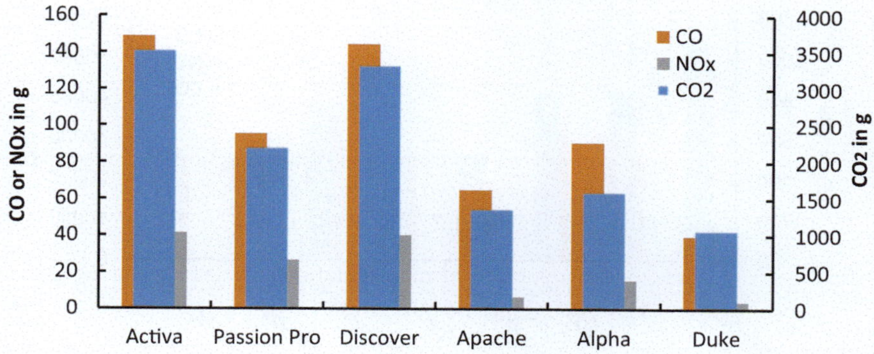

Fig. 2 Emission from two-wheelers for the faculty of engineering and technology

considered. Similar procedure has been followed for the calculation of emission by cars as well and the bar chart made to represent the data is shown in Fig 3. Using the data for individual faculties, overall emissions were calculated as depicted in Fig. 4.

As shown in Fig. 4, total CO_2 emission for the Faculty of Engineering and Technology from all vehicles, two-wheeler and cars, coming and going via all the three entrances is 53867.70 grams. For the same faculty, total CO emission is 1023.28 grams and total NO_X emission is 194.21 grams. For the Faculty of Arts, emission of CO_2, CO, and NO_X, respectively are 42459.94 grams, 819.44 grams, and 150.15 grams and similarly for other locations shown in the chart. For the whole campus, constituting all eight faculties inside, net emission of CO_2 is 279036.17 gram, net emission of CO is 5260.25 gram, and finally net emission of NO_X is 972.23 gram in a normal working day.

Overall parking framework for the AMU campus may constitute:

1. Locating the places of requirement

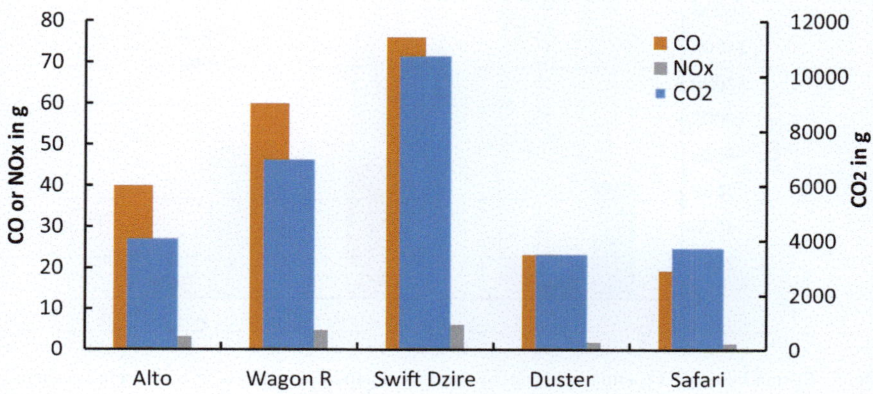

Fig. 3 Emission from cars for faculty of engineering and technology

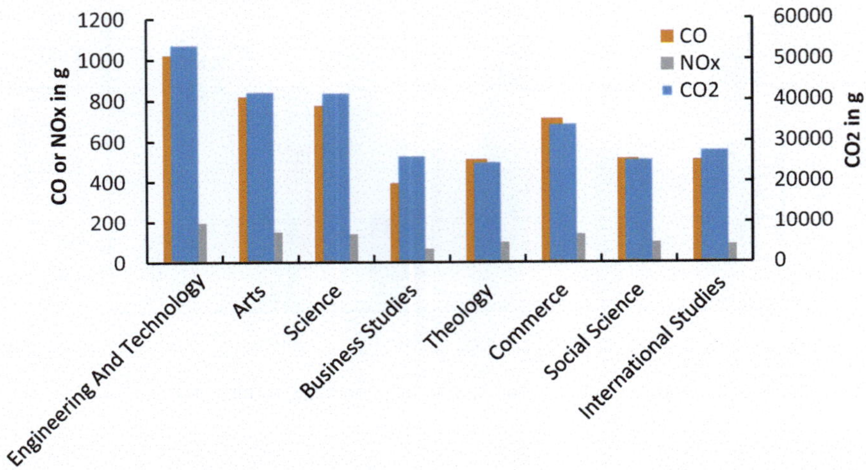

Fig. 4 Total emission from different faculties

2. Cost-effective system installation for smart parking.

To minimize emissions, parking facilities are supposed to be established/constructed at locations where the vehicle has to cover the least distance. Through utilizing optimization techniques and considering commuters' convenience, following places are identified for multistorey parking setup as outlined in Fig. 1. The Location '**X**' represents parking near the engineering faculty that is supposed to accommodate commuters coming from '**B**' and want to go to engineering faculty '**D**', faculty of arts '**E**', faculty of science '**F**', faculty of social science '**J**' and faculty of theology '**H**'. The location '**X**' will also facilitate parking for vehicles coming to Engineering faculty '**D**' from the entrances '**A**' and '**C**' (Fig. 5).

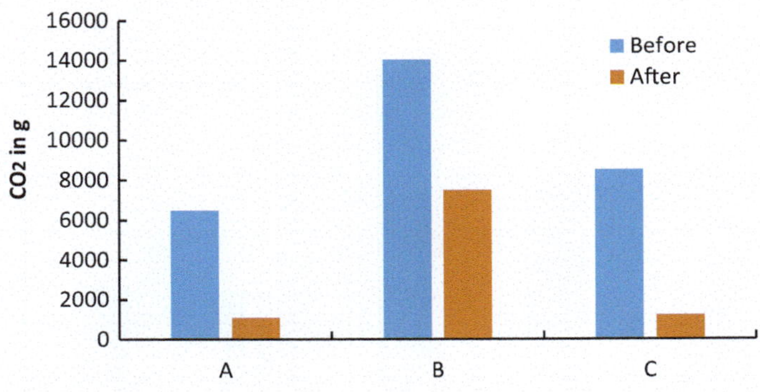

Fig. 5 Comparison of CO_2 emission (in g) by cars coming to faculty of arts from different entrances

In a similar manner, a multistorey parking **'Y'** is proposed near Science faculty **'F'** that will accommodate all the vehicles coming to Faculty of Commerce **'I'**, Faculty of Science **'F'** and vehicles from **'A'** to **'H'**, **'E'** and **'J'**. Finally the third parking lot **'Z'** is to be established near the Faculty of International Studies that will serve all the vehicles coming towards Faculty of International studies **'K'** and Faculty of Business Studies **'G'**. It will also help commuters coming from entrance **'C'**, **'E'**, and **'J'**.

2.2 Potential Effects of Relocating Parking Lots

Upon relocating parking locations, significant amount of pollutants emission can be reduced. For example, CO_2 emitted by cars coming from **'A'** to Arts faculty **'E'** can be reduced from 6475.2 to 1079.2 g which is equivalent to 83% reduction as shown in Fig. 6. Similarly, cars coming from **'B'** to **'E'** show about 47% reduction in CO_2 emissions and from **'C'** to **'E'** same figure goes about 87%. Overall reduction in CO_2 emissions from cars coming to Arts faculty is approximately 66%. Considering the whole university campus, by updating parking locations, total CO_2 emissions come down to 176113.32 g with approximately 37% of reduction. Similar reduction can be seen in the case of CO and NO_X emissions as shown in Figs. 7, 8, and 9. CO emissions were reduced from 5260.25 to 3290.12 g so were NO_X emissions from 972.23 to 604.43 g.

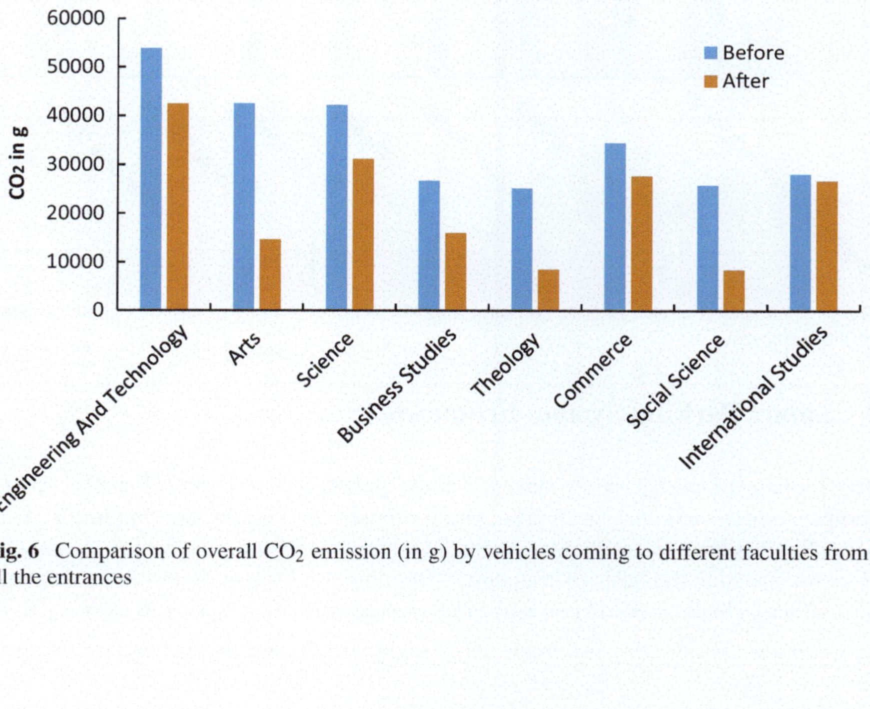

Fig. 6 Comparison of overall CO_2 emission (in g) by vehicles coming to different faculties from all the entrances

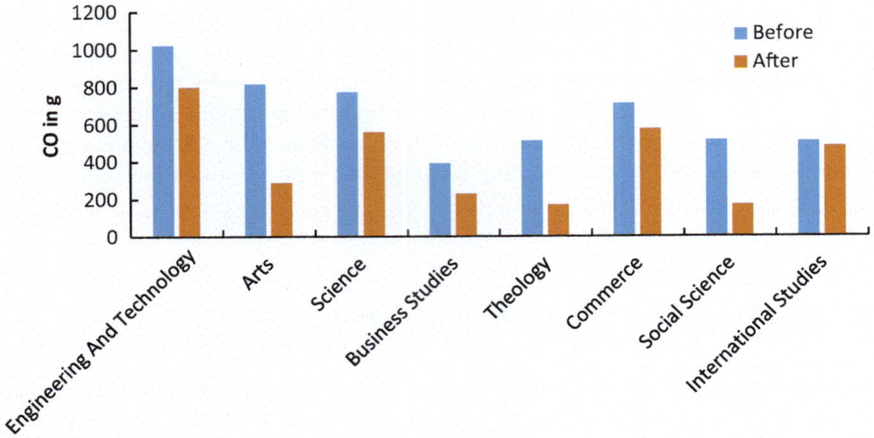

Fig. 7 Comparison of overall CO emission (in g) by vehicles coming to different faculties from all the entrances

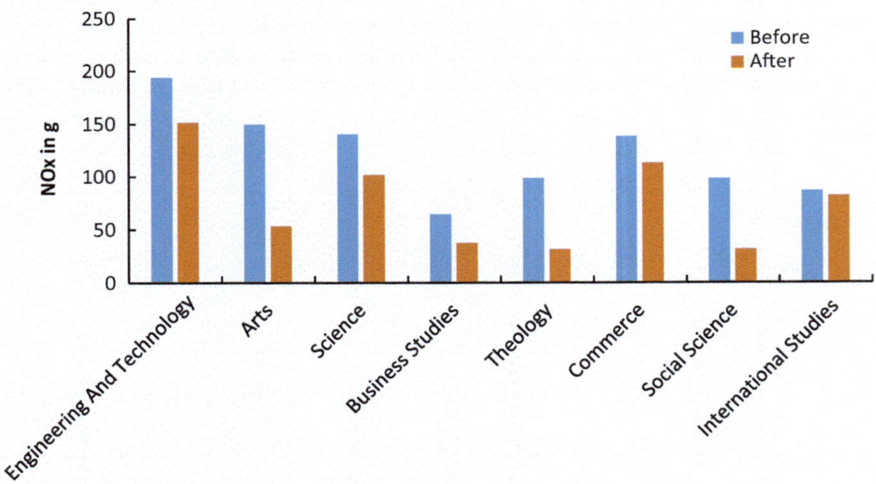

Fig. 8 Comparison of overall NO_x emission (in g) by vehicles coming to different faculties from all the entrances

3 Smart Parking System Framework

The proposed smart parking system for the pollution and congestion-free AMU campus is depicted in Fig. 7. The idea proposed by (Telles and Meduri 2017), SParkSys Vision, to let vehicle owners know where to park on the basis of available slots was became the basis of conceptual framing. This is simple to implement and cost effective having reasonable real-time accuracy. For the purpose, Raspberry Pi-3

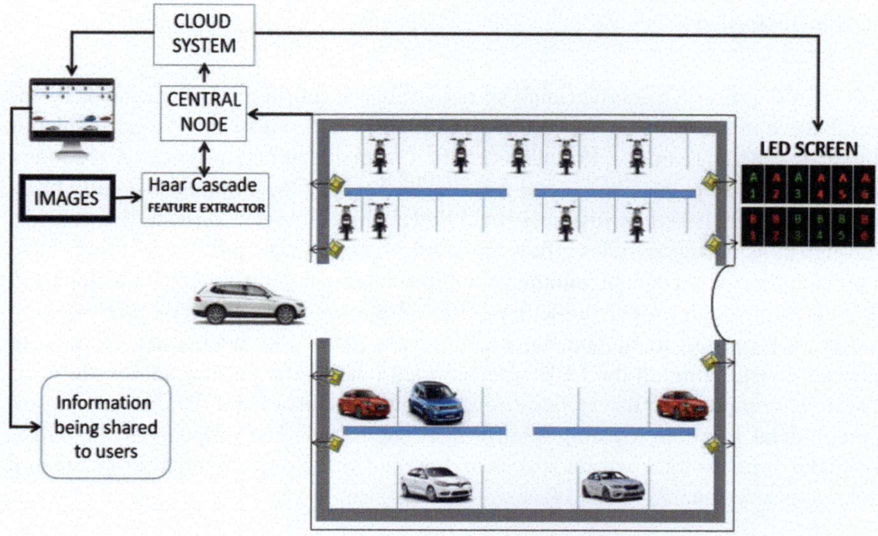

Fig. 9 Proposed system architecture

model B accompanied with a Pi Camera and ARM Cortex-A53 processor can be used to collect inputs from the parking area. The camera will take images that will be processed in the server to harvest required data like number of spaces available and their locations. The parking spots can be assigned names using numeric and alphabets. The status of each space will be sent to the central node from camera nodes where required information of the surrounding will be detected using a machine learning algorithm (MLA) that will help identifying empty and occupied spaces. To enhance cost optimization, a single camera node may be utilized to take images of multiple parking lots, and identify empty and occupied spaces after training the data on Haar Cascade algorithm. After collecting data at each node, the information will be sent to network and stored in a cloud server from where information can be shared/accessed. To filter duplicate data, each node will share data within nodes and with central node. Then the data will be shared from cloud server to LED screens at the entrance of each parking lot. Vacant spots can be displayed on the screen by its unique number preferably in green colour and occupied ones in red colour. A mobile application can also be developed to provide information to commuters before they come to the parking facility. In case one forgets where the vehicle is parked and to enhance security, an OCR technology is embedded to the overall system to screen vehicle number from the images taken and will share information via mobile application.

4 Conclusion

This chapter presents a detailed analysis of current transportation system and different pollutants emissions by motor vehicles coming in and out of the Aligarh Muslim University (AMU) campus. Pollutants emitted from different two-wheelers (scooters and bikes) and four-wheelers (cars) entering and exiting the campus on a daily basis were calculated based on distance travelled from entrances to desired locations. Through considering vehicles inflow and outflow, appropriate parking locations have been identified and a decent amount of pollution emissions (up to 37%) was found to be reduced through proper implementation. Moreover, an IoT-based smart parking system is proposed to update daily commuters about the availability of parking spaces by displaying on the LED screens fitted outside the parking facility/lots. To ensure its competency, use of one-to-many camera sensors and data processing on a methodical machine learning algorithm is suggested. The proposed system also facilitates users to trace vacant spaces and parked vehicle locations through using a mobile application.

Acknowledgements The authors gratefully acknowledge the overall support provided by the Director and all other staff of Centre of Advanced Research in Electrified Transportation (CARET), recognized and supported by the Department of Heavy Industries (DHI), Government of India as a Centre of Excellence for Electric Vehicle Charging Infrastructure.

References

Jariwala HJ, Syed HS, Pandya MJ, Gajera YM (2017) Noise pollution & human health: a review. Indoor Built Environ 1–4

Khanna A, Anand R (2016) IoT based smart parking system. In: 2016 international conference on internet of things and applications (IOTA). IEEE, pp 266–270

Ling X, Sheng J, Baiocchi O, Liu X, Tolentino ME (2017) Identifying parking spaces & detecting occupancy using vision-based IoT devices. In: 2017 global internet of things summit (GIoTS). IEEE, pp 1–6

Lookmuang R, Nambut K, Usanavasin S (2018) Smart parking using IoT technology. In: 2018 5th international conference on business and industrial research (ICBIR). IEEE, pp 1–6

Sadhukhan P (2017) An IoT-based E-parking system for smart cities. In: 2017 international conference on advances in computing, communications and informatics (ICACCI). IEEE, pp 1062–1066

Shinde S, Patil A, Chavan S, Deshmukh S, Ingleshwar S (2017) IoT based parking system using Google. In: 2017 international conference on I-SMAC (IoT in social, mobile, analytics and cloud) (I-SMAC). IEEE, pp 634–636

Telles E, Meduri P (2017) SParkSys: a framework for smart parking systems. In: 2017 international conference on computational science and computational intelligence (CSCI). IEEE, pp 1396–1399

Wang M, Dong H, Li X, Song L, Pang D (2017) A novel parking system designed for smart cities. In: 2017 Chinese Automation Congress (CAC). IEEE, pp 3429–3434

Yang LE, Hoffmann P, Scheffran J (2017) Health impacts of smog pollution: the human dimensions of exposure. The Lancet Planetary Health 1(4):e132–e133

Zhang J, Liu Y, Cui LL, Liu SQ, Yin XX, Li HC (2017) Ambient air pollution, smog episodes and mortality in Jinan, China. Sci Rep 7(1):1–8
https://www.who.int/news/item/02-05-2018-9-out-of-10-people-worldwide-breathe-polluted-air-but-more-countries-are-taking-action
https://www.who.int/health-topics/air-pollution#tab=tab_1
http://mospi.nic.in/statistical-year-book-india/2015
https://www.amu.ac.in/fct.jsp
https://shaktifoundation.in/wp-content/uploads/2017/06/WRI-2015-India-Specific-Road-Transport-Emission-Factors.pdf
https://www.transportpolicy.net/standard/india-light-duty-emissions/
https://www.transportpolicy.net/standard/india-motorcycles-emissions/

Chapter 18
Levelized Cost of Sustainable Electricity Production and Storage in India

Asif Pervez, Jahangir Chauhan, and Irfan Ali

Abstract This study examines various technical and financial determinants of levelized cost of electricity production and storage in India based on different technologies. Descriptive research design is adopted in this study. Data are obtained from reports of the Central Electricity Authority (C.E.A.), Central Electricity Regulatory Commission (CERC), Government of India and India stats database. Raw data are analysed and have been presented using tables and figures. The study's findings show significant reductions in LCOE of Solar P.V. and LCOS of battery between 2021 and 22 and 2029 and 30. It was concluded that the conventional and non-conventional energy system could be combined in hybrid energy systems with battery storage and pumped storage to supply consumer energy demands consistently.

Keywords Levelized cost of electricity · Battery storage · Pumped storage · Conventional and non-conventional energy sources · India

1 Introduction

Electricity is crucial for the development of an economy, especially an emerging economy like India. Inaccessibility to electricity is a significant challenge for a country's social and economic progress (Dinkelman 2011; Stern et al. 2019). Despite being the seventh-largest economy of the world in 2018, India has a meagre per capita electricity consumption of 1122 kWh in 2017 compared to many other developed and developing nations globally. Although various reforms were initiated after 1991

A. Pervez (✉)
Centre for Distance and Online Education, Jamia Millia Islamia, New Delhi, India
e-mail: asifpervez10@jmi.ac.in

J. Chauhan
Department of Commerce, Aligarh Muslim University, Aligarh, India
e-mail: jchauhan.cm@amu.ac.in

I. Ali
Department of Statistics and Operations Research, Aligarh Muslim University, Aligarh, India
e-mail: irfii.st@amu.ac.in

I. Ali et al. (eds.), *Computational Modelling in Industry 4.0*,
https://doi.org/10.1007/978-981-16-7723-6_18

and by Electricity Act 2003, the Indian electricity system's financial health is still an area of concern (Agrawal et al. 2017). The problem of India's power sector and its prospect has been studied and documented in Bhattacharyya (1994), where it is asserted that there are widespread energy shortages due to the high demand for electricity that outstripped the supply. The 7th Sustainable development Goal (SDGs) of the United Nations is determined to give sustainable and clean energy for the social and economic development of all (United Nation 2015). Economic growth and energy-related issues for developing nations under the SDGs have been studied and documented in (Nomani et al. 2017, Gupta et al. 2018, Ali et al. 2020, and Modibbo et al. 2020). A developing nation's energy policy should aim to provide uninterrupted energy in a reasonable, just, efficient, and environmentally caring manner (Narula 2014). The initiatives for India's renewable energy sector under 'National action plan on climate change' has been intensively reviewed by Chandel et al. (2016), according to them, the power sector contributes about half of the total carbon emission in the world due to its fast fossil fuels depletion. India in its Intended Nationally Determined Contribution to UNFCCC intended to reduce the carbon emissions by 33–35% and to increase non-fossil fuel-based electric power installed capacity up to 40% of total installed capacity by 2030 (UNFCCC 2015). Therefore, the Government of India needed to promote renewable energy. The government has declared a target for installing 175GW of renewable energy capacity by 2022 (Government of India 2015) and an installed capacity of 275 GW by 2026–27 (CEA 2018). However, the likelihood of adverse environmental effects of India's renewable energy sources have studied Abbasi and Abbasi (2000), and they concluded that the impact could be as strongly negative as compared to the conventional sources of energy.

The current status of solar power policies has studied by Rathore et al. (2019), and the motivating factors for investing in solar rooftop PV also analysed and finally, a summary of barriers to the solar rooftop PV growth and development are provided. Central Electricity Authority has also considered grid battery storage and pumped storage in their optimal generation mix for 2029–3 (CEA 2018) to complement Renewable Energy generation. The cost of battery energy storage systems has been decreasing with advancements in technology, and it may help absorb more Renewable Energy into the electricity system. Therefore, the conventional and non-conventional energy systems could be combined in hybrid energy systems with battery storage and pumped storage to supply consumer energy demand consistently; (see Fig. 1).

1.1 Current Scenario of the Electricity Sector in India

Total Installed Capacity of electricity in India as on 31.01.2019 was 349,288.2 MW, which comprises of 45,399.2 MW from Hydro, 223,027.34 MW from Thermal, 74,081.66 MW from R.E.S, and 6780 MW from Nuclear (CSO 2019).

Figure 2 shows the installed capacity of different sources from 2009 to 2019. India's electricity sector has been traditionally dominated by a thermal and hydro

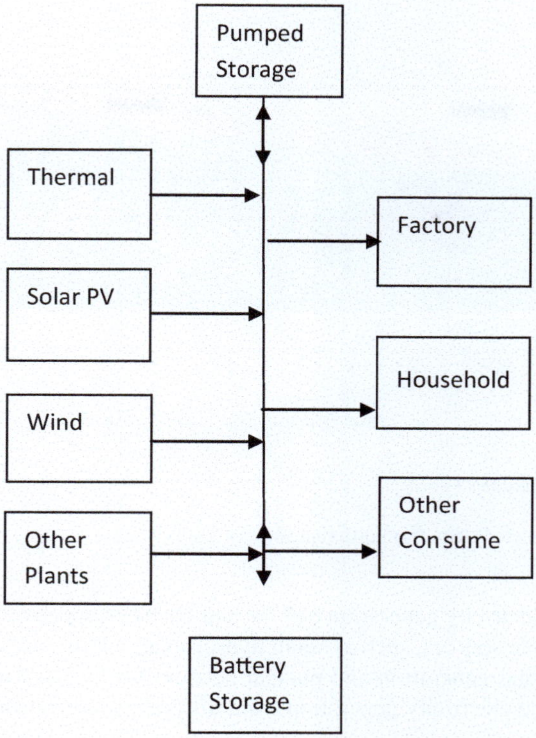

Fig. 1 Hybrid electricity system with storage

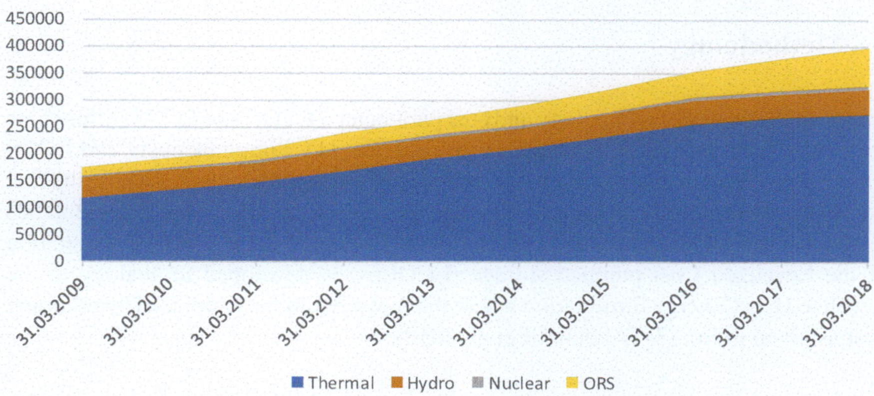

* *ORS means Other Renewable Sources*

Fig. 2 Installed electricity generation capacity of different sources

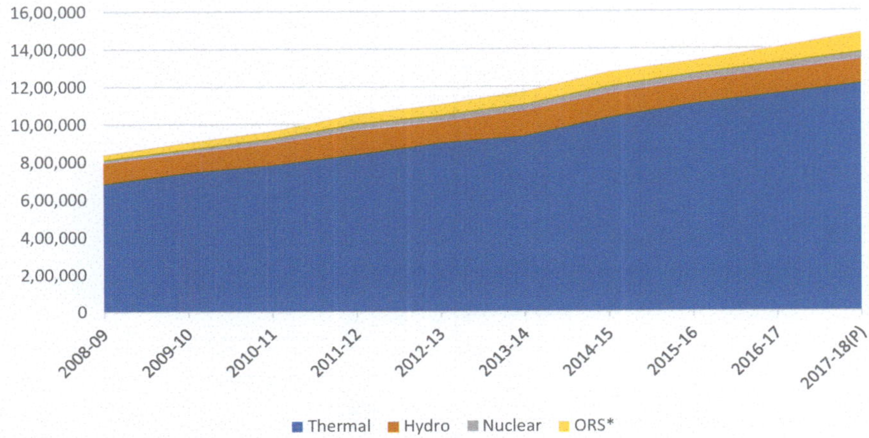

(P)-Provisional
* *ORS means Other Renewable Sources*

Fig. 3 Generation of electricity from different sources

generation with electricity generation of 77% and 10.7%, respectively, during 2018–19. Other renewable sources, such as small hydro, solar, wind, and biomass, account for 9.3% of the total generation and nuclear account for 2.7% of total generation. Figure 3 shows the electricity generation from different sources from 2009 to 2019 (CEA 2019).

2 Methodology

Based on the available technology and cost parameters, LCOE and LCOS estimations are made based on LCOE Formula as given in Eq. (1). Simultaneously, the LCOS form is developed in analogy to the Levelized Cost Energy formulation where the fuel cost has been replaced by charging cost and generation electricity has been replaced by the discharged electricity as given in Eq. (2) (Belderbos et al. 2016). Formulation of the Levelized Cost pumped storage (LCOEps) is developed in analogy to the Levelized Cost Energy formulation where the fuel cost has been replaced by pumping cost as given in Eq. (3) (Abdellatif et al. 2018).

$$\text{Levelizsed cost of energy (LCOE)} = \frac{\sum\left[\frac{(I_t+OM_t+F_t)}{(1+r)^t}\right]}{\sum\left[\frac{E_t}{(1+r)^t}\right]} \quad (1)$$

$$\text{Levelizsed cost of storage (LCOS)} = \frac{\sum\left[\frac{(I_t+OM_t+C_t)}{(1+r)^t}\right]}{\sum\left[\frac{DE_t}{(1+r)^t}\right]} \quad (2)$$

$$\text{Levelizsed cost of pumped storage plant (LCOEps)} = \frac{\sum\left[\frac{(I_t+OM_t+P_t)}{(1+r)^t}\right]}{\sum\left[\frac{E_t}{(1+r)^t}\right]} \quad (3)$$

where symbols used in the Eqs. (1) to (3) have their usual definitions that is $\boldsymbol{I_t}$ is the initial investment (Capital cost), $\boldsymbol{OM_t}$ is the operating and maintenance costs, $\boldsymbol{F_t}$ is the fuel cost, $\boldsymbol{C_t}$ is the battery charging cost, $\boldsymbol{P_t}$ is the pumping cost, E_t is the system energy yield, $\boldsymbol{DE_t}$ is the discharged electricity, $\boldsymbol{t}$ is number of years, and $\boldsymbol{r}$ is for a discount rate.

3 Levelized Cost of Electricity

In this section, cost parameters, technical parameters, and the levelized cost of conventional and non-conventional sources have been given in Tables 1, 2, 3, 4, 5 and 6.

Table 1 Cost parameters of conventional sources of electricity

Technology	Hydro	Coal (Pithead)	Coal (Load Centred)	Gas	Nuclear (LRW)	Nuclear (PHWR)	References
Capital Cost(Lakh/MW)	1170	785	760	523	1900	1170	CSO (2019),CERC (2014)
Operation & maintenance (Lakh/MW)	29.25	18	18	28.61	20	20	CSO (2019)
Fuel cost (Rs/Kwh)	0	2.59	2.59	2.70	0.85	0.85	CSO (2019)
Escalation of O&M (%)	3	3	3	3	3	3	Assumed
Escalation of fuel cost per year (%)	0	2.5	2.5	2	0	0	CSO (2019)

Table 2 Technical parameters of conventional sources of electricity

Technology	Hydro	Coal (Pithead)	Coal (Load Centred)	Gas	Nuclear (LRW)	Nuclear (PHWR)	References
Construction time (Years)	8	4	4	4	6	6	CSO (2019)
Plant life (Years)	35	25	25	25	30	30	CSO (2019)
PLF/CUF (%)	35	60	60	30	68	68	CSO (2019), Instat (2019)
Auxiliary consumption (%)	0.7	6.5	6.5	2.5	10	10	CSO (2019)

Table 3 LCOE of conventional sources of electricity

Technology	Hydro	Coal (Pithead)	Coal (Load Centred)	Gas	Nuclear (LRW)	Nuclear (PHWR)
LCOE (Rs/Kwh)	4.64	5.40	5.35	5.96	4.77	3.48

Table 4 Cost parameters of non-conventional sources of electricity

Technology	Solar PV	Wind	Biomass	Small hydro	References
Capital Cost(Lakh/MW)	450	600	570	760	CSO (2019)
Operation & maintenance(Lakh/MW)	5.625	6	11.4	19	CSO (2019)
Fuel cost (Rs/Kwh)	0	0	7	0	CSO (2019)
Escalation of O&M (%)	3	3	3	3	Assumed
Escalation of fuel cost per year (%)	0	0	2	0	CSO (2019)

Table 5 Technical parameters of non-conventional sources of electricity

Technology	Solar PV	Wind	Biomass	Small hydro	References
Construction time (Years)	0.5	1.5	3	5	CSO (2019))
Plant life (Years)	25	25	20	35	CSO (2019)
PLF/CUF (%)	19	29	30	45	
Auxiliary consumption (%)	1	0.5	8	0.7	CSO (2019)

Table 6 LCOE of non-conventional sources of electricity

Technology	Solar PV	Wind	Biomass	Small hydro
LCOE (Rs/Kwh)	2.98	2.52	11.70	2.34

3.1 Cost and Technical Parameters of Conventional Sources

Thermal electricity sources have the highest LCOE among the conventional electricity sources, followed by nuclear and hydro. However, Nuclear (PHWR) has a low LCOE. This high LCOE of thermal indicates low Plant load Factor, increase in the cost of coal, and high operating and maintenance cost.

3.2 Cost and Technical Parameters of Non-Conventional Sources

Biomass sources have the highest LCOE among the non-conventional sources, followed by solar, wind, and small hydro. High LCOE of Biomass results from low Plant load factor (30%) due to non-availability of fuel and high cost of baggage. At the same time, lower LCOE of solar and wind is due to an increase in their Plant load Factor, improvement in the technology, and a decrease in the solar panel cost.

Figure 4 shows a reduction in LCOE of solar PV plants as Capital costs of solar plants is expected to reduce from Rs 4.5 Crores in 2021–22 to Rs 4.1 Crores in the year 2029–30 (CSO 2019). Therefore, levelized cost of solar is expected to decrease from Rs 2.99 per kwh in 2021–22 to 2.76 in 2029–30, other factors assuming to be constant. Figure 5 depicted LCOE of solar PV plants at different capacity utilization factor as CUF of solar PV is expected to increase in coming years. Levelized cost of solar is expected to reduce from Rs 2.99 per Kwh in 2021–22 at 19% CUF to 1.89 at 30% CUF, other factors assuming to be constant.

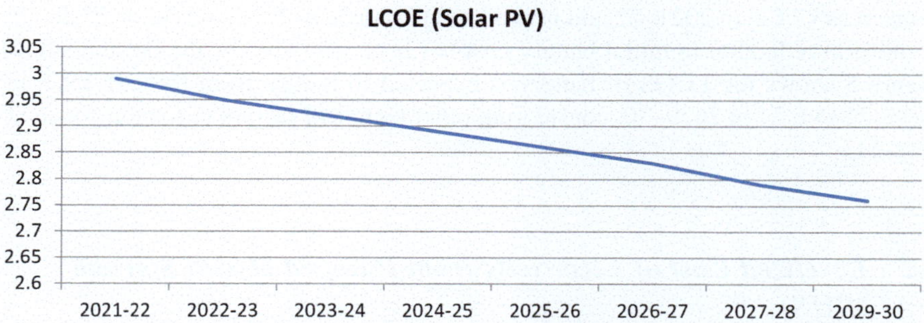

Fig. 4 Projected levelized cost of electricity production from solar PV

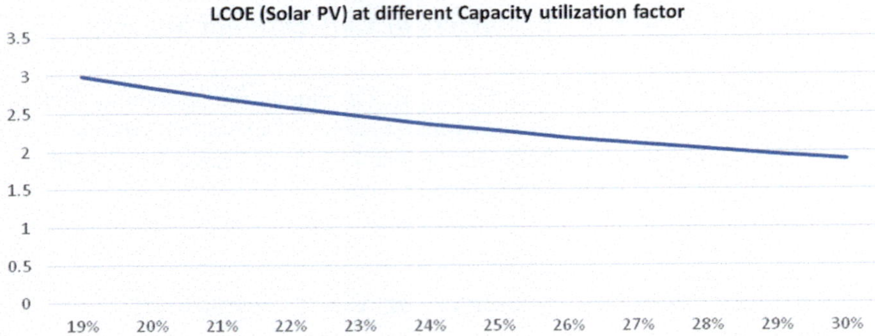

Fig. 5 LCOE (Solar PV) at different Capacity utilization factor

4 Levelized Cost of Energy Storage

In this section, the levelized cost of battery storage and pumped hydro storage have been calculated and discussed.

4.1 Battery Storage

In Table 7, assuming number of cycles (charging/discharging events) as 365, a life of 10 years, a battery storage degradation rate of 1% per year (Comello and Reichelstein 2019) a 9% cost of capital, an 85% round-trip efficiency, the corresponding Levelized cost of storage is Rs 9.36 per kWh for 2021–22. The cost of battery energy storage is taken as ₹7 Cr/ in 2021–22 and is expected to reduce to ₹4.3 Cr/MW in 2029–30. A uniform reduction in initial battery cost has been assumed for the present study. Figure 5 shows the LCOS of battery is expected to reduce from Rs 9.36/kWh in 2021–22 to 6.24 in 2029–30, considering other technical factors to remain constant (Fig. 6).

4.2 Levelized Cost of Electricity from Pumped Hydro Storage (PHS)

PHS is crucial for India's commitments for 275 GW renewable energy installation by 2027. Around 2612 MW of PHS is working in the country, and another 3145 MW is under construction. For further progress and exploiting the potential of PHS, at least US $20 billion of investment is required in the coming years (Buckley and Shah 2019). Pumping cost is the cost of electricity that is used to lift up water from the lower reservoir to the upper reservoir during off-peak times. The pumping cost is assumed

Table 7 Cost and technical parameters for battery storage

Parameters	Battery storage	References
Installed power storage (MW)	1.0	
Initial investment (Rs Lakh/MW)	700	CSO (2019)
Operating & Maintenance (Rs lakh/MW)	2%	CSO (2019)
Charging cost (Rs/Kwh)	0	Assumed
Utilization of usable storage capacity	100%	CSO (2019)
Number of cycle per year	365	Comello and Reichelstein (2019)
Degradation in storage capacity	1%	Comello and Reichelstein (2019)
Auxiliary consumption	12%	CSO (2019)
Round trip efficiency	88%	Assumed
Residual value (Rs Lakh/MW)	0	Assumed
Construction time (Years)	0.5	CSO (2019)
Depth of discharge	100%	Assumed
Life (years)	10	CSO (2019)
Discount rate	9%	CSO (2019)
LCOS (Rs/Kwh)	9.36	Calculated

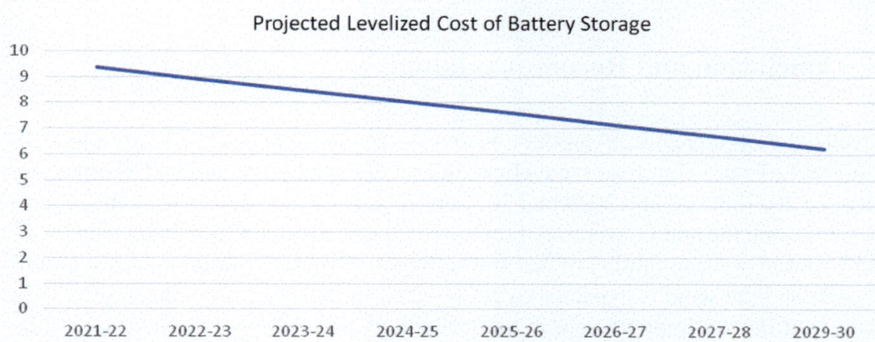

Fig. 6 Projected levelized cost of battery storage

Table 8 Cost and technical parameters of pumped storage plant

Parameters	Pumped storage	References
Capital Cost(Rs Lakh/MW)	1287	CSO (2019)
Operation & maintenance (In Percentage)	2.5	CSO (2019)
Pumping Cost (Rs/Kwh)	0	Assumed
Construction Time (Years)	8	CSO (2019)
Plant Life (Years)	35	CSO (2019)
PLF/CUF (In percentage)	35	
Auxiliary consumption (In percentage)	0.7	CSO (2019)
Round trip efficiency (In percentage)	75	Abdellatif et al. (2018)
Discount rate (In percentage)	9	CSO (2019)
LCOE (Rs/Kwh)	7.51	Calculated

to be zero for the present study understanding supplying electricity for pumping from grid-connected renewable power plants not generating electricity during peak times or the spinning reserve of thermal power plants during the off-peak time (Abdellatif et al. 2018). The corresponding levelized cost of energy for PHS is 7.51 Rs /Kwh (Table 8).

5 Conclusion and Recommendations

Levelized cost of electricity production and storage in India based on different technologies indicates significant reductions in LCOE of Solar PV and LCOS of battery between 2021 and 22 and 2029 and 30. Currently, Biomass sources have the highest LCOE among the non-conventional sources of electricity due to low Plant load factor (30%) due to non-availability of fuel and high cost of bagasse. Simultaneously, LCOE of solar and the wind is expected to get reduced further due to the increase in their Plant load Factor, improvement in the technology and decrease in the cost of solar panel and wind plant. PHS is crucial for India's commitments for renewable energy installation by 2027. However, The corresponding levelized cost of energy for PHS is high. The conventional and non-conventional energy system could be combined in hybrid energy systems with battery storage and pumped storage to consistently supply consumer energy demands. Policymakers should analyse different pathways oriented to the sustainable development of electricity system while developing a cost optimization model for the electricity system integrating renewable energy sources and electricity storage system and developing strategies for achieving SDGs related to clean electricity production.

Declaration of Interests The authors declare that they have no known competing financial interests or personal relationships that could have appeared to influence the work reported in this paper.

References

Abbasi SA, Abbasi N (2000) The likely adverse environmental impacts of renewable energy sources. Appl Energy 65(1–4):121–144
Abdellatif D, AbdelHady R, Ibrahim AM, Abu El-Zahab E (2018) Conditions for Economic competitiveness of pumped-storage hydroelectric power plants in Egypt. Renew Wind Water Solar 5(1):2
Agrawal A, Kumar A, Rao TJ (2017) Future of Indian power sector reforms: electricity amendment bill 2014. Energy Policy 107:491–497
Ali I, Modibbo UM, Chauhan J, Meraj M (2020) An integrated multi-objective optimization modelling for sustainable development goals of India. Environ Dev Sustain:1–21. https://doi.org/10.1007/s10668-020-00745-7
Belderbos A, Delarue E, D'haeseleer W (2016) Calculating the levelized cost of electricity storage. In: Proc IAEE, pp 19–22
Bhattacharyya SC (1994) An overview of problems and prospects for the Indian power sector. Energy 19(7):795–803
Central Electricity Authority (CEA) (2018) National Electricity Plan, CEA, New Delhi
Central Electricity Authority (CEA) (2019) Draft Report on optimal Generation Capacity Mix For 2029–30
Central Electricity Regulatory Authority (2014) Petition No. 303/GT/2014. http://www.cercind.gov.in/2015/orders/SO303-GT-2014.pdf
Central Statistics Office (2019) Energy Statistics 2019, New Delhi
Chandel SS, Shrivastva R, Sharma V, Ramasamy P (2016) Overview of the initiatives in the renewable energy sector under the national action plan on climate change in India. Renew Sustain Energy Rev 54:866–873
Comello S, Reichelstein S (2019) The emergence of cost-effective battery storage. Nat Commun 10(1):1–9
Dinkelman T (2011) The effects of rural electrification on employment: new evidence from South Africa. Am Econ Rev 101(7):3078–3108
Government of India (2015) Working towards climate justice
Gupta S, Fügenschuh A, Ali I (2018) A multi-criteria goal programming model to analyze the sustainable goals of India. Sustainability 10(3):778
Indiastat (2019) Sector/category wise plant load factor in India
IRENA (2015) Case studies: battery storage
Modibbo UM, Ali I, Ahmed A (2020) Multi-objective optimization modelling for analysing sustainable development goals of Nigeria: agenda 2030. Environ Dev Sustain:1–35.https://doi.org/10.1007/s10668-020-01022-3
Narula K (2014) Is sustainable energy security of India increasing or decreasing? Int J Sustain Energ 33(6):1054–1075

Nomani MA, Ali I, Fügenschuh A, Ahmed A (2017) A fuzzy goal programming approach to analyse sustainable development goals of India. Appl Econ Lett 24(7):443–447
Rathore PKS, Chauhan DS, Singh RP (2019) Decentralized solar rooftop photovoltaic in India: on the path of sustainable energy security. Renew Energy 131:297–307
Stern DI, Burke PJ, Bruns SB (2019) The impact of electricity on economic development: a macroeconomic perspective
Buckley T, Shah K (2019) Pumped Hydro Storage in India, Getting the right plans in place to achieve a lower cost low carbon electricity market. Institute of energy economics and financial analysis
UNFCCC (2015) India's intended nationally determined contribution: working towards climate justice
United Nations (2015) Sustainable development goals

Chapter 19
Application of Principal Component Analysis and Analytical Hierarchical Process in Surface Water Quality Assessment in Hatta Catchment, Emirate of Dubai, UAE

Izrar Ahmed

Abstract United Arab Emirates is a water-scarce country in terms of availability of conventional water resources. Water experts always explore sustainable demand–supply management through Integrated Water Resources Management. With this preview, a detailed surface water quality monitoring programme was carried out to assess its suitability for intended use. Statistical treatment of large water quality dataset helps in understanding of various processes involved in chemical alteration of water resources. Principal Component Analysis (PCA) is data reduction technique attempts to explain the correlation between the observations in terms of the underlying factors which are not directly observable. The results showed that all samples could be analysed by three main components, which accounted for 85.86% of the total variance. PCA technology identified important water quality parameters and revealed that the agricultural activities and domestic discharges are the main causes of water pollution in the study area. Analytic Hierarchical Process (AHP) is a decision support tool widely used in management of natural resources. It compares all possible pairs of criteria and determines most suitable criterion to develop irrigation water quality index. Study infers that water lies in good to moderate suitable classes. Any negative change in the form of qualitative or quantitative surface water resources would directly affect the shallow groundwater. This necessitates the urgent requirement to protect surface water from pollution threats to help integrated water resources management.

Keywords Surface water quality · Analytical Hierarchical Process (AHP) · Irrigation water quality · Principal Component Analysis (PCA)

I. Ahmed (✉)
Department of Geology, Aligarh Muslim University, Aligarh 202002, India

Formerly Soil and Water Resources Specialist, Environment Department, Dubai Municipality, Emirate of Dubai, Dubai, UAE

I. Ali et al. (eds.), *Computational Modelling in Industry 4.0*,
https://doi.org/10.1007/978-981-16-7723-6_19

1 Introduction

Managing water scarcity is a major problem in the arid regions, continuously increasing demand makes sustainable development of water resources a challenging task (Murad et al. 2007; Abdalla and Chan 2017; Alsharhan and Rizk 2020). Water Planners are constantly working to develop UAE's Integrated Water Resources Management strategy (Dawoud 2013; Ahmed et al. 2019). Water quality monitoring is an important part of almost all water resources management policies. Chemical analyses of water provide crucial information in relation to water quality status, which enables efficient Demand Side Management (DSM) decision-making. Use of statistical methods in Water quality data helps in providing key information. There are numerous researches presenting quality status of the water resources (Umar and Ahmed 2007; Umar et al. 2009; Nazzal et al. 2014, 2015). From simple statistical operations, e.g. range, distribution, deviation, outliers, missing values, correlations to complex statistical treatment, e.g. Principal Component Analysis help the researchers in many ways. Bivariate plots and graphical presentation do need certain statistical treatment of dataset. This helps in revealing evolution trends, source genesis, spatial distribution, impact caused due to landuse activities, and suitability for specific criteria.

Water quality analyses have two main objectives, i.e. finding genesis of the water and evaluating suitability for any specific usage. It is important to adopt suitable statistical methodology when analysing water quality data to draw relevant conclusion. Multivariate statistical methods have been widely applied in interpretation of multiparameter large water quality dataset. Cluster analysis (CA) and factor analysis (FA) are used for water quality analysis which helps in chemical evolution of water types (Singh et al. 2004a, b). It has been used to characterize water quality for spatial and temporal changes due to natural or manufactured environments (Singh et al. 2006; Bhat et al. 2013).

Core water quality findings and complex processes involved in it are of little interest for policy makers hence water planners require to generate a Water Quality Index (WQI) based on which prompt decision can be taken. Quality status of surface water is a dynamic process due to its vulnerability to receive contaminants. Therefore, the water quality index should be capable of addressing pertaining variables as well as suitability for specific usage. In view of this, water quality data is further classified into relevant criteria to establish a unique water quality index employing Analytical Hierarchical Process (AHP). AHP was developed in the 1970s by Thomas L. Saaty and has since been widely used for unbiased decision-making (Bhushan and Rai, 2004). AHP is multi-criteria programming for decision-making for large number of variables or criteria based on the prioritization.

The present study performed to serve as a baseline for surface water resources, inferring genesis and contamination trends using 25 water quality parameters from the rain-fed surface water reservoirs. The study investigates the current status of water pollution; carryout PCA for the spatial and temporal changes of water quality and finding possible pollution sources and using AHP technique to develop a suitable WQI.

2 Materials and Methods

2.1 Description of the Study Area

Hatta, a mountainous region, is being developed as a peri-urban excellent centre for adventure sports and tracking without compromising its traditional and historical values. Rainfall is highly unpredictable in terms of frequency and magnitude. A mean annual rainfall of 125.6 mm and a mean temperature of 27.8 °C characterize the area. Historically, Hatta's inhabitants do traditional farming by growing mainly dates, fruits, vegetables, fodder, etc. Traditionally, surface water remained the source of irrigation. With the advent of water extraction technology, the mode of irrigation shifted towards the groundwater resources. Surface water is stored in dam's reservoirs, which increases aesthetic values and help recharging groundwater. This can also serve as source of water supply in emergency time.

2.2 Water Resources

Hatta catchment is one of the east coast catchments that form one of the major five hydrological zones in the UAE. The entire region generates surface water runoff during rainy seasons. The annual rainfall in the northern areas is relatively higher, i.e. 150 mm/year, compared to the middle and southern parts of the country that receive about 110 mm/year (Sherif et al. 2011). Groundwater remains the most promising source for agricultural works. Rainfall induces surface water storage which in turn lead to quick groundwater level fluctuations and increase in baseflow in wadi channels. Falaj, an ancient subsurface water supply network rejuvenates aftermath of the rain events. This is all indicative of hydrodynamic connectivity between surface water and groundwater. Falaj is subsurface horizontal channel receiving groundwater from up gradient and all along the channel, which is available at the tail end to be used for the irrigation in downstream region.

2.3 Sampling and Analytical Techniques

Sixteen samples were collected from all surface water repositories including 04 water reservoirs and the Falaj water during the year 2019 sampling campaign (Fig. 1). Representative surface water samples were collected on quarterly frequency

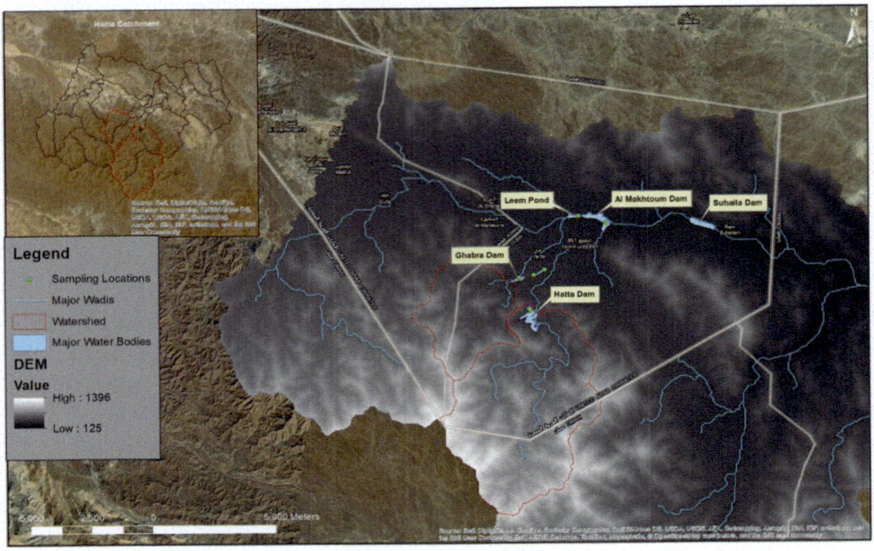

Fig. 1 Base map showing Hatta Catchment and water sampling locations

from each location following standard sampling procedures and protocols (USGS 2018). Samples were preserved and transported to Dubai Central Laboratory (DCL). Samples were analysed for pH, EC, TDS, Turbidity, Alkalinity, major elements (Ca, Na, K, Mg, Cl, SO_4, HCO_3, NO_3), Trace elements (Fe, Zn, Cu, Co, Cr, As, Ni, Cr, Pb, Mn), pathogens (Total Coliform, E. Colie) using standard analytical techniques (APHA 2005).

Sampling results are tested for analytical inaccuracies for electro neutrality condition using the below equation.

$$IB = \frac{\sum Cations - \sum Anions}{\sum Cations + \sum Anions} \times 100$$

IB is ion balance error in percentage, cations and anions are denoted by sum of all cations and anions, respectively, expressed in meq/L and IB error should be less than 10% (Domenico and Schwartz 1990).

Interpretation of the results was done using Min, Max, mean, median, Standard deviation, kurtosis, skewness, and coefficient of variation were calculated using statistical technique.

Mean is the average of all the numbers and then divide by the total number. Median gives the middle values and mode is the value occurring maximum number of times. Standard deviation is a measure of variability of the sample. However, kurtosis means the degree of flatness and skewness denotes the symmetry of the data. The coefficient of variation is a measure of relative variability of the samples. Principal Component analysis (PCA) was performed using SPSS 25. Analytical hierarchical Process (AHP) is used for assigning relative weights of the selected criteria.

3 Results and Discussions

3.1 *Surface Water Quality Evaluation*

Several factors like topography, geology, landuse, seasonal variations, aridity index, and rainfall can influence the surface water quality. Chemical alterations of surface water begin soon after it starts flowing as runoff over different surfaces interacting with anthropogenic influences, landuse activities of the catchment, soil, and bare rock interactions. The nature and general conditions at reservoir conditions have great influence on the water's chemistry. The water quality analysis is used to identify the processes affecting the surface water quality of the study area. Seasonal changes in the water flow rate and biological activity contribute to differences in the chemical composition of surface water (Augustyn et al. 2012). Table 1 shows the chemical concentration of selected variables in the water samples.

The groundwater has low to moderate EC values ranging from 366 to 2774 μS/cm and average value of 730.04 μS/cm. pH is ranging from 7.6 to 10 with an average value 8.8, indicative of slight alkaline nature. The possible reason for high pH may be diurnal temperature change. The photosynthesis often raises respiration as a result CO_2 is extracted from the water which in turn may raise pH levels. The photosynthesis decreases in night time, leading to fall in pH values as respiring organisms add CO_2 to the water.

Like the elevated pH values, water is characterized by high alkalinity and hardness with an average value of 294.3 and 413.8, respectively. The general dominance of cations is Na > Mg > Ca > K while anions are in the order of HCO_3 > Cl > SO_4 > NO_3.

Dominance of Na^+ and Cl^- in water is a characteristic feature of most arid regions (Ahmed et al. 2019; Nazzal et al. 2014). Bicarbonate seems to acquire from recent biochemical reactions after rainfall events (Umar et al. 2009). Ca^{2+} and Mg^{2+} are found naturally in surface and ground water, and the only other elements that occur in greater abundance are Na^+ and Ca^{2+} cations. Na^+, Mg^{2+}, and Cl^- concentrations in ground and surface waters increase as those elements are washed out from bedrock. The sediments collected in wadi channel confluences to main dam reservoirs. Reservoir water gets mineralized through periodic evaporation and dilute after episodic runoffs.

Na^+ and Cl^- exhibit large variations and so is the standard deviation. Likewise, Ca^{2+}, Mg^{2+}, and $SO_4{}^{2-}$ show high variation and high standard deviations. First inference, which can be made from these observations, is that the several factors influence the water quality of the study.

Table 1 General statistical analysis of surface water quality

Parameter	n	Min	Max	Mean	Median	Sd	Skew	Kurtosis
pH	16	7.64	10.06	8.77	8.68	0.81	0.39	−1.12
Alk	12	116	410	294.33	377.00	126.45	−0.43	−2.06
Tur	16	0.49	30	6.74	3.30	9.00	2.07	3.40
EC	16	366	2774	1275.00	1500.00	730.04	0.45	−0.33
TDS	16	175.44	1919.09	887.71	1079.82	507.93	0.17	−0.63
K	16	0.56	6.07	4.17	4.55	1.54	−1.18	1.29
Na	16	9.17	260	100.77	109.25	76.27	0.47	−0.40
Ca	16	11.06	38.12	23.73	26.74	8.97	−0.01	−1.63
Mg	16	29.15	148	83.95	94.30	36.44	0.01	−0.97
Cl	16	20.4	1133	282.51	286.00	287.79	1.94	4.60
SO_4	16	47	332	92.31	56.40	94.04	2.44	4.67
HCO_3	16	21	500	293.25	325.00	195.03	−0.19	−1.92
F	16	0	1.8	0.21	0.10	0.43	3.93	15.57
NO_3	16	0	16	6.81	6.30	6.63	0.09	−2.12
Zn	16	0.009	0.0357	0.01	0.01	0.01	3.99	15.95
Fe	16	0.049	0.1727	0.06	0.05	0.03	4.00	16.00
Ni	16	0.00031	0.0167	0.00	0.00	0.00	3.30	11.97
Mn	15	0.00049	0.0089	0.00	0.00	0.00	3.41	12.35
Cu	16	0.009	0.02	0.01	0.01	0.00	3.86	15.17
As	16	0.00049	0.0022	0.00	0.00	0.00	0.74	−1.17
Ba	16	0.009	0.0489	0.02	0.01	0.01	0.81	−0.78
Cr	16	0.0023	0.0125	0.01	0.01	0.00	0.02	−2.13
T Coli	16	0	2421	1315.06	1327.00	1096.24	−0.10	−2.02
E Colie	16	0	2421	301.25	2.00	632.32	2.92	9.12

The striking feature emerged out from the statistical analysis is major elements vary widely evident from the high standard deviation. Trace elements show highest Skewness and kurtoses. Skewness is a measure of symmetry of data which in other words also the absence of symmetry. A distribution is called symmetric if it appears the same to the left and right from the centre, i.e. having a 0 value. On the other hand, the kurtosis parameter is a measure of the combined weight of the tails relative to the rest of the distribution.

A correlation statistics of the sampled data brings interesting hidden facet into the front (Table 2). TDS correlated well with most major elements having good correlations with K, Na, Mg, and Cl, moderate correlation with Ca, SO4, and Ba. The good correlation implies that the increase in total dissolved salts in a function of these elements. While TDS is a collective indicator of water chemistry, individual correlations often provide good inferences. Na correlates well with Cl and Mg while Ca correlates well with HCO3 and NO3. These individual correlations show basic

Table 2 Pierson Correlation coefficient of important parameters

	TDS	K	Na	Ca	Mg	Cl	SO_4	HCO_3	F	NO_3	Zn	Fe	Ni	Ba	Cr	Tcoli
TDS	1.00															
K	**0.74**	1.00														
Na	**0.98**	**0.73**	1.00													
Ca	*0.60*	0.49	0.59	1.00												
Mg	**0.99**	**0.78**	**0.98**	0.66	1.00											
Cl	**0.91**	*0.61*	**0.92**	0.28	**0.88**	1.00										
SO4	*0.59*	0.38	*0.65*	0.05	0.57	**0.82**	1.00									
HCO_3	0.36	0.38	0.26	*0.69*	0.37	−0.04	−0.52	1.00								
F	0.16	−0.13	0.16	0.08	0.14	0.10	−0.08	0.20	1.00							
NO_3	0.44	0.31	0.36	**0.73**	0.45	0.08	−0.42	**0.95**	0.23	1.00						
Zn	0.11	−0.19	0.12	0.04	0.09	0.06	−0.13	0.20	**0.99**	0.22	1.00					
Fe	0.11	−0.19	0.12	0.03	0.09	0.06	−0.12	0.19	**0.99**	0.21	1.00	1.00				
Ni	−0.21	−0.19	−0.20	−0.19	−0.23	−0.13	−0.01	−0.22	0.16	−0.20	0.15	0.15	1.00			
Ba	0.57	0.43	0.47	0.61	0.57	0.46	0.36	0.30	−0.15	0.36	−0.22	−0.22	−0.15	1.00		
Cr	0.49	0.35	0.40	**0.73**	0.49	0.12	−0.38	**0.96**	0.32	**0.97**	0.32	0.31	−0.26	0.36	1.00	
Tcoli	0.49	0.42	0.40	**0.75**	0.51	0.16	−0.22	**0.84**	0.31	**0.86**	0.28	0.27	−0.28	0.58	**0.87**	1.00

affinity of the cations towards the anions counterpart and also indicative of similar source genesis (Umar et al. 2009). Cr good correlations with Ca (0.73), NO_3 (0.97), and HCO_3 (0.96) is another striking feature of the statistical analysis which is indicating a common genesis of these parameters. Na:Cl shares good bonding (0.92), Cl:SO_4 shares good correlation (0.82) but Na:SO_4 shares moderate correlation (0.62) designating a different evolution mechanism and source genesis.

Weathering of limestone, dolomite, and sulphate-containing sediments such as gypsum and anhydrite contain Mg-Ca-SO_4. Large quantities of magnesium are released to groundwater from sedimentary rocks, mainly dolomite. In a complex geological set up consisting of ophiolites, ultramafic rocks, chert, limestone, and quaternary-alluvium are highly disturbed and seems to acquire complex relationship with the water quality that is difficult to trace out.

Piper diagram is widely used plot for describing water quality to help in understanding the sources of the dissolved constituents in water (Piper 1944). The major ions were plotted in the Piper Diagram to understand the main water facies occurring in the region (Fig. 2). The cationic triangle show significant presence of Mg ions followed by Na ions while the anionic triangle show dominance of HCO_3 ions followed by Cl-ions. Dam 1 is represented by Mg-HCO_3 type water. Two samples

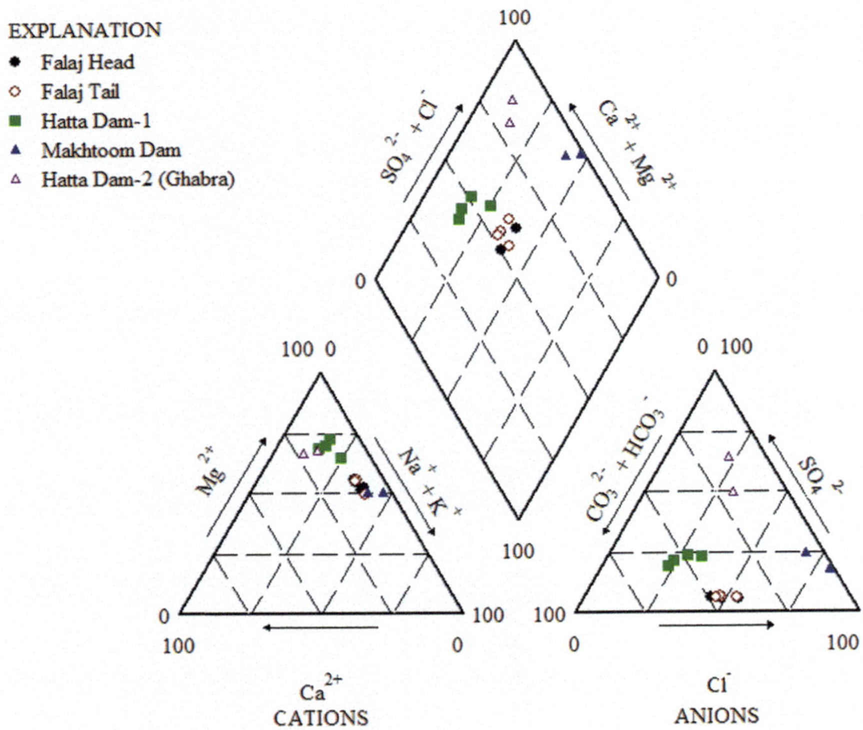

Fig. 2 Piper trilinear diagram for chemical classification of water

from Dam 1 and Falaj water are classified as mixed type water. This striking similarity makes a relationship between Dam 1 and Falaj water. Dam 2 has unique chemical signature Mg-Cl_2 type water. Dam 3 (Makhtoom Dam) is altering its chemical characteristics from Mg-Ca-Cl_2 to mixed type water.

There is a potential health risk due to presence of pathogens in surface water which may enter into the groundwater environment. Pathogens especially in Falaj water are of the major concerns. The major sources of pathogens are sewage effluents, animal waste, and some wildlife species dwelling in surface water or the upper catchment. These pathogens are most common causative factor of intestinal diseases like typhoid, paratyphoid, salmonellosis, cholera. Escherichia Coli are true indicators of fecal contamination because they are only found in feces (WHO 2011). Their presence show recent encroachment as well as gap in monitoring activity.

3.2 Principal Component Analysis/Factor Analysis

PCA is used for identification of the factors that influence water resources and solution for pollution problems (Reghunath et al., 2002; Simeonov et al. 2003). Factor analysis is applied to reduce the dimensionality of interrelated variables. The Principal components are data reduction techniques through linear combinations of the original variables and the eigenvectors. The main benefit of PCA is that it attempts to explain the correlation between the observations in terms of the underlying factors, which are not directly discernable. In the present study PCA of factor analysis was estimated.

In a natural resources environment correlations exist among multi-indicators. PCA converts large inter-correlated indicators into a smaller set of indicators that are uncorrelated variables (Jianqin et al. 2010). The correlation of PCs and original variables is designated by loadings. Scores are the individual transformed observations (Wunderlin et al., 2001). The classification of the factor loadings as 'strong', 'moderate' and 'weak', corresponding to absolute loading values of >0.75, 0.75–0.50, and 0.50–0.30, respectively (Liu et al. 2003). It is imperative to note that loading reflects the relative importance of a variable within the component and does not reflect the importance of the component.

Principle component analysis shows three main components on a rotated component matrix (Table 3). All 3 components are distinct and seem to develop acquiring distinct chemical processes (Fig. 3). PC-1 is represented by HCO_3, NO_3, Ca, Cr, Total coliform, and E Coli. This group shows recent influence by near surface biochemical reaction and anthropogenic influence. This also infers that the meteoric characteristics are not completely obliterated. PC-2 contains common parameters, e.g. EC, Na, Cl, Mg, SO_4, and K, which represent major ions chemistry of the water. PC-3 is represented by heavy metals, e.g. Zn, Fe, F, and Mn indicating a common genesis of these heavy metals.

Table 3 Principle component analysis showing 3 main components

Rotated component matrix[a]			
	Component		
	1	2	3
HCO_3	**0.978**		
NO_3	**0.947**		
pH	−0.934		
Cr	**0.928**		
T COLI	**0.859**		
Ca	**0.741**	0.419	
E COLIE	**0.598**		
Cu	−0.457		
Cl		**0.977**	
EC		**0.956**	
Na		**0.953**	
TDS	0.329	**0.933**	
Mg	0.360	**0.927**	
SO_4	−0.512	**0.831**	
K	0.398	**0.706**	
Zn			**0.988**
Fe			**0.987**
F			**0.981**
Mn			**0.978**

Extraction Method: Principal Component Analysis. Rotation Method: Varimax with Kaiser Normalization. Rotation converged in 5 iterations

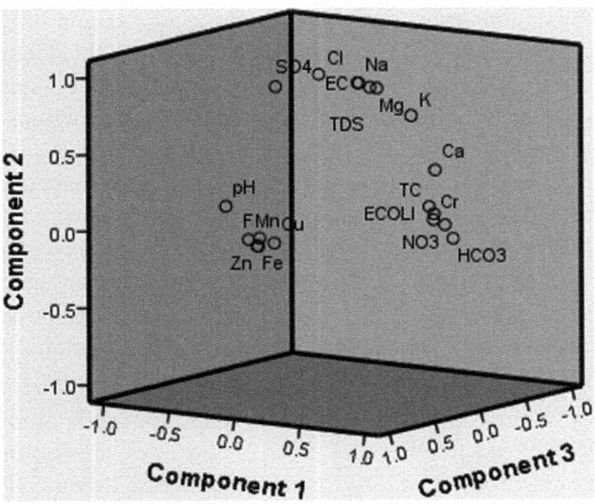

Fig. 3 Components plot in rotated space

3.3 Cluster Analysis

Geographical influences and spatial similarity in surface waters can be detected through Hierarchical Cluster analysis (CA) represented as a dendrogram as shown in (Fig. 4) grouping all the 16 samples into 2 statistically significant clusters. The clustering procedure generated two groups of sites in a logical way. The sites in these groups have similar characteristic features and sources of contamination.

Cluster 1 (Subgroup-1: sample ID 1, 2, 4, 5, 6, 7, 8 Subgroup-2: 3, 13, 14), Cluster 2 (9, 10, 11, 12, 15, 16). Cluster 1 is mainly represented by Falaj water which is basically groundwater and hence tends to be different than surface water characteristics. A small number of samples from Makjtoom dam also fall in this cluster.

Cluster 2 is mainly represented by waters of Dam 1 and Dam 2. These samples are least altered due to the geographical isolation from urban activities and falling in the upper reaches of the catchment.

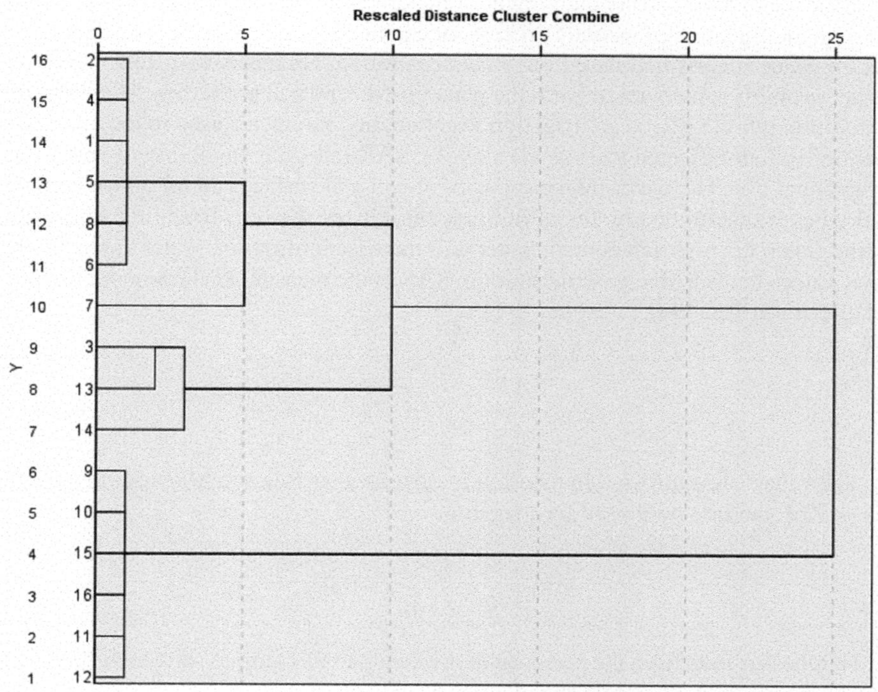

Fig. 4 Classification of water samples on Dendrogram using ward linkage

4 Developing Irrigation Water Quality Index

One of main objectives of water quality monitoring programme is analysing the suitability for specific usage. Monitoring of water sources determines the probable parameters that indicate the type of pollution and its possible genesis. Water pollution is generally caused by waste disposal sewage mixing, animal excrements, storage of waste, animal manure, and artificial fertilizers (Fridrich et al. 2014). Principally, the existing surface water usage defines the criteria for developing the irrigation Water Quality index. Several studies have documented important parameters used to assess irrigation water quality (Zaman et al. 2018). In recent years, Analytic Hierarchy Process (AHP), developed by Saaty (1987, 2004), has been used in environmental management initiatives (Pramanik 2016). It combines and uses pairwise matrices to compare all possible pairs of criteria to determine the maximum priority (Bozdag 2015). The present study used six criteria that influence irrigation water quality, i.e. Electrical conductivity (EC), Sodium Adsorption Ratio (SAR), Magnesium Ratio (MR), Total hardness (TH), Chloride (Cl), and Kelley Ratio (KR).

Presence of salts affects the growth of the crops by limiting the uptake of water through modification of osmotic process, or chemically by metabolic reactions. Irrigation water is classified based on Na concentration. Na reacts with soil to reduce its permeability which intern harm the plant growth and soil properties. The Sodium adsorption ratio (SAR) is an irrigation water quality parameter used in the management of sodium-affected soils (USDA 1954). SAR indicates suitability of water for irrigational use. The water intake capacity of soil and surface run off above certain soil types is determined by the infiltration capacity of the soil. Irrigating land with saline water, i.e. high salt content water will increase infiltration. Water having high SAR values tends to decrease infiltration. SAR is the measure of the amount of Na^{+} relative to Ca^{2+} and Mg^{2+}.

$$SAR = \frac{Na}{\sqrt{\frac{Ca+Mg}{2}}}$$

The values are expressed in meq/l. It is calculated as follows. Magnesium hazard was <50% can be safely used for irrigation.

$$\mathrm{MR} = \frac{\mathrm{Mg}}{\mathrm{Ca} + \mathrm{Mg}} \times 100$$

Hardness results from the presence of divalent metallic cations, of which Calcium and Magnesium are the most abundant in surface water

$$\mathrm{H_T} = 2.5\mathrm{Ca} + 4.1\,\mathrm{Mg}$$

Na, Cl, and Boron from soil or water accumulate in a sensitive crop to concentrations high enough to reduce yields. Chloride is selected to check specific toxicity.

Kelley's index is calculated by Na measured against Ca and Mg (Kelley et al. 1940). A Kelley 's index of greater than 1 indicates an excess level of Na in water. Values are expressed in meq/L.

$$\text{KI} = \frac{\text{Na}}{\text{Ca} + \text{Mg}}$$

4.1 Analytical Hierarchical Process

The AHP is an important decision support tool in natural resource management studies, solving complex problem hierarchically. It examines each level of the hierarchy individually utilizing pairwise matrices to compare all possible pairs of criteria. The AHP tool also determines which criterion has the highest priority (Saaty 1980).

A standard scale was used to describe the relative influence of parameters, where score 1 denotes equal influence of parameters and score 9 denotes extreme influence of a parameter on groundwater recharge compared to the other parameters (Table 4). The application of AHP starts with disintegrating into a hierarchy of different criteria which can analysed easily independently. The decision-makers can use different alternatives by making pairwise comparisons for each of the chosen criteria (Saaty 2008).

Study selected 6 criteria EC, SAR, MAR, TH, Cl, and KR having distinct influence on irrigation water quality. Each criterion reflects a different aspect of water quality. These criteria are given unique weights, which is assigned with good understanding of the water quality and use pattern. Applying relative class rate scale given in Table 5, average relative weights of the selected criteria are developed (Table 6).

Water quality criteria are divided into subgroups and assigned respective ratings (Ayers and Wescot 1994) (Table 6). Considering the selection of each parameter or index on irrigation water quality rate subgroups as 1, 2, and 3 designating no harmful effect, moderate effect, and harmful effect.

Irrigation suitability index of the study area is obtained as suitability score (IW),

$$\text{IW} = \sum_{i=1}^{n} R_i \times W_i$$

where Ri is the reclassified rank of individual water quality criterion, Wi is the weight of the individual criterion obtained from AHP, and n is the total number of criteria. The irrigation suitability represents samples into 3 main categories, i.e.

Table 4 AHP relative class rate scale according to Saaty (1980)

Importance	Equal	Weak	Moderate	Moderate plus	Strong	Strong plus	Very strong	Very, Very strong	Extreme
Scale	1	2	3	4	5	6	7	8	9
	1/9, 1/8	1/7, 1/6	1/5, 1/4	1/3, 1/2	1, 2	3, 4	5, 6	7, 8	9
	← Less important				More important →				

Table 5 Average relative weights assigned to selected 6 criteria

Criteria	Weight	Consistency check
EC	0.312	OK 1%
SAR	0.279	
Cl	0.202	
KR	0.106	
TH	0.063	
MR	0.038	

Table 6 Ratings of the selected criteria

Irrigation criteria	Range	Water class/Restriction	Rank
EC	<0.7	None	1
	0.7–3	Slight to Mod	2
	>3	Severe	3
SAR	<3	None	1
	3–9	Slight to Mod	2
	>9	Severe	3
Cl	<140	None	1
	140–350	Slight to Mod	2
MAR	>350	Severe	3
	<50	Suitable	1
	>50	Unsuitable	3
KR	<1	Suitable	1
	>1	Unsuitable	3
TH	<75	Soft	1
	75–150	Mod	2
	150–300	Hard	2
	>300	Very hard	3

suitable (IW = 1.00 – 1.33), moderately suitable (IW = 1.34 – 2.33), or unsuitable (IW = 2.34 – 3.00). IWS estimates of water samples are presented in Table 7.

5 Conclusions and Recommendations

Most countries of the Arabian region are falling in arid and semi-arid areas, therefore, facing severe pressures due to limited water resources. Water scarcity is expected to increase with increased per capita water use associated with life style and high pace of development. Based on WHO (2003) criteria for recreational use, Hatta surface water qualifies in very good category with the exception of Falaj water. Falaj water

Table 7 Showing Irrigation Water Suitability (IWS) Classes

	IWS-season	IWS-location	Class-location	IWS-overall	Class-overall
HD1-Q-1	1.139	1.139	Suitable	1.66125	Mod suitable
HD1-Q-2	1.139				
HD1-Q-3	1.139				
HD1-Q-4	1.139				
HD2-Q-1	1.418	1.418	Mod suitable		
HD2-Q-2	1.418				
HD2-Q-3	Dry				
HD2-Q-4	Dry				
MD-Q-1	2.197	2.197	Mod suitable		
MD-Q-2	2.197				
MD-Q-3	Dry				
MD-Q-4	Dry				
HFH-Q1	1.716	1.716	Mod suitable		
HFH-Q2	1.716				
HFH-Q3	1.716				
HFH-Q4	1.716				
HFT-Q1	1.918	1.83625	Mod suitable		
HFT-Q2	1.716				
HFT-Q3	1.995				
HFT-Q4	1.716				

is polluted with fecal pollution and hence categorized as fair to very good at the upstream side and poor to good at the downstream side. This indicates that the Falaj water is getting polluted while enroute to discharging point. Water quality is mainly influenced by geology of the area, evaporation, landuse pattern, and anthropogenic factors.

Dam 1 is falling in Suitable Irrigation Water Quality class. Slightly higher SAR values in Dam 2 make it into 'moderately suitable class' for irrigation. Relatively high EC and Cl values make Dam 3 water into 'moderately suitable class' for irrigation. Relatively high EC, Cl, and hardness values makes Fallaj water into 'moderately suitable class' for irrigation.

Pollution of surface water is a problem itself and that may be threat to fresh groundwater resources as both are found hydrodynamically connected. Apart from that it can find ways to enter the food chain directly or indirectly posing a risk to human health. It is evident that with few exceptions, surface water is not used for anything except recreational usage. This makes the problem of bacteriological contamination less problematic. Although, contaminants presented in the surface water may enter the groundwater resources. This highlights the need of restricting human interactions with the surface water bodies which may be protected through several options while fenced boundaries wherever possible may be one such option.

It is further recommended to adopt controlling measures to prevent microbiological contamination of surface water especially within the Falaj. Bats and other wildlife should be prevented to enter into the Falaj's ducts. Makhtoom dam should be cleaned for sediments and salt encrust that have been deposited after multiple drying panning of past rain events. Makhtoom dam can also be investigated for artificial rainwater recharge through recharge shafts. It is recommended to continue keep track of surface water quality monitoring programme. Statistical treatment and graphical presentation help in better understanding of processes involved in chemical alteration of waters.

References

Abdalla O, Chen M (2017) Special issue on water resources in arid areas. Arab J Geosci 10:6. https://doi.org/10.1007/s12517-016-2797-8

Ahmed I, Tariq N, Al Muhery A (2019) Hydrochemical characterization of groundwater to align with sustainable development goals in the Emirate of Dubai UAE. Environ Earth Sci 78:44. https://doi.org/10.1007/s12665-018-8030-1

Alsharhan AS, Rizk ZE (2020) Water resources and water demands in the UAE. In: Water resources and integrated management of the United Arab Emirates. World Water Resources, vol 3. Springer, Cham. https://doi.org/10.1007/978-3-030-31684-6_27

APHA (2005) Standard methods for the examination of water and wastewater. APHA, Washington, DC, USA, 2005. ISBN 978-0875532356

Augustyn L, Kaniuczak J (2012) Stanek-Tarkowska J (2012) Selected physicochemical and chemical properties of surface water of Wisloka for consumption. Ecol Eng 28:7–20 [in Polish]

Ayers A, Wescot D (1994) Water quality for agriculture. FAO irrigation and drainage paper 29. Rome, Italy, pp 1–130

Bhat SA, Meraj G, Yaseen S, Bhat AR, Pandit AK (2013) Assessing the impact of anthropogenic activities on spatiotemporal variation of water quality in Anchar lake, Kashmir Himalayas. Int J Env Sci V(3)5: 1625–1640

Bhushan N, Rai K (2004) The analytic hierarchy process. strategic decision making: applying the analytic hierarchy process, pp 11–21

Bozdag A (2015) Combining AHP with GIS for assessment of irrigation water quality in Çumra Combining AHP with GIS for assessment of irrigation water quality in Çumra irrigation district (Konya), Central Anatolia Turkey. Environ Earth Sci 73:8217–8236

Dawoud MA (2013) The development of integrated water resources information management system in arid regions. Arab J Geosci 6:1601–1612. https://doi.org/10.1007/s12517-011-0449-6

Domenico PA, Schwartz FW (1990) Physical and chemical hydrogeology. Wiley, New York, NY, USA, 824p

Fridrich B, Krcmar D, Dalmacija B, Molnar J, Peši C, Kragulj (2014) Impact of wastewater from pig farm lagoons on the quality of local groundwater. Agric Water Manag 135:40–53. https://doi.org/10.1016/j.agwat.2013.12.014

Jianqin M, Jingjing G, Xiaojie L (2010) Water quality evaluation model based on principal component analysis and information entropy: application in Jinshui River. J Resour Ecol V(1)3:249–252

Kelley WP, Brown SM, Leibig GI Jr (1940) Chemical effects of saline irrigation water on soils. Soil Sci 49:95–107

Liu CW, Lin KH, Kuo YM (2003) Application of factor analysis in the assessment of groundwater quality in a blackfoot disease area in Taiwan. Sci Total Environ V-313(1–3):77–89. https://doi.org/10.1016/S0048-9697(02)00683-6
Murad AA, Al Nuaimi H, Al Hammadi M (2007) Comprehensive assessment of water resources in the United Arab Emirates (UAE). Water Resour Manage 21:1449–1463. https://doi.org/10.1007/s11269-006-9093-4
Nazzal Y, Ahmed I, Al-Arifi NSN et al (2014) A pragmatic approach to study the groundwater quality suitability for domestic and agricultural usage, Saq aquifer, northwest of Saudi Arabia. Environ Monit Assess 186:4655–4667. https://doi.org/10.1007/s10661-014-3728-3
Nazzal Y, Ahmed I, Al-Arifi NSN et al (2015) (2015) A combined Hydrochemical-statistical analysis of Saq aquifer, northwestern part of the Kingdom of Saudi Arabia. Geosci J 19:145–155. https://doi.org/10.1007/s12303-014-0016-8
Piper AM (1944) A graphic procedure in the geochemical interpretation of water-analyses. EOS Trans Am Geophys Union 25(6):914–928
Pramanik MK (2016) Site suitability analysis for agricultural land use of Darjeeling district using AHP and GIS techniques Model. Earth Syst Environ 2:1–22
Reghunath R, Murthy T, Raghavan B (2002). The utility of multivariate statistical techniques in hydrogeochemical studies: an example from Karnataka, India. Water Res V-36(10):2437–2442
Saaty TL (1980) The analytic hierarchy process: planning, priority setting, resource allocation. McGraw, New York, NY, USA, p 281
Saaty R (1987) The analytic hierarchy process-What it is and how it is used. Mathl Model 9:161–176
Saaty TL (2004) Decision making—the analytic hierarchy and network processes (AHP/ANP). J Syst Sci Syst Eng 13:1–35
Saaty TL (2008) Decision making with the analytic hierarchy process. Int J Serv Sci 1:83
Sherif MM, Mohamed MM, Shetty A, Almulla M (2011) Rainfall-runoff modeling of three wadis in the northern area of UAE. J Hydrol Eng 16(1):10–20
Simeonov V, Stratis J, Samara C, Zachariadis G, Voutsa D, Anthemidis A, Sofoniou M, Kouimtzis T (2003) Assessment of the surface water quality in Northern Greece. Water Res 37(17):4119–4124
Singh KP, Malik A, Mohan D, Sinha S (2004b) Multivariate statistical techniques for the evaluation of spatial and temporal variations in water quality of Gomti River (India): a case study. Water Res 38(18):3980–3992
Singh KP, Malik A, Singh VK, Basant N, Sinha S (2006) Multi-way modeling of hydro-chemical data of an alluvial river system—a case study. Anal Chim Acta 571(2):248–259
Singh K, Malik A, Mohan D, Sinha S (2004) Multivariate statistical techniques for the evaluation of spatial and temporal variations in water quality of Gomti River (India)-a case study. Water Res V-38(18):3980–3992
Umar R, Ahmed I (2007) Hydrochemical Characteristics of Groundwater in parts of Krishni-Yamuna Basin, Muzaffarnagar district, U.P. J Geol Soc India 69:989–995. Springer
Umar R, Ahmed I, Alam F, Khan MMA (2009) Hydrochemical characteristics and seasonal variations in groundwater quality of an alluvial aquifer in parts of Central Ganga Plain, Western Uttar Pradesh, India. Env Earth Sci 58:1295–1300. Springer
USDA (1954) Handbook No. 60, United States Department of Agriculture (USDA). Washington, DC, US Government Printing Office, pp 1–160
WHO (2003) World Health Organization. Guidelines for safe recreational water environments. Volume 1, Coastal and fresh waters, pp200
WHO (2011) Enterohaemorrhagic Escherichia coli (EHEC) Fact sheet N◦ 125. http://www.who.int/mediacentre/factsheets/fs125/en/
Wunderlin D, Diaz M, Ame M, Pesce S, Hued A, Bistoni M (2001) Pattern recognition techniques for the evaluation of spatial and temporal variations in water quality, a case study: Suquia river basin (Cordoba Argentina). Water Res 35(12):2881–2894
Zaman M, Shahid SA, Heng L (2018) Irrigation water quality. In: Guideline for salinity assessment, mitigation and adaptation using nuclear and related techniques. Springer, Cham. https://doi.org/10.1007/978-3-319-96190-3_5

The manufacturer's authorised representative in the EU is Springer Nature Customer Service Centre GmbH, Europaplatz 3, 69115 Heidelberg, Germany. If you have any concerns regarding our products, please contact ProductSafety@springernature.com

Printed and bound by CPI Group (UK) Ltd, Croydon, CR0 4YY
30/09/2026
02225233-0007